LOW CARBON ENERGY TECHNOLOGIES IN SUSTAINABLE ENERGY SYSTEMS

LOW CARBON ENERGY TECHNOLOGIES IN SUSTAINABLE ENERGY SYSTEMS

Edited by

GRIGORIOS L. KYRIAKOPOULOS

School of Electrical and Computer Engineering, Electric Power Division, Photometry Laboratory, National Technical University of Athens, Athens, Greece

Academic Press is an imprint of Elsevier
125 London Wall, London EC2Y 5AS, United Kingdom
525 B Street, Suite 1650, San Diego, CA 92101, United States
50 Hampshire Street, 5th Floor, Cambridge, MA 02139, United States
The Boulevard, Langford Lane, Kidlington, Oxford OX5 1GB, United Kingdom

Notices

Knowledge and best practice in this field are constantly changing. As new research and experience broaden our understanding, changes in research methods, professional practices, or medical treatment may become necessary.

Practitioners and researchers must always rely on their own experience and knowledge in evaluating and using any information, methods, compounds, or experiments described herein. In using such information or methods they should be mindful of their own safety and the safety of others, including parties for whom they have a professional responsibility.

To the fullest extent of the law, neither the Publisher nor the authors, contributors, or editors, assume any liability for any injury and/or damage to persons or property as a matter of products liability, negligence or otherwise, or from any use or operation of any methods, products, instructions, or ideas contained in the material herein.

Library of Congress Cataloging-in-Publication Data
A catalog record for this book is available from the Library of Congress

British Library Cataloguing-in-Publication Data
A catalogue record for this book is available from the British Library

ISBN: 978-0-12-822897-5

For information on all Academic Press publications
visit our website at https://www.elsevier.com/books-and-journals

Publisher: Joe Hayton
Acquisitions Editor: Graham Nisbet
Editorial Project Manager: Naomi Robertson
Production Project Manager: Nirmala Arumugam
Designer: Mark Rogers

Typeset by Thomson Digital

Contents

Part 3 Conclusions and future research

Contributors

Dmitry V. Antonenkov
Department of Power Supply Systems of Enterprises, Novosibirsk State Technical University, Novosibirsk, Russian Federation

G. Arabatzis
Democritus University of Thrace, Department of Forestry and Management of the Environment and Natural Resources, Pantazidou, Orestiada, Greece

Sofia Asonitou
University of West Attica, Aegaleo, Athens, Greece

Aldo Barrueto
Department of Electrical Engineering, Universidad Tecnica Federico Santa María (UTFSM), Santiago, Chile

Felisa Cordova
Faculty of Engineering, Universidad Finis Terrae, Santiago, Chile

Efi Drimili
Laboratory of Technology and Policy of Energy and Environment, School of Science and Technology, Hellenic Open University, Parodos Aristotelous, Patra, Greece

Zoe Gareiou
Laboratory of Technology and Policy of Energy and Environment, School of Science and Technology, Hellenic Open University, Parodos Aristotelous, Patra, Greece

Pandora Gkeka-Serpetsidaki
Renewable and Sustainable Energy Lab (ReSEL), Technical University of Crete, School of Environmental Engineering, University Campus, Chania, Greece

Theodore Kalyvas
Laboratory of Technology and Policy of Energy and Environment, School of Science and Technology, Hellenic Open University, Parodos Aristotelous, Patra, Greece

Vasileios C. Kapsalis
National Technical University of Athens, School of Mechanical Engineering, Section of Industrial Management and Operational Research, Athens, Greece

Evangelia Karasmanaki
Department of Forestry and Management of the Environment and Natural Resources, Democritus University of Thrace, Orestiada, Greece

Grigorios L. Kyriakopoulos
School of Electrical and Computer Engineering, Electric Power Division, Photometry Laboratory, National Technical University of Athens, Athens, Greece

S.V. Leontopoulos
General Department, University of Thessaly, Larissa, Greece

Vadim Z. Manusov
Department of Power Supply Systems of Enterprises, Novosibirsk State Technical University, Novosibirsk, Russian Federation

Antonio Parejo
Department of Electronic Technology, Escuela Politécnica Superior, University of Seville, Seville, Spain

Hans Rother
Enel Distribución Chile S.A., Santiago, Chile

Sarat Kumar Sahoo
A Constituent College of Biju Patnaik Technological University, Parala Maharaja Engineering College, Department of Electrical Engineering, Govt. of Odisha, Berhampur, Odisha, India

Antonio Sanchez-Squella
Department of Electrical Engineering, Universidad Tecnica Federico Santa María (UTFSM), Santiago, Chile

Dhruv Shah
Mukesh Patel School of Technology Management & Engineering (MPSTME), Mumbai Campus, India

Evgenia Y. Sizganova
Department of Electrotechnical Complexes and Systems, Siberian Federal University, Krasnoyarsk, Russian Federation

Denis B. Solovev
Far Eastern Federal University, Engineering School, Vladivostok, Russian Federation; Vladivostok Branch of Russian Customs Academy, Vladivostok, Russian Federation

Georgios Tsantopoulos
Department of Forestry and Management of the Environment and Natural Resources, Democritus University of Thrace, Orestiada, Greece

Theocharis Tsoutsos
Renewable and Sustainable Energy Lab (ReSEL), Technical University of Crete, School of Environmental Engineering, University Campus, Chania, Greece

Fernando Yanine
Faculty of Engineering, Universidad Finis Terrae, Santiago, Chile

Miltiadis Zamparas
School of Science and Technology, Hellenic Open University, Patra, Greece

Efthimios Zervas
Laboratory of Technology and Policy of Energy and Environment, School of Science and Technology, Hellenic Open University, Parodos Aristotelous, Patra, Greece

Preface

A new era of energy systems is enabled by the integration of low carbon energy technologies and novel energy production schemes. In this context, the adoption of low carbon energy technologies brings new insights of wider socio-economic and environmental interest. This book also delivers new and flexible responses to crucial social attitudes and technological challenges in complementing novel energy systems of environmental sensitivity in the energy markets of the future. I am very pleased to deliver this book title that offers an in-depth analysis of such sustainable energy systems. This book merits its attractiveness as it is one of the first initiatives worldwide analyzing the multifaceted impacts of this kind of research on individuals under a wide geographical coverage and containing a pluralistic thematic corpus. I am grateful to all the contributors to this book for their intellectual research, insightful propositions, and creative arguments. The relevant book structure is aligned around the following three pillars:

Part 1: Introduction and fundamentals

At this introductory part of the book, emphasis is paid to the presentation of technologies that are both energy intensive and socio-culturally important, further representing a wide spectrum of environmental sensitivity energy schemes. Selected topics in this section include:

- The role of resource recovery technologies in reducing the demand of fossil fuels and conventional fossil-based fertilizers;
- Increasing efficiency of mining enterprises power consumption; and
- The contribution of energy crops to biomass production.

Part 2: Examining low carbon energy technologies and their contribution as sustainable energy systems

In this part of the book, selected, sophisticated research and case studies of sustainable energy systems, in both theoretical and applicability contexts of analysis, are presented. Selected topics include:

- Public attitudes toward the major renewable energy types in the last five years: A scoping review of the literature;
- Understanding willingness to pay for renewable energy among Europeans during the period 2010–20;
- Linking energy homeostasis, exergy management, and resiliency to develop sustainable grid-connected distributed gen-

eration systems for their integration into the distribution grid by electric utilities;

- Smart energy systems: The need to incorporate homeostatically controlled distributed generation systems to the electric power distribution industry: An electric utilities' perspective;
- Grid-tied distributed generation with energy storage to advance renewables in the residential sector: Tariffs analysis with energy sharing innovations;
- Integrating green energy into the grid: How to engineer energy homeostaticity, flexibility, and resiliency in energy distribution and why should electric utilities care;
- Multi energy systems of the future;
- Bibliometric analysis of scientific production on energy, sustainability, and climate change;
- Public acceptance of renewable energy sources;
- Sustainable site selection of offshore wind farms using GIS-based multi-criteria decision analysis and analytical hierarchy process. Case study: Island of Crete (Greece); and
- Accounting and sustainability.

Part 3: Conclusions and future research

In this part of the book, the concluding chapter titled "Should Low Carbon Energy Technologies be Envisaged in the Context of Sustainable Energy Systems?" offers a literature collection and an argumentative synthesis, debated on the socio-economic and environmental aspects of sustainable energy systems. The theoretical background and the key aspects were approached at an integrative discussion of complementary research strategies including: operability of low carbon energy technologies within complex transitioning of energy economies, technical aspects of exploitation, models and simulations, R&D, infrastructure, as well as efficiency improvements prospected. Besides, constraints, barriers, drivers, and challenges of scaling-up RES-based projects to real energy markets worldwide, they have been systematically signified, through the pluralistic literature collected and the argumentation developed.

I want to warmly thank the professional staff at Elsevier for their qualitative work that made this book possible. I hope you to find the contents of the book fascinating and its reading attractive. If you would like any further information on this book, I am at your disposal.

Dr. Grigorios L. Kyriakopoulos
Guest Editor

1

Introduction and fundamentals

1

The role of resource recovery technologies in reducing the demand of fossil fuels and conventional fossil-based mineral fertilizers

Miltiadis Zamparas
School of Science and Technology, Hellenic Open University, Patra, Greece

Chapter outline

Low Carbon Energy Technologies in Sustainable Energy Systems. http://dx.doi.org/10.1016/B978-0-12-822897-5.00001-8

1 Introduction

1.1 Urban wastewater and energy resource recovery

Urban wastewater is characterized as wastewater generated from domestic activities or as a mixture of wastewater generated by household, industrial, and rainwater outflows [1,2]. Urban wastewater is considered as a hazardous material that has to be disinfected to support public health and protect the environment. In European Union, more than half of the population lives in agglomerations or more than 150,000 population equivalent (PE), generating a daily amount of 41.5 million m^3 of wastewater. Besides, an annual portion of 2.4% (counts for 1 billion m^3) of treated domestic wastewater discharges contains reusable nutrients, organic carbon, lipids, and biosolids. The collection and the treatment of urban wastewater in EU are made in the form of a mixed gray- and black-water (named as storm-water, combined sewer system). Even though certain portions of wastewater effluents remain unexploited, a portion of urban wastewater is retrievable while using Nature-Based Solutions (NBS) [3–5]. Typical plants of treating unsegregated wastewater include reclaimed fertigation and irrigation waters, P-rich sludge, biopolymers, alginates, materials and energy, in terms of biogas, biofuel, electricity, and heat (Fig. 1.1).

NBS for resource recovery from wastewater are involving mature technologies—mainly constructed wetlands and algae ponds—up to the advanced biological methods of rotating

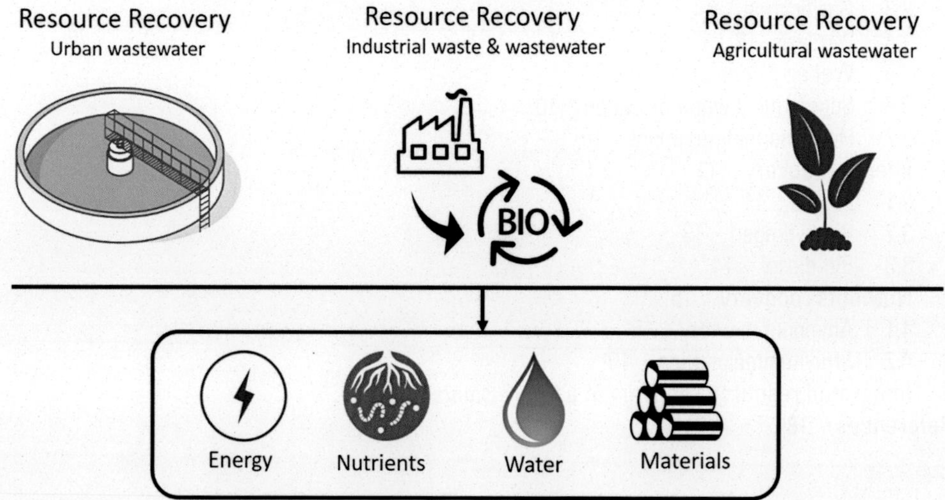

Figure 1.1. Resource streams and opportunities for recovery. Author own study.

biological contactors, aerobic granulation, and anaerobic reactors [6]. The broad range of recoverable products includes commonly complementary products are biogas from primary and secondary sludge and recovered water for agricultural, industrial, residential, and urban uses as well as for recharge of groundwaters. Plant biomass is transformable to biogas and digestate as fertilizers, bioethanol through sugar fermentation, and biochar via pyrolysis, or treated for the fabrication of pulp-paper and bioplastics. Bio-oil is generated by handling biomass under anoxic conditions of high temperature, biohydrogen by steam reformation of bio-oils, and photolysis of water catalyzed by specific microalgae species. Commercial uses of algae biomass are that of feed production and high-value chemicals [3,6].

1.2 The global demand of P-fertilizers and the need of nutrient recovery

As "waste" is generally considered the flow of resources, such waste is intended for ending disposal in sewer networks, containing rich-quantities of resources such as nutrients (N, P, K), organic, water, and minerals. In European Union, the generation of 3.6 Mt N, 1.7 Mt P, and 1.3 Mt K regards the portion of its citizens' excreta. Nevertheless, in particular, the excessive consumption of European fertilizers is reaching at 11 Mt N, 2.9 Mt P, and 2.5 Mt K of fertilizers [7].

The primary source of P, phosphate rock (PR), it is non-renewable, while recently there has been uncertainty due to reduced supply and the rising prices in the international market [8–11]). It is forecasted that global demand will suppress supplies since the production rate of phosphorus fertilizers will decrease while the readily available phosphorus resources are constantly depleted (Fig. 1.2). Besides, almost 90% of the globally estimated reserves of phosphate rock are deposited in Morocco, Iraq, China, Algeria, and Syria, which arouse a matter of food security for other nations [9,10]. The seemingly steady applications of P-rock fertilizers have reached its limits; thus, a sustainable way of utilization is necessary. Besides, resource recovery technologies are promisingly proven to be alternative methods in which the phosphorus stock is utilized as fertilizer by enriching the soil conditioners.

Nutrient recycling from WWTPs also includes nutrients' recycling thus positively impacting on ecosystem, while decreasing the extensive use of conventional fossil-based fertilizers and, therefore, reducing water and energy consumption [15,16].

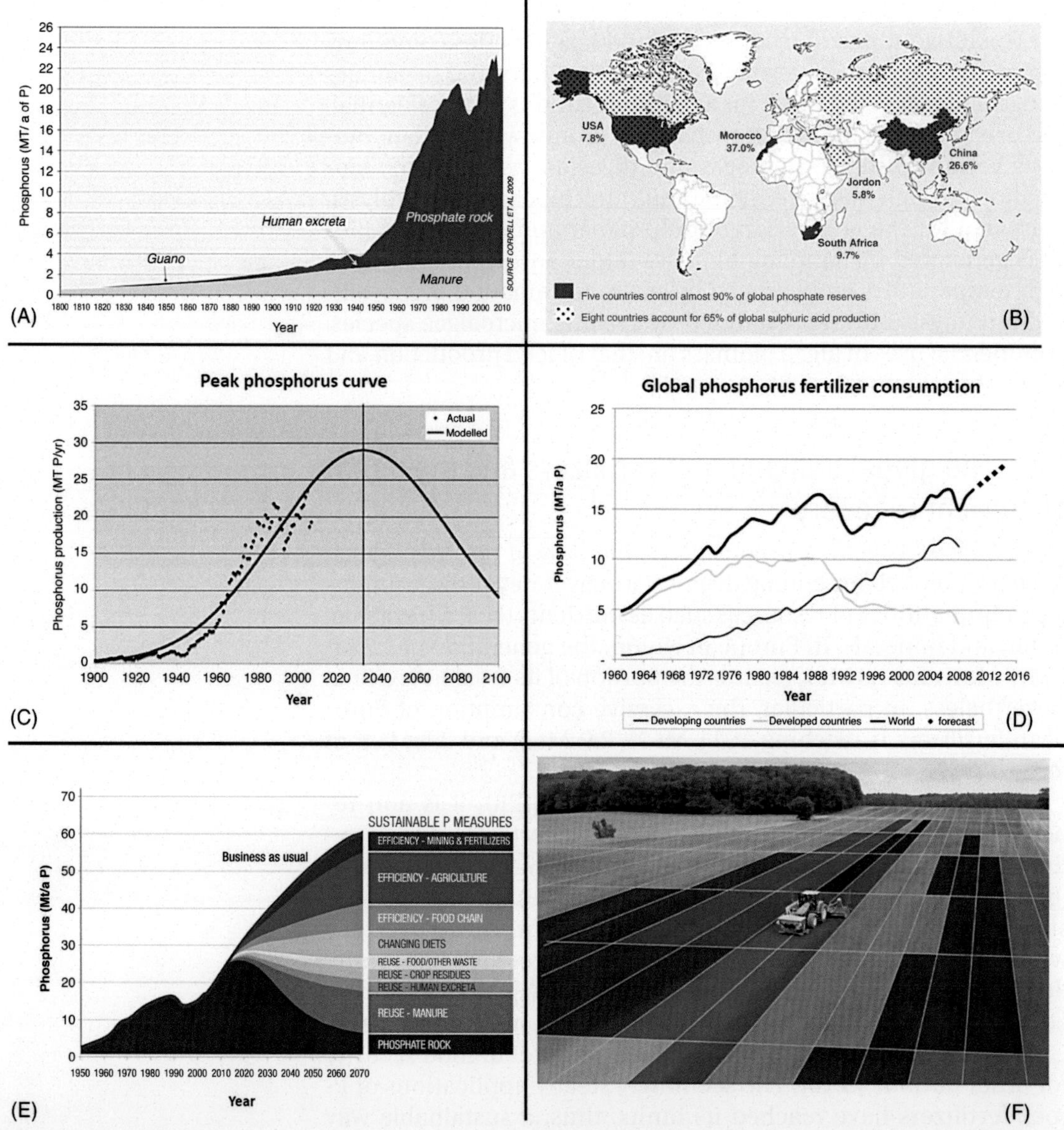

Figure 1.2. (A) Historical sources of phosphorus fertilizers (1800–2010), (B) Global contribution of P-Rock, (C) The phosphorus curve peak denotes, in a similar way to oil, that global phosphorus reserves are prone to peak, after which a sharp reduction of production it is reported, (D) Global phosphorus fertilizer consumption between 1961 and 2006 (in million tons P), (E) Sustainable phosphorus measures: efficiency, recycling, and changing diets, (F) Precision farming by jointly applying fertilizer with actual soil nutrient needs. The color indicates soil nutrient demand regarding collected data from satellite, field sensors, and farmers' knowledge. From Refs. [12–14].

2 Methods for energy and resource recovery

Sludge management is mostly proven to be a challenging activity of wastewater treatment plants (WWTPs) that sustains high-water content, low dewaterability, and stringent regulation concerning the reuse or the disposal of the generated sludge. WWTPs are also scheduled to ensure those eco-friendly processes that control the volumes of sludge for disposal and its bioenergy conversion (Fig. 1.3). Besides, the energy recovery of sludge comprises its transition to biogas, syngas, and bio-oil, toward the end-uses of electricity, mechanical energy, and heat [18].

2.1 Anaerobic digestion

Waste management methods are linked to energy recovery, thus, positively controlling the global waste and optimizing resources through the energy generation from renewables. Such energy conversion key technologies are highlighted in Fig. 1.4, showing the conversion routes of sludge to syngas, liquid fuel, chemicals, heat, and/or electricity [20].

Anaerobic digestion is a low cost applied biological conversion process providing organic waste without reduction to its high calorific value of the produced biogas (combination of methane

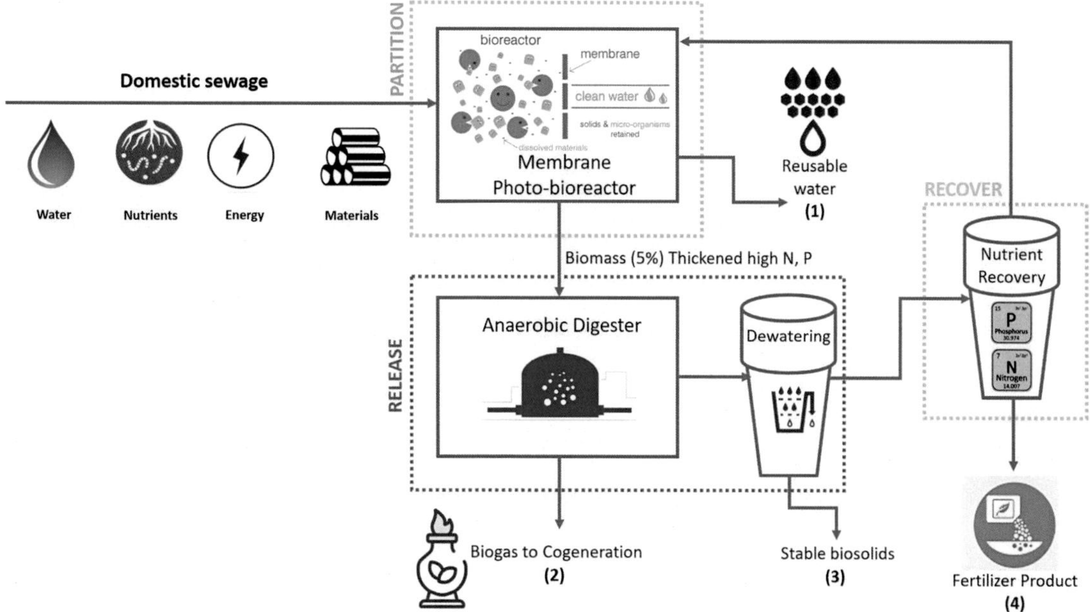

Figure 1.3. Partition-release-recover concept (PRR) as a platform for energy recovery from domestic wastewater. Modified from Ref. [17].

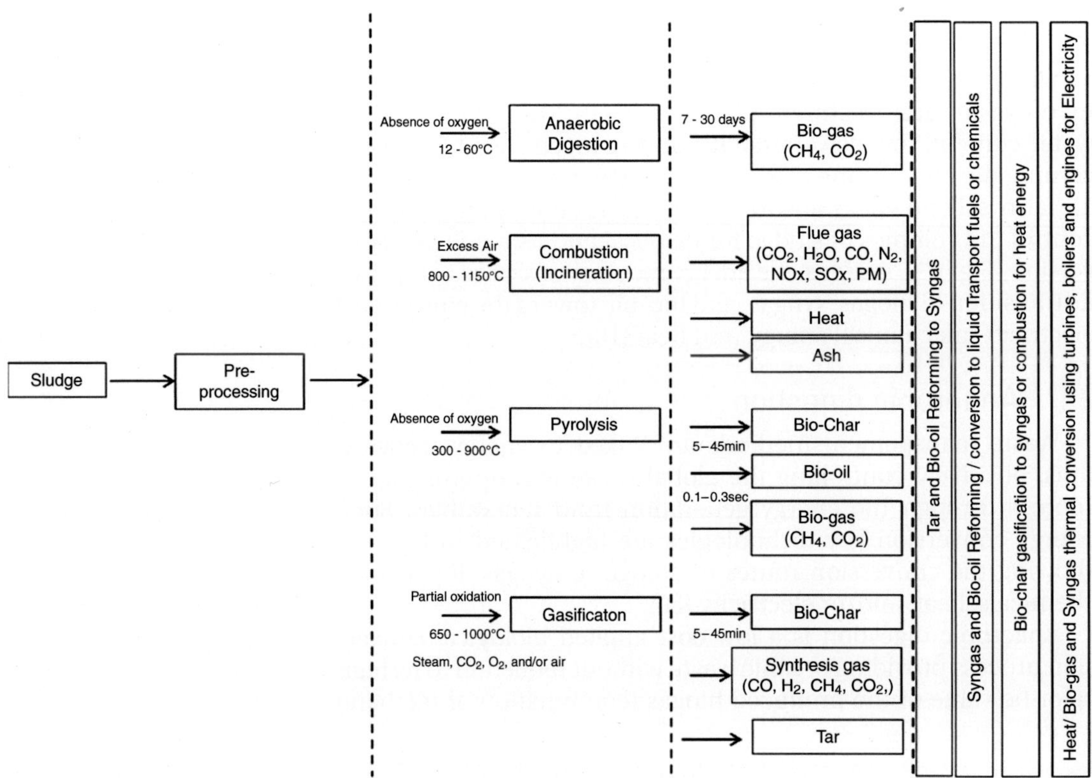

Figure 1.4. Potential sludge-to-energy recovery routes [19].

and carbon dioxide), even though its high moisture content. Biogas conversion to heat and electricity can be achieved through cogenerating of thermal reactors. The typical procedural conditions of biochemical processes are that of an inert environment at mesophilic temperature range for sludge stabilization, while the processing residual matter can be used for agricultural purposes [20–23]).

2.2 Incineration and co-incineration

Sludge incineration is achieved by the organic compounds' oxidation at elevated temperatures. In this process, the biosolids' burning can be achieved through supplying excess quantity of air to the combustion chamber toward the production of carbon dioxide and water, having only the by-product of ash inert material. This inert material can be further upcycled as building material, or be disposed [24–26] (Fig. 1.5).

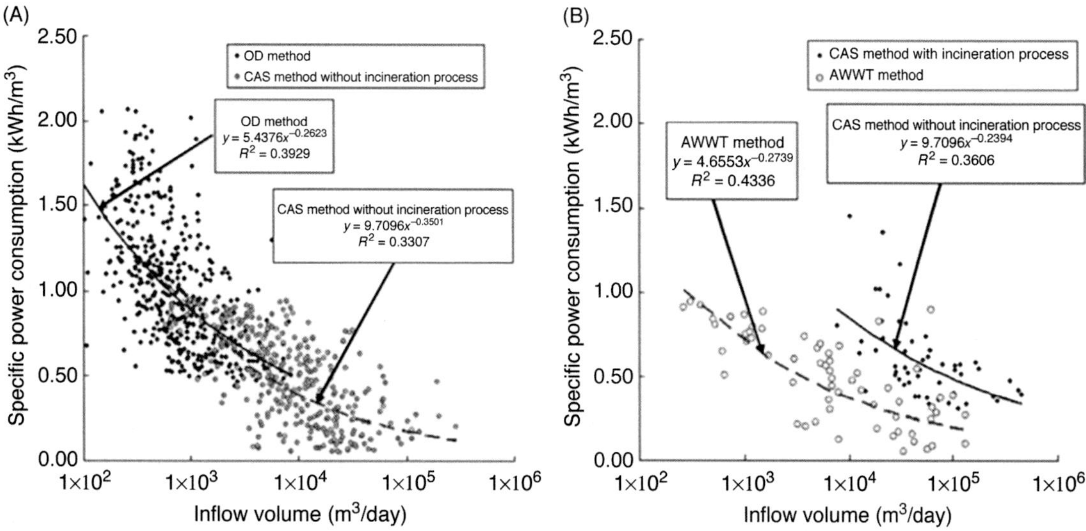

Figure 1.5. The joint profiles of energy consumption and inflow volume of (A) OD and CAS treatment systems without incineration process and (B) CAS treatment systems in Japan, co-representing incineration process and advanced (AWWT) wastewater treatment [27].

2.3 Gasification

The gasification process includes the transformation of dried sludge in ash and, subsequently, in combustible gases at the temperature of 1000°C under the conditions of limited amounts of oxygen. In the gasification process, heat can be used for power generation and heat-demanding processes, while it also produces syngas (also named synthetic gas). At the pressure range of 0.6–2.6 MPa and temperature range of 1400–1700°C, gasification is achieved by the utilization of pure oxygen oxidant [28].

2.4 Pyrolysis

The method of sludge pyrolysis can innovatively manage sludge by the thermal treatment of 350–500°C, while the energy needs are met from reduced oxygen and high-pressure environment. Sludge transformation contains the production of indicative products, such as pyrolysis oils and combustible gases [19,29].

2.5 Wet air oxidation

Wet oxidation is a method of chemical oxidation of sludge by adding oxygen at a high temperature, 150–330°C, and high pressure, 6–20 MPa. The ZIMmerman PROcess (ZIMPRO) is dating back to 1960s in the Netherlands, being among the oldest processes that are based on wet oxidation technology. Sludge's organic matter can be oxidized to produce water, carbon dioxide, along with easily biodegradable organic agents [30].

2.6 Supercritical water oxidation

Supercritical water oxidation (SCWO) is a method of wastewater treatment, which follows high temperature (600°C) and pressure (25 MPa), and it is proven to be an effective method for sludge disintegration. In this method, the carbon and hydrogen from organic and biological components are oxidized to CO_2 and H_2O; while other elements, such as nitrogen, sulfur, and phosphorus form N_2O, SO_4^{2-}, and PO_4^{3-}, respectively. In this logic, the conversion of organic chlorides to heavy metals and Cl is accompanied by their oxidization to the relevant oxides. However, the fact that the SCWO is a costly method, there is an advantageous cost-effective balance; the value from this method along with the sludge volume decrease can reach an excess value of 90% recovery of energy, coagulants, and phosphate, and thus, offsetting the operational cost of the SCWO method [31–33].

2.7 Hydrothermal treatment

The method comprises heating the sludge in the water phase at the temperature range of 150–450°C in anoxic conditions. Sludge hydrolysis during hydrothermal treatment resulting in the production and accumulation of high quantities of dissolved organic compounds in the liquid phase. Besides, protein hydrolysis is transformed into amino acids, while lipid hydrolysis produces fatty acids, and the hydrolysis of fiber material and hydrocarbons produces low molecular hydrocarbon compounds, such as sugars. The interest of these compounds resides to the fact that these are not only carbon resources for biogas production, but can also help in the advancement of the denitrification process of wastewaters and biological removal of phosphorus. Moreover, hydrothermally treated sewage sludge (SS) in the form of liquid residue contains the elemental nutrients of N, P, and K, thus, proven especially effective and marketable fertilizer [20,34–36] (Table 1.1).

Table 1.1 Evaluation profile of the methods for sludge processing.

Method	Technical adaptability in raw and digested sludge	Ecosystems' protection and sustainability	Economic scalability	Energy efficiency	Technological maturity/ industrial scalability
Anaerobic digestion					
Combustion					
Pyrolysis					
Gasification					

Scale of weighted indicators: highly positive fairly positive low positive.
Source: Author's own study, data collected from Ref. [19].

3 Energy recovery

3.1 Biogas

The main compositional allocation of applying the anaerobic digestion method to SS, results in the production of biogas that comprises of 60%–70% methane (CH_4), 30%–40% carbon dioxide (CO_2), as well as trace amounts, such as nitrogen (N), hydrogen (H_2), and hydrogen sulfide (H_2S).

Anaerobic digesters are generating methane gas that is considered as the vital source of energy at an urban WWTP in the form of methane production to power gas engines and electrical-thermal energy for onsite treatment plants. A typical WWTP electricity cost is reaching almost 80% of the overall functioning cost, while half of this cost can be recovered through methane production for energy purposes [37,38]. A typical paradigm of biogas technology is that of the bio-terminator 24/85, which is a technology of mesophilic anaerobic digestion that has been developed after the research conducted on Total Solids Solution at the University of Louisiana, United States. This process was proven especially effective to destroy an amount of 85% TS after the reactor retention time for 24 h or less. Similarly, an installed capacity of 3.785-m^3 power plant in the year 2005 operated for 5 months at Baton Rouge, Louisiana [37,39]. The system removed an amount of 93% VS at 2 days' hydraulic retention time.

Another commercial method is that of "Columbus Advanced Biosolids Flow-through Thermophilic Treatment" (CBFT3). This method contains the modification of thermophilic anaerobic digestion using plug flow reactor engines that covered 40%–50% of the plant electricity production needs. The reported total energy efficiency of the process ranges from 68% to 83% [40,41]. Physico-chemical, thermo-mechanical, and biological pretreatment steps, as that of microwave (MW) heating, ultrasonication, ozonation, use of liquid jets, and wet oxidation, all can be assessed in terms of biogas production, energy balance, quantity of sludge produced, and pricing [20,41]. Several countries deployed a full developmental scale of pretreatment technologies, including thermal (Cambi and BioThelys), physical-chemical (MicroSludge), mechanical (Lysatec GmbH), and ultrasonic (CROWN) [42–49].

Biogas is considered as an excellent fuel for numerous applications that can be used to a wide range of applications developed for natural gas. Particularly, applications of biogas are heat and steam production, generation/cogeneration of electricity, energy for transportation machinery, and as fuel in gas vehicles, as well as for chemicals' production.

The Combined Heat and Power (CHP) application denoted that if CHP is installed at 544 wastewater treatment facilities of about 340 MW to support the energy needs for 261,000 homes. It has been estimated that an amount of 2.3 million MT of carbon dioxide emissions can offset if existing WWTP of capacity treatment of over 5 million gallons/day of anaerobic digestion, they are installing energy recovery facilities to support high energy security and to lower greenhouse gas emissions through less intensive use of fossil fuels [50,51].

As aforesaid, important utilities of sludge-derived biogas as biofuel is associated with transportation purposes. In such case of the Swedish Henriksdal treatment plant, it is producing and selling biogas to Stockholm's bus company, running more than 30 biogas-fueled buses. In the United States, the deployment of various energy recovery techniques are well established and they include electro-mechanical energy production and heat recovery through biogas anaerobic-waste-sludge generation [52,53] or, similarly, the hydrogen yield from methane for energy generation with liquefied carbonate fuel at the King County, Washington's South treatment plant. It has also been reported that the utility of grease from restaurant trap haulers—mainly composed by energy-rich compounds, such as fats, sugars, and carbohydrates—they can undergo co-digestion with SS in Watsonville, California at more than 50% biogas yield. Therefore, grease can form a suitable substrate for biogas production during anaerobic digestion of sludge, while dewatered SS can produce fuel charcoal for the thermal power and electricity generation upon the syngas production through SS pyrolysis [54]. Biogas harvesting from SS is proven to be highly efficient method of resources' recovery in China. The annual stock of methane generation by all sewage-sludge feedstocks was assessed at 720 million cubic meters. In the United Kingdom, specific government programs have been proposed for energy recovery, aiming at electricity generation of 20% from renewable sources by 2020, while in the reference year 2005, the profile of energy recovery from renewables was achieved by combustion (10.8%) and biogas production (4.2%), respectively [52,55,56].

3.2 Bio-hydrogen

Hydrogen is known as a sturdy energy carrier with a large energy capacity per unit mass and having the advantage of not emitting CO_2 throughout combustion [57]. The prevalent hydrogen technologies can use fossil fuels and consume large amounts of energy, having a high-level carbon footprint, including reform of fossil gas steam (50% of global production), petroleum reform (30%), and underground gasification of coal (18%). Therefore, the efficient production of hydrogen requires to be focused on ecologically friendly and economic technologies [58], while bioprocesses are proven to be among the most promising approaches in hydrogen production, autotrophic (e.g., biofotolysis) and heterotrophic (e.g., photofermentation and dark fermentation). However, dark fermentation (DF) is considered as the only capable system to simultaneously fulfill the two goals of waste management and energy recovery, such as agricultural waste and wastewater are used as feedstock [59,60]. DF functioning contains the process in which carbohydrate-rich substrates are processed into simpler organic compounds by anaerobic bacteria with simultaneous hydrogen output [61–63].

Hydrogen is a favorable source of energy to fossil fuels. Hydrogen has high energy (122 kJ/g), which is two- to sevenfold greater than hydrocarbon fuel. The main research interest is whether thermochemical treatments for wet SS could produce hydrogen-rich fuel gas. Therefore, examination of the prospects to generate hydrogen-rich fuel gas by thermochemical technologies includes drying, pyrolysis, and gasification. Pyrolysis of wet SS (1000°C), linked with high heating levels, increases the production of hydrogen [64]. Furthermore, the pyrolysis from wet sludge can create a gaseous liquid that contains a significant portion of hydrogen [65]. Then, the high moisture content of SS procures elevated temperatures and a steam-rich environment, can subsequently lead to volatile organic compounds' steam reformation and solid char's gasification, contributing to hydrogen production. High moisture level can increase the quantities of CO_2, CH_4, and H_2 and decrease that of CO [66,67].

3.3 Bio-diesel

Biodiesel is a carbon-neutral source of energy that selectively substitutes fossil fuels in a range of applications, such as in the transportation sector, which counts for 23% of global greenhouse gasses [68]. Biodiesel is valued as a clean option while it can be used in the current engines, alongside in the production and supply network without necessitating significant changes. The majority of biodiesel (~95%) is currently provided from the transesterification of edible vegetable oils, such as canola, palm, rapeseed, and soybean, which are professedly first-generation biofuels. However, biodiesel production from non-edible oils, which are second-generation biofuels, is gaining vivid interest, especially regarding the contentious disputes of the related food versus fuel antagonism for land and water [69]. Besides, public SS is gaining traction worldwide as a lipid feedstock since such high energy lipids comprise of phospholipids, mono-, di-, and tri- glycerides and free fatty acids in producing biodiesel from sludge lucrative [70,71].

In the relevant literature [71,72], it has been stressed out that by integrating lipid extraction and transesterification processes in 50% of all current municipal WWTP in the United States, will produce 1.8 billion gallons of biodiesel, which counts for ~0.5% of the annual nationwide needs for oil diesel. Nowadays, it has been estimated that the average production expense is 3.11 US$/gallon of biodiesel, whereas lowering its pricing at or below the price of current petro-diesel [73–75].

It is noteworthy that in a South Korean case study, it has been reported that the price of lipids extracted from the SS could be about $0.03/L lower than all the current feedstocks for biodiesel production. These researchers showed that biodiesel production using SS lipids is cost effective due to low cost of this type of feedstock and the substantially high yields of oil. The SS yield of oil, counting of 980,000 L/ha/year, is higher than that of microalgal and soya oil, 446 and 2200 L/ha/year, respectively. Economically, the Korean Government charged US$ 58.3/ton wet SS for disposal, while the drying cost is US$ 57.17. This dried SS is commonly delivered to the manufacturer of the coal power station and the value (for a one-ton of dried SS) is set as 10 US$. Therefore, the cheapest choice would be the biodiesel production from SS-extracted lipids. This policy is compatible to a proven cost-effective and reliable process of biodiesel conversion [20,76–78].

4 Nutrients recovery

In the context of a worldwide constriction of resources availability, the recovery of nutrients from WWTPs should be taken into account. In the case of nitrogen, bio-electrochemical system (BES) has been demonstrated as a promising technology for recovering nitrogen, either in form of ammonia or struvite [79,80].

4.1 Ammonia recovery

Ammonium is a widely used fertilizer, and it has been conventionally obtained by the energy intensive Haber-Bosch process. The energy costs associated to both Haber-Bosch process and nitrification-denitrification treatments could be minimized if ammonia was recovered from WWTPs and used as a fertilizer. Following this objective, several studies have evaluated the recovery of ammonia from wastewater and, more specifically, from urine using BES [79]. In order to recover ammonia, BES can be operated either as MFC [81] or microbial electrolysis cell (MEC) [81,82]. The recovery of ammonia using BES is based on the charge neutrality principle and requires the use of a cation exchange membrane. The electron flow between anode and cathode implies that cations (like ammonium NH_4^+) are forced to diffuse from the anode to the cathode compartment through the membrane. Once in the cathode, its high pH allows ammonium to be converted into ammonia (NH_3), which is highly volatile, and thus can be easily recovered through stripping [83,84].

4.2 Struvite precipitation

Struvite is a crystalline solid composed of magnesium, ammonia, and phosphate at equimolar concentrations ($MgNH_4PO_4 \cdot 6H_2O$). The precipitation of struvite in WWTPs would not only allow nitrogen, but also, phosphorus recovery [85]. Phosphorus is an essential fertilizer, but it has been estimated that accessible phosphorus reserves could be depleted in the next 50 years. Hence, its recovery has become a research priority. In order to precipitate struvite, alkaline conditions are required, which are conventionally imposed by chemical additions [86]. In BES, the reducing processes occurring in cathodes generate alkaline conditions in this compartment. Besides cathode basification is seen as a drawback in most of BES applications, it is worthy for struvite recovery. The cathodic alkaline conditions can be provoked using different BES configurations. Accordingly, struvite has been recovered in a plethora of BES as air-cathode MFCs [87], or single chamber [88] and double-chamber MEC [89,90].

Bioelectrochemical systems possesses a realistic economic potential, while, future research can be directed to determine those technological hurdles related to methane production in anodes, hydrogen losses and electrochemical losses. Developmental perspectives of BESs are cross-disciplinary and they are highlighting the need for truly integrated process at all scales. In terms of energy production, BES can be used for direct electricity production or for an indirect energy recovery through hydrogen or methane production. Both applications have a promising future. The removal or recovery of nutrients in BES has presented promising results. In terms of treatment, it could improve the economic viability of the current WWTP. Specifically, WWTPs can adopt the principles of circular economy and support the upcycling recovery of (otherwise disposable) nutrients, giving a second chance of their utility and adding value to the whole BES. Ideally, the utility of a BES for sewage treatment can couple energy production and nutrient recovery (or removal) [91,92] (Fig. 1.6).

Nutrient recovery from wastewater is attracting the scientific interest in alignment with industrial communities, regulators, and public-interest groups. Many technologies and processes are being developed for the recovery of products such as ammonium nitrate, ammonium sulfate, ammonium water, bio-struvite, calcium phosphate, hydroxyapatite, phosphoric acid, potassium phosphate, struvite, white phosphorus, etc. Many of these processes require specific assets and well-trained operators. The numbers of installations worldwide for nutrient recovery is increasing rapidly as technological advances are occurring at a fast pace. Nevertheless, it continues to challenge the cost effectiveness of these processes, while providing a reasonable end route and launch of these

Figure 1.6. Crystal Green struvite produced by the Pearl process developed by Ostara. From Ref. [93].

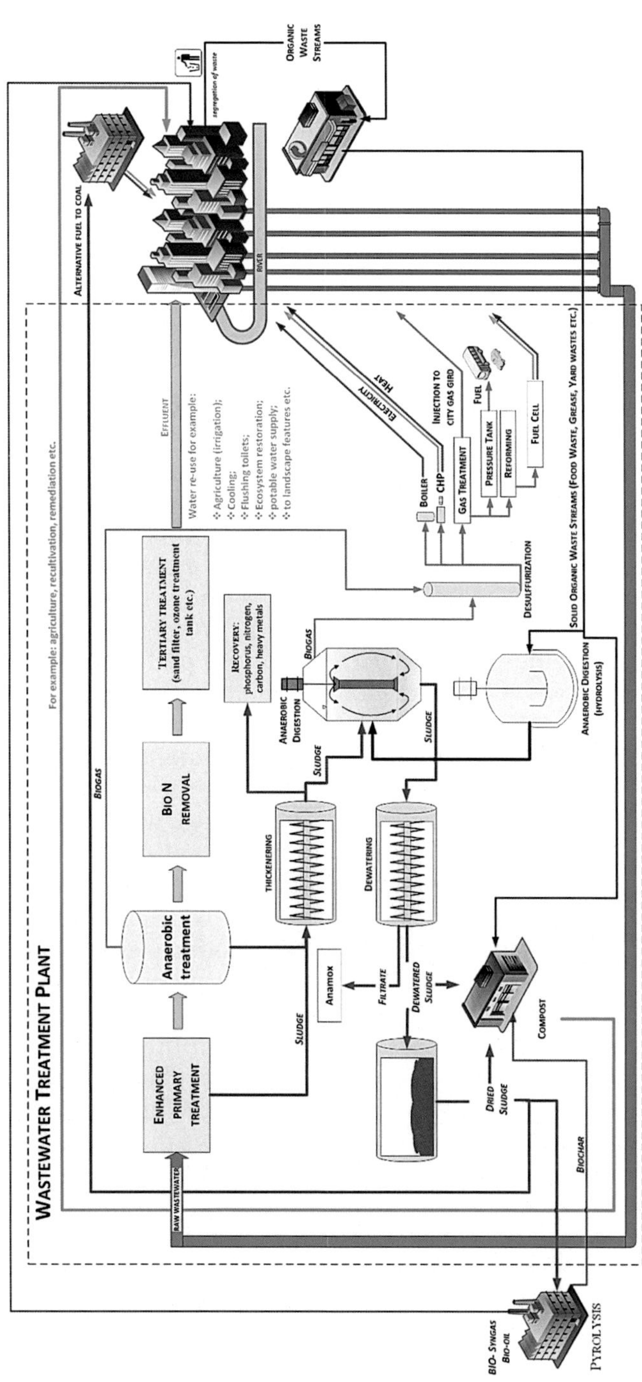

Figure 1.7. Wastewater treatment plant in a future smart city [16].

products to well developed marketplace through a stable supply chain. The attractiveness of these processes to the water industry is still very much driven by reducing problems down stream (e.g., struvite precipitation in pipes) or reducing the recirculation of nutrients within the WWTP and consequent reduce the pollutant load to the secondary treatment. Government and national agencies worldwide are striving for innovation, while bringing the circular economy to practice within the wastewater industry. Therefore, they need to take an active role in order to initiate and support supply chains and markets for the recovered/upcycling products.

5 Integrated resource recovery in a future smart city

Circular economy (CE) is targeted on the optimization of circle approach functions by recovery resources. Moving toward more CE can help to deliver assumptions of the resource effectiveness agenda formed under the Europe 2020 Strategy for smart, comprehensive, and green growth. The model of smart cities is realistic and balanced pattern of social, economic, and environmental, for sustainable growth. Eco-cities worldwide reveal a common aim, "to improve the well-being of people via integrated urban planning that totally uses the benefits of natural resources, protecting and supporting these assets for forthcoming generations" [16]. Nowadays, wastewater treatment plant is considered as a vital component of SMART city. Fig. 1.7 illustrates the integrated wastewater system model in which WWTPs not only handle wastewater with efficacy that make available sewage recycle but also produce energy and generate fertilizers [16,94]. Progressively more cities worldwide are applying SMART models in their zone. For instance, the City of Borås, in Sweden has established a plan in which WWTP will be placed along with the regional power plant supplying renewable fuel for the city. The goal of this recycling concept is to use the energy of the city's wastewater flows making a municipality exempt from fossil fuels [16].

References

[1] E. Eriksson, K. Auffarth, M. Henze, A. Ledin, Characteristics of grey wastewater, Urban Water 4 (2002) 85–104 https://doi.org/10.1016/S1462-0758(01)00064-4.

[2] A.L. Thebo, P. Drechsel, E.F. Lambin, K.L. Nelson, A global, spatially-explicit assessment of irrigated croplands influenced by urban wastewater flows, Environ. Res. Lett. 12 (2017) https://doi.org/10.1088/1748-9326/aa75d1.

[3] K. Krauze, I. Wagner, From classical water-ecosystem theories to nature-based solutions—Contextualizing nature-based solutions for sustainable city, Sci. Total Environ. 655 (2019) 697–706 https://doi.org/10.1016/j.scitotenv.2018.11.187.
[4] R. Lafortezza, J. Chen, C.K. van den Bosch, T.B. Randrup, Nature-based solutions for resilient landscapes and cities, Environ. Res. 165 (2018) 431–441 https://doi.org/10.1016/j.envres.2017.11.038.
[5] M. van den Bosch, Ode Sang, Urban natural environments as nature-based solutions for improved public health—A systematic review of reviews, Environ. Res. 158 (2017) 373–384 https://doi.org/10.1016/j.envres.2017.05.040.
[6] J. Kisser, M. Wirth, B. De Gusseme, M. Van Eekert, G. Zeeman, A. Schoenborn, et al. A review of nature-based solutions for resource recovery in cities, Blue-Green Syst. 2 (1) (2020) 138–172 https://doi.org/10.2166/bgs.2020.930.
[7] O.F. Schoumans, F. Bouraoui, C. Kabbe, O. Oenema, K.C. van Dijk, Phosphorus management in Europe in a changing world, Ambio. 44 (2015) 180–192 https://doi.org/10.1007/s13280-014-0613-9.
[8] D. Cordell, A. Rosemarin, J.J. Schröder, A.L. Smit, Towards global phosphorus security: A systems framework for phosphorus recovery and reuse options, Chemosphere 84 (6) (2011) 747–758 https://doi.org/10.1016/j.chemosphere.2011.02.032.
[9] S.M. Jasinski, Phosphate rock: statistics and information, USGS Mineral Commodities Summaries (2015).
[10] S.M. Jasinski, Phosphate Rock 2016. Mineral Commodity Summaries, US, Geologic. Survey (2016).
[11] C.M. Mehta, W.O. Khunjar, V. Nguyen, S. Tait, D.J. Batstone, Technologies to recover nutrients from waste streams: A critical review, Critic. Rev. Environ. Sci. Technol 45 (4) (2015) 385–427 https://doi.org/10.1080/10643389.2013.866621.
[12] D. Cordell, S. White, Peak phosphorus: Clarifying the key issues of a vigorous debate about long-term phosphorus security, Sustainability 3 (2011) 2027–2049 https://doi.org/10.3390/su3102027.
[13] D. Cordell, J.O. Drangert, S. White, The story of phosphorus: Global food security and food for thought, Global Environ. Change. 19 (2009) 292–305 https://doi.org/10.1016/j.gloenvcha.2008.10.009.
[14] D. Cordell, S. White, Sustainable phosphorus measures: strategies and technologies for achieving phosphorus security, Agronomy 3 (2013) 86–116 https://doi.org/10.3390/agronomy3010086.
[15] J. Bohdziewicz, E. Neczaj, A. Kwarciak, Landfill leachate treatment by means of anaerobic membrane bioreactor, Desalination 221 (2008) 559–565 https://doi.org/10.1016/j.desal.2007.01.117.
[16] E. Neczaj, A. Grosser, Circular economy in wastewater treatment plant–challenges and barriers, Proceedings 2 (2018) 614 https://doi.org/10.3390/proceedings2110614.
[17] D.J. Batstone, T. Hülsen, C.M. Mehta, J. Keller, Platforms for energy and nutrient recovery from domestic wastewater: A review, Chemosphere 140 (2015) 2–11 https://doi.org/10.1016/j.chemosphere.2014.10.021.
[18] V.K. Tyagi, S.L. Lo, Sludge: A waste or renewable source for energy and resources recovery?, Renew. Sustain. Energy Rev. 25 (2013) 708–728 https://doi.org/10.1016/j.rser.2013.05.029.
[19] J. Oladejo, K. Shi, X. Luo, G. Yang, T. Wu, A review of sludge-to-energy recovery methods, Energies 12 (2019) 60 https://doi.org/10.3390/en12010060.
[20] V.K. Tyagi, S.L. Lo, Energy and resource recovery from sludge: full-scale experiences, Environ. Mater. Waste: Resour. Recovery Pollut. Prevent. (2016) 221–244 https://doi.org/10.1016/B978-0-12-803837-6.00010-X.

[21] D.J. Batstone, B. Virdis, The role of anaerobic digestion in the emerging energy economy, Curr. Opin. Biotechnol. 27 (2014) 142–149 https://doi.org/10.1016/j.copbio.2014.01.013.

[22] J.B. Holm-Nielsen, T. Al Seadi, P. Oleskowicz-Popiel, The future of anaerobic digestion and biogas utilization, Bioresour. Technol. 100 (2009) 5478–5484 https://doi.org/10.1016/j.biortech.2008.12.046.

[23] A.S. Stillwell, D.C. Hoppock, M.E. Webber, Energy recovery from wastewater treatment plants in the United States: A case study of the energy-water nexus, Sustainability 2 (2010) 945–962 https://doi.org/10.3390/su2040945.

[24] J. Havukainen, M. Zhan, J. Dong, M. Liikanen, I. Deviatkin, X. Li, M. Horttanainen, Environmental impact assessment of municipal solid waste management incorporating mechanical treatment of waste and incineration in Hangzhou, China, J. Clean. Product. 141 (2017) 453–461 https://doi.org/10.1016/j.jclepro.2016.09.146.

[25] H.k.A. Jeswani, Incineration of municipal solid waste, Waste Manage (2016) 191–205 https://doi.org/10.1016/B978-0-12-207690-9.50021-0.

[26] H.K. Jeswani, A. Azapagic, Assessing the environmental sustainability of energy recovery from municipal solid waste in the UK, Waste Manage. 50 (2016) 346–363 https://doi.org/10.1016/j.wasman.2016.02.010.

[27] Y. Gu, Y. Li, X. Li, P. Luo, H. Wang, Z.P. Robinson, et al. The feasibility and challenges of energy self-sufficient wastewater treatment plants, Appl. Energy 204 (2017) 1463–1475 https://doi.org/10.1016/j.apenergy.2017.02.069.

[28] M. Jaeger, M. Mayer, The Noell conversion process—a gasification process for the pollutant-free disposal of sewage sludge and the recovery of energy and materials, Water Sci. Technol. 41 (2000) 37–44 https://doi.org/10.2166/wst.2000.0140.

[29] B. Cieślik, P. Konieczka, A review of phosphorus recovery methods at various steps of wastewater treatment and sewage sludge management. The concept of "no solid waste generation" and analytical methods, J. Clean. Product. 142 (2017) 1728–1740 https://doi.org/10.1016/j.jclepro.2016.11.116.

[30] E. Neyens, J. Baeyens, A review of thermal sludge pre-treatment processes to improve dewaterability, J. Hazard. Mater. 98 (2003) 51–67 https://doi.org/10.1016/S0304-3894(02)00320-5.

[31] P. Kritzer, E. Dinjus, An assessment of supercritical water oxidation (SCWO): Existing problems, possible solutions and new reactor concepts, Chem. Eng. J. 83 (2001) 207–214 https://doi.org/10.1016/S1385-8947(00)00255-2.

[32] L. Qian, S. Wang, D. Xu, Y. Guo, X. Tang, L. Wang, Treatment of municipal sewage sludge in supercritical water: A review, Water Res. 89 (2016) 118–131 https://doi.org/10.1016/j.watres.2015.11.047.

[33] D. Xu, S. Wang, J. Zhang, X. Tang, Y. Guo, C. Huang, Supercritical water oxidation of a pesticide wastewater, Chem. Eng. Res. Des. 94 (2015) 396–406 https://doi.org/10.1016/j.cherd.2014.08.016.

[34] A. Hammerschmidt, N. Boukis, E. Hauer, U. Galla, E. Dinjus, B. Hitzmann, et al. Catalytic conversion of waste biomass by hydrothermal treatment, Fuel 90 (2011) 555–562 https://doi.org/10.1016/j.fuel.2010.10.007.

[35] A. Jain, R. Balasubramanian, M.P. Srinivasan, Hydrothermal conversion of biomass waste to activated carbon with high porosity: A review, Chem. Eng. J. 283 (2016) 789–805 https://doi.org/10.1016/j.cej.2015.08.014.

[36] S. Paul, A. Dutta, Challenges and opportunities of lignocellulosic biomass for anaerobic digestion, Resourc. Conserv. Recycl. 130 (2018) 164–174 https://doi.org/10.1016/j.resconrec.2017.12.005.

[37] N. Bachmann, Sustainable biogas production in municipal wastewater treatment plants, IEA Bioenergy (2015).

[38] R. Cano, S.I. Pérez-Elvira, F. Fdz-Polanco, Energy feasibility study of sludge pretreatments: A review, Appl. Energy 149 (2015) 176–185 https://doi.org/10.1016/j.apenergy.2015.03.132.
[39] C. Burnett, A.P. Togna, A. Bohner, Y. Yang, N. Warmate, P.M. Paul Sutton, A Simple But Efficient High Performance Anaerobic Digester System, Proc. Water Environ. Feder. 2007 (2012) 1290–1297 https://doi.org/10.2175/193864707787975769.
[40] H. Monteith, Y. Kalogo, L. Fillmore, F. Schulting, C. Uijterlinde, S. Kaye, Resource recovery from wastewater solids: a global review, Proc. Water Environ. Feder. 2008 (2012) 7808–7826 https://doi.org/10.2175/193864708788808573.
[41] D. Puyol, D.J. Batstone, T. Hülsen, S. Astals, M. Peces, J.O. Krömer, Resource recovery from wastewater by biological technologies: Opportunities, challenges, and prospects, Front. Microbiol. 7 (2017) https://doi.org/10.3389/fmicb.2016.02106.
[42] M. Abu-Orf, T. Goss, Comparing thermal hydrolysis processes (CAMBI™ and EXELYS™) for solids pretreatmet prior to anaerobic digestion, Proc. Water Environ. Feder. (2014) https://doi.org/10.2175/193864712811693272.
[43] J. Chauzy, D. Crétenot, A. Gillbert, Operation of the new BIOTHELYS facility at Saumur, France, Eur. Biosolids (2007).
[44] Julien Chauzy, D. Cretenot, A. Bausseron, S. Deleris, Anaerobic digestion enhanced by thermal hydrolysis: First reference BIOTHELYS® at Saumur, France, Water Pract. Technol. 3 (2008) https://doi.org/10.2166/wpt.2008.004.
[45] D. Gary, R. Morton, C.-C. Tang, R. Horvath, The effect of the microsludge ™ treatment process on anaerobic digestion performance, Proc. Water Environ. Feder. (2012) https://doi.org/10.2175/193864707788116455.
[46] E. Neczaj, A. Grosser, Biogas production by thermal hydrolysis and thermophilic anaerobic digestion of waste-activated sludge, Industr. Municipal Sludge: Emerg. Concerns Scope Resour. Recover. (2019) 741–781 https://doi.org/10.1016/B978-0-12-815907-1.00031-3.
[47] M. Saha, C. Eskicioglu, J. Marin, Microwave, ultrasonic and chemo-mechanical pretreatments for enhancing methane potential of pulp mill wastewater treatment sludge, Bioresour. Technol. 102 (2011) 7815–7826 https://doi.org/10.1016/j.biortech.2011.06.053.
[48] R. Stephenson, S. Laliberte, P. Hoy, D. Britch, Full scale and laboratory scale results from the trial of microsludge at the joint water pollution control plant at Los Angeles County, Water Pract. 1 (2012) 1–13 https://doi.org/10.2175/193317707x243391.
[49] US Environmental ProtectionMunicipal solid waste generation, recycling, and disposal in the united states: facts and figures for 2010, AMBRA GmbH (2010) https://doi.org/EPA-530-F-14-001.
[50] S. Kelly, B. Forbes, W. Miles, M. White, Reducing SSOs without breaking the bank, WEFTEC 2012-85th Annual Technical Exhibition and Conference 2012 (2012) https://doi.org/10.2175/193864712811726734.
[51] D. Ott, Guide for municipal wet weather strategies: Overview of how the collection system fits into a holistic approach. 86th Annual Water Environment Federation Technical Exhibition and Conference, WEFTEC 2013 (2013) https://doi.org/10.2175/193864713813726830.
[52] P. Jones, A. Salter, Modelling the economics of farm-based anaerobic digestion in a UK whole-farm context, Energy Policy 62 (2013) 215–225 https://doi.org/10.1016/j.enpol.2013.06.109.
[53] A. Salter, C.J. Banks, Establishing an energy balance for crop-based digestion, Water Sci. Technol. 59 (2009) 1053–1060 https://doi.org/10.2166/wst.2009.048.

[54] M.P. Bailey, Show preview: A new flowmeter for biogas applications, Chem. Eng. (United States) (2017).
[55] C.J. Banks, A.M. Salter, M. Chesshire, Potential of anaerobic digestion for mitigation of greenhouse gas emissions and production of renewable energy from agriculture: Barriers and incentives to widespread adoption in Europe, Water Sci. Technol. 55 (2007) 165–173 https://doi.org/10.2166/wst.2007.319.
[56] Andrew Salter, C. Banks, Anaerobic digestion: overall energy balances—parasitic inputs & beneficial outputs presented, Sustain. Organic Resour. Partner. Adv. Biol. Process. Organ. Energy Recycling (2008).
[57] I.K. Kapdan, F. Kargi, Bio-hydrogen production from waste materials, Enzyme Microbial Technol. 38 (2006) 569–582 https://doi.org/10.1016/j.enzmictec.2005.09.015.
[58] A. Asghar, A.A.A. Raman, W.M.A.W. Daud, Advanced oxidation processes for in-situ production of hydrogen peroxide/hydroxyl radical for textile wastewater treatment: A review, J. Clean. Product. 87 (2015) 826–838 https://doi.org/10.1016/j.jclepro.2014.09.010.
[59] P.C. Hallenbeck, C.Z. Lazaro, E. Sagir, Biohydrogen, Comprehen. Biotechnol. 3 (2019) 128–139 https://doi.org/10.1016/B978-0-444-64046-8.00155-5.
[60] K. Seifert, R. Zagrodnik, M. Stodolny, M. Łaniecki, Biohydrogen production from chewing gum manufacturing residue in a two-step process of dark fermentation and photofermentation, Renew. Energy 122 (2018) 526–532 https://doi.org/10.1016/j.renene.2018.01.105.
[61] A. Ghimire, L. Frunzo, F. Pirozzi, E. Trably, R. Escudie, P.N.L. Lens, et al. A review on dark fermentative biohydrogen production from organic biomass: Process parameters and use of by-products, Appl. Energy 144 (2015) 73–95 https://doi.org/10.1016/j.apenergy.2015.01.045.
[62] J.X.W. Hay, T.Y. Wu, J.C. Juan, J. Md. Jahim, Biohydrogen production through photo fermentation or dark fermentation using waste as a substrate: Overview, economics, and future prospects of hydrogen usage, Biofuels Bioproduct. Biorefin. 7 (2013) 334–352 https://doi.org/10.1002/bbb.1403.
[63] T. Laurinavichene, D. Tekucheva, K. Laurinavichius, A. Tsygankov, Utilization of distillery wastewater for hydrogen production in one-stage and two-stage processes involving photofermentation, Enzyme Microbial Technol. 110 (2018) 1–7 https://doi.org/10.1016/j.enzmictec.2017.11.009.
[64] M.C. Samolada, A.A. Zabaniotou, Comparative assessment of municipal sewage sludge incineration, gasification and pyrolysis for a sustainable sludge-to-energy management in Greece, Waste Manage. 34 (2014) 411–420 https://doi.org/10.1016/j.wasman.2013.11.003.
[65] P. Manara, A. Zabaniotou, Towards sewage sludge based biofuels via thermochemical conversion—A review, Renew. Sustain. Energy Rev. 16 (2012) 2566–2582 https://doi.org/10.1016/j.rser.2012.01.074.
[66] A. Raheem, V.S. Sikarwar, J. He, W. Dastyar, D.D. Dionysiou, W. Wang, et al. Opportunities and challenges in sustainable treatment and resource reuse of sewage sludge: A review, Chem. Eng. J. 337 (2018) 616–641 https://doi.org/10.1016/j.cej.2017.12.149.
[67] S.S.A. Syed-Hassan, Y. Wang, S. Hu, S. Su, J. Xiang, Thermochemical processing of sewage sludge to energy and fuel: Fundamentals, challenges and considerations, Renew. Sustain. Energy Review. 80 (2017) 888–913 https://doi.org/10.1016/j.rser.2017.05.262.
[68] A.E. Atabani, A.S. Silitonga, I.A. Badruddin, T.M.I. Mahlia, H.H. Masjuki, S. Mekhilef, A comprehensive review on biodiesel as an alternative energy resource and its characteristics, Renew. Sustain. Energy Review. 16 (2012) 2070–2093 https://doi.org/10.1016/j.rser.2012.01.003.

[69] A.E. Atabani, A.S. Silitonga, H.C. Ong, T.M.I. Mahlia, H.H. Masjuki, I.A. Badruddin, et al. Non-edible vegetable oils: A critical evaluation of oil extraction, fatty acid compositions, biodiesel production, characteristics, engine performance and emissions production, In Renew. Sustain. Energy Review. 18 (2013) 211–245 https://doi.org/10.1016/j.rser.2012.10.013.
[70] T.M. Mata, A.A. Martins, N.S. Caetano, Microalgae for biodiesel production and other applications: A review, Renew. Sustain. Energy Review. 14 (2010) 217–232 https://doi.org/10.1016/j.rser.2009.07.020.
[71] M. Olkiewicz, A. Fortuny, F. Stüber, A. Fabregat, J. Font, C. Bengoa, Effects of pre-treatments on the lipid extraction and biodiesel production from municipal WWTP sludge, Fuel 141 (2015) 250–257 https://doi.org/10.1016/j.fuel.2014.10.066.
[72] M. Olkiewicz, A. Fortuny, F. Stüber, A. Fabregat, J. Font, C. Bengoa, Evaluation of different sludges from WWTP as a potential source for biodiesel production, Proced. Eng. 42 (2012) 634–643 https://doi.org/10.1016/j.proeng.2012.07.456.
[73] D.M. Kargbo, Biodiesel production from municipal sewage sludges, Energy Fuels 24 (2010) 2791–2794 https://doi.org/10.1021/ef1001106.
[74] M. Kumar, P. Ghosh, K. Khosla, I.S. Thakur, Biodiesel production from municipal secondary sludge, Bioresour. Technol. 216 (2016) 165–171 https://doi.org/10.1016/j.biortech.2016.05.078.
[75] J.A. Melero, R. Sánchez-Vázquez, I.A. Vasiliadou, F. Martínez Castillejo, L.F. Bautista, J. Iglesias, et al. Municipal sewage sludge to biodiesel by simultaneous extraction and conversion of lipids, Energy Conversion Manage. 103 (2015) 111–118 https://doi.org/10.1016/j.enconman.2015.06.045.
[76] M. Husseien, A.A. Amer, A. El-Maghraby, N.A. Taha, Availability of barley straw application on oil spill clean up, Int. J. Environ. Sci. Technol. 6 (2009) 123–130 https://doi.org/10.1007/BF03326066.
[77] J. Kim, H.J. Kim, S.H. Yoo, Public value of marine biodiesel technology development in South Korea, Sustainability (Switzerland) 10 (2018) 4252 https://doi.org/10.3390/su10114252.
[78] B.H. Um, Y.S. Kim, Review: A chance for Korea to advance algal-biodiesel technology, J. Industr. Eng. Chem. 15 (2009) 1–7 https://doi.org/10.1016/j.jiec.2008.08.002.
[79] H. Wang, H. Luo, P.H. Fallgren, S. Jin, Z.J. Ren, Bioelectrochemical system platform for sustainable environmental remediation and energy generation, Biotechnol. Adv. 33 (2015) 317–334 https://doi.org/10.1016/j.biotechadv.2015.04.003.
[80] F. Zhang, J. Li, Z. He, A new method for nutrients removal and recovery from wastewater using a bioelectrochemical system, Bioresour. Technol. 166 (2014) 630–634 https://doi.org/10.1016/j.biortech.2014.05.105.
[81] R.A.A. Meena, R. Yukesh Kannah, J. Sindhu, J. Ragavi, G. Kumar, M. Gunasekaran, et al. Trends and resource recovery in biological wastewater treatment system, Bioresour. Technol. Rep. 7 (2019) 100235 https://doi.org/10.1016/j.biteb.2019.100235.
[82] L. Rago, Y. Ruiz, J.A. Baeza, A. Guisasola, P. Cortés, Microbial community analysis in a long-term membrane-less microbial electrolysis cell with hydrogen and methane production, Bioelectrochemistry 106 (2015) 359–368 https://doi.org/10.1016/j.bioelechem.2015.06.003.
[83] M. Badia-Fabregat, L. Rago, J.A. Baeza, A. Guisasola, Hydrogen production from crude glycerol in an alkaline microbial electrolysis cell, Int. J. Hydrogen Energy 44 (2019) 17204–17213 https://doi.org/10.1016/j.ijhydene.2019.03.193.
[84] D. Cecconet, M. Devecseri, A. Callegari, A.G. Capodaglio, Effects of process operating conditions on the autotrophic denitrification of nitrate-contami-

nated groundwater using bioelectrochemical systems, Sci. Total Environ. 613-614 (2018) 363–371 https://doi.org/10.1016/j.scitotenv.2017.09.149.

[85] N.Y. Acelas, E. Flórez, D. López, Phosphorus recovery through struvite precipitation from wastewater: effect of the competitive ions, Desalin. Water Treat. (2015) 2468–2479 https://doi.org/10.1080/19443994.2014.902337.

[86] O. Krüger, K.P. Fattah, C. Adam, Phosphorus recovery from the wastewater stream—necessity and possibilities, Desalin. Water Treat. (2016) 15619–15627 https://doi.org/10.1080/19443994.2015.1103315.

[87] J. You, J. Greenman, C. Melhuish, I. Ieropoulos, Electricity generation and struvite recovery from human urine using microbial fuel cells, J. Chem. Technol. Biotechnol. 91 (2016) 647 https://doi.org/10.1002/jctb.4617.

[88] M. Liao, Y. Liu, E. Tian, W. Ma, H. Liu, Phosphorous removal and high-purity struvite recovery from hydrolyzed urine with spontaneous electricity production in Mg-air fuel cell, Chem. Eng. J. 391 (2019) 123517 https://doi.org/10.1016/j.cej.2019.123517.

[89] T. Pepè Sciarria, G. Vacca, F. Tambone, L. Trombino, F. Adani, Nutrient recovery and energy production from digestate using microbial electrochemical technologies (METs), J. Clean. Product. 208 (2019) 1022–1029 https://doi.org/10.1016/j.jclepro.2018.10.152.

[90] P. Zamora, T. Georgieva, A. Ter Heijne, T.H.J.A. Sleutels, A.W. Jeremiasse, M. Saakes, et al. Ammonia recovery from urine in a scaled-up microbial electrolysis cell, J. Power Source. 356 (2017) 491–499 https://doi.org/10.1016/j.jpowsour.2017.02.089.

[91] J.M. Lema, S. Suarez, Innovative wastewater treatment & resource recovery technologies: impacts on energy, Econ. Environ. Water Intell. Online. (2017) https://doi.org/10.2166/9781780407876.

[92] E.M. Sander, B. Virdis, S. Freguia, Bioelectrochemical nitrogen removal as a polishing mechanism for domestic wastewater treated effluents, Water Sci. Technol. J. Int. Assoc. Water Pollut. Res. 76 (2017) 3150–3159 https://doi.org/10.2166/wst.2017.462.

[93] European Sustainable phosphorus Platform. Avalilable from: https://phosphorusplatform. eu/scope-in-print/news/681-struvite-sales-success-in-uk

[94] M. Papa, P. Foladori, L. Guglielmi, G. Bertanza, How far are we from closing the loop of sewage resource recovery? A real picture of municipal wastewater treatment plants in Italy, J. Environ. Manage. 198 (2017) 9–15 https://doi.org/10.1016/j.jenvman.2017.04.061.

2

Increasing efficiency of mining enterprises power consumption

Vadim Z. Manusov[a], Dmitry V. Antonenkov[a], Evgenia Y. Sizganova[b], Denis B. Solovev[c,d]

[a]Department of Power Supply Systems of Enterprises, Novosibirsk State Technical University, Novosibirsk, Russian Federation; [b]Department of Electrotechnical Complexes and Systems, Siberian Federal University, Krasnoyarsk, Russian Federation; [c]Far Eastern Federal University, Engineering School, Vladivostok, Russian Federation; [d]Vladivostok Branch of Russian Customs Academy, Vladivostok, Russian Federation

Chapter outline

1 Significance

The concept of energy efficiency implies the rational use of energy resources. Various measures can be taken for this, including but not limited to replacing obsolete equipment, using secondary energy resources, creating conditions for lower energy consumption, and controlling energy consumption. The implementation of these measures is financially beneficial because lower resource consumption leads to a decrease in the services-associated costs.

For any organization, the starting point for the development of an energy-efficiency program is audit, as it shows a time-fixed picture of the current state of energy efficiency for analysis of the use of energy resources, its costs, identification of bottlenecks for rational use of resources, and development of a program of energy-saving measures and projects.

Low Carbon Energy Technologies in Sustainable Energy Systems. http://dx.doi.org/10.1016/B978-0-12-822897-5.00002-X

Due to the steady growth trend in energy prices, including electricity, the issue of energy savings is very relevant and important for mining enterprises. The rated capacity of electrical equipment in a modern electrical profile of the mining enterprise tops 20 MW, which is comparable in capacity to large industrial enterprises. Unlike industrial enterprises, the electrical profile of a mining enterprise is dispersed over a large territory and changes its structure over time.

One of the most important analytical procedures for rank analysis—interval estimation of a parametric ranked distribution—can serve as a method for determining, as part of an electrical profile of a mining enterprise, units that have been abnormally consuming electricity and therefore require a priority energy inspection. Energy-saving goals coincide with other goals of the mining enterprise, such as improving the environmental situation, increasing the efficiency of energy supply systems, etc.

2 The degree of elaboration of the issue

The laws of the development of technology consisting of individual elements, and wildlife consisting of individual beings, have much in common [1,2]. Therefore, it seems possible to describe the electrical profile of mining enterprises on the basis of cenological patterns and rules as an object of technocenosis.

In the study of the properties of individual beings and communities of living organisms in 1877, Karl August Möbius introduced the concept of "biocenosis." Biocenosis is a set of living organisms that live in a particular area, where environmental conditions determine its species composition. The term "technocenosis" and the cenological approach, where technocenosis is defined as a community of all products, including all populations, limited in space and time, were proposed in 1974 by B.I. Kudrin [3,4].

Similar studies of such complexity are considered in other areas of science [3,4], in a study of various systems: physical (inorganic world [5]), biological (organic world), technical, informational, and social.

In biology, a quantitative study of species by repeatability apparently began with the work by S. Gartside (1928). Without making a mathematical interpretation [6], he argued that despite the large number of zooids that characterize some species, a large number of species are represented by a relatively small number of zooids, this is confirmed by C. Darwin [7].

R. Fisher proposed a logarithmic series to describe the distribution of the number of species by repeatability; with suitable approximation, it can be applied to a wide range of values [8]. Previously, A. Corbet obtained a hyperbolic dependence, which R. Fisher considered as the limiting case [6]. Attempts to describe the experimental data by other distribution laws are known. Preston [9] proposed a lognormal law to describe the distribution of species by the number of zooids. MacArthur [10] investigated a number of models based on simple logical premises, later they were widely quoted and refined. And although MacArthur acknowledged his approach as imperfect [11], his merit lies in abandoning a purely descriptive method. Waterson [12] proposed three models of biocenosis and showed a relationship between the number of species and the number of zooids. In all these models, there is undoubtedly a dependence qualitatively described by S. Gartside.

From the methodological viewpoint, information systems are considered completely in the works [13–17]. There are many varieties of Zipf's law and many supporting examples (the distributions and laws of Mandelbrot, Lotka, Pareto, Bradford, Yule, Willis, Estoup). The universal nature of Zipf's law for science studies, computer science, linguistics, and other areas related to human behavior (information systems), has caused many hypotheses attempting to explain the revealed pattern.

Zipf [18] explained the stability found in the analysis of the vocabulary by the fact that a person always economizes on own efforts. This point of view goes back to physics (the principle of least action). The linguistic formulation of this principle is as follows: human communication consists of the speaker's desire to be understood and the listener's—to understand with a minimal effort. This representation may be useful in a substantive study of the quantitative results of human intellectual activity [17]. In thermodynamic models of energy and entropy, there is an analogue, too: the concept of complexity [13], the number of letters in a word [19], and the efforts required for publication [17]. Also promising is the possibility of a formal representation of Bose-Einstein statistics to describe the distribution of elements in a stable system [20].

A probabilistic and statistical interpretation of Zipf's law has become widespread: the existence of invariant properties of elements that form an integrity, which can be distinguished, but whose determination is difficult, is assumed. It is this method that is the basis of the dictionaries [16].

Thus, biological and information systems are described using a methodologically identical approach: the distribution by repeatability of elements that differ from each other. The construction of technocenoses is similar to the formation of biocenoses, the patterns of creating texts, arrays of scientific publications, and other aggregates described by the Zipf's law. The drawn conclusion logically follows from the nature of system research: technocenosis is a set of products with relations and connections between them, which forms a certain hierarchically organized entirety. Like other natural systems—biological and informational, it has a common, systemic content, which can be described by the distribution of species by repeatability.

For each product as a species, the limit values of all physico-chemical factors can be compiled, resulting in formation of an n-dimensional volume—an ecological niche. The ecological niche of A type is the volume in n-dimensional space determined by the minimum and maximum values of environmental factors that ensure the survival of the species. The ecological niche is conceived in the framework of a single four-dimensional spatio-temporal macrocosmic structure, that is, it is fixed in space and time. But the ecological niche is not a place in the Euclidean space, rather it is a description model (a similar concept is accepted in biology [21,22]).

A number of scientific studies [23–27] have shown the need to implement scientifically based approaches in the issue of determining the energy saving potential. In the works [26,28–32], the main procedures of this method are: creating a database, interval estimation, forecasting and rationing. It is proposed to supplement these with another key procedure—exponentiation [33,34]. Exponentiation is determining the resource's integral volume by which the resource consumption should be reduced for a given analyzed period of time, without affecting the normal functioning of systems [26,31,32]. The method of optimal power management implies exponentiation as a procedure for determining the energy saving potential, by the value of which the energy consumption should be reduced without affecting the normal functioning of electric consumers in a given time interval [26,31,32]. The concept of energy-saving potential is a key element in the exponentiation procedure. The energy-saving potential (or the potential of an energy-saving system) that is the absolute difference between the energy consumption in a technocenosis without implementing the saving procedure calculated over a period of time, on the one hand, and energy consumption, which corresponds to the lower boundary of the variable confidence interval, on the other. Power consumption is calculated as inseparable, from zero to infinity, under the corresponding curve ranked parametric distribution. The unit of calculation is either the curve obtained for the empirical values of electricity consumption or the lower boundary of the confidence interval variable. The time interval of the calculation is determined, on the one hand, by the depth of the database on past electricity consumption, which serves as a base for the confidence interval variable, and, on the other hand, the necessary horizon for modeling the potential in the future [26,30,31,35]. The proposed approach is fundamentally different from the traditional approach, when the energy-saving potential is understood as the sum of the differences between the actual energy consumption and some hypothetical value of the energy consumption for each electrical installation and achieves the best indicators of energy efficiency.

In paper [36], it was found that the power consumption of the electrical profile of a mining enterprise (using coal mining enterprise as an example) is an element of a technocenological type system (technocenosis). This finding was used to reason the feasibility of a technocenological approach to solving the problem of determining the energy-saving potential of an electrical profile. It was established that the empirical distribution of power consumption belongs to the type of stable ranked parametric H-distributions, for the purpose of forecasting these distributions on the basis of extrapolation of the parameters. Based on the foregoing, it follows that the use of a rank analysis algorithm to determine the energy-saving potential will allow taking into account the particular features of a mining enterprise with its individual administrative properties as an object of a technocenological type.

3 Theoretical part

In this article, the subject of the research is to determine the potential value for energy saving of electricity consumption, which will be used at a given time interval to reduce the power consumption of a mining enterprise unit (e.g., coal mining) without compromising its normal functioning.

Energy-saving potential is the absolute difference obtained for the estimated time period between the power consumption of a mining enterprise unit (technocenosis) without implementing energy-saving procedures, on the one hand, and power consumption corresponding to the lower boundary of the variable confidence interval, on the other.

To emphasize the fundamental difference discussed earlier, a new system potential is proposed—the one that is calculated using Zipf distributions known from rank analysis, as follows:

$$\Delta W_1 = \int_0^\infty W(r)dr - \int_0^\infty W_1(r)dr, \tag{2.1}$$

where ΔW_1 is the energy saving potential of mining enterprise; $W(r)$ is the approximation curve obtained for empirical values of power consumption of units; $W_1(r)$ is the lower boundary of the variable confidence interval, obtained on the basis of database processing; and r is the rank of mining enterprise unit.

As a data source, a database on electricity consumption (earlier data) is used. Interval estimation is carried out in order to determine the boundaries of a variable confidence interval for the exponentiation vector. The lower boundary of the variable confidence interval is a hyperbolic curve obtained by approximating the lower boundaries of the 95% confidence intervals calculated for each of the ranks of the ranked parametric distribution.

The construction of a confidence interval based on the values of ranked power consumption allows one to take into account the systemic influence of technocenosis on units and vice versa, as well as the multiple mutual influence of units on each other. The analysis performed for a significant number of dissimilar objects of various technocenoses from various areas made it possible to confirm the earlier assumption about the normal distribution of energy consumption values within ranks, which enables one to construct a confidence interval for each rank based on empirical data for a number of time intervals [37,38].

As is known, if Θ^* serves as an estimate of an unknown parameter Θ, then the interval, which covers an unknown parameter with a given reliability γ is called confidence interval $[\Theta^* - \delta, \Theta^* + \delta]$ [37,38]:

$$P\,[\Theta^* - \delta < \Theta < \Theta^* + \delta] = \gamma, \tag{2.2}$$

where δ is the accuracy of assessment.

The absolute difference, obtained by modeling for the estimated time period between the power consumption of a mining enterprise unit (technocenosis) without implementing energy-saving procedures, on the one hand, and power consumption corresponding to the lower boundary of the variable confidence interval, on the other. The electricity consumption of a mining enterprise (technocenosis) is calculated as an integral ranging from zero to infinity under the curve of ranked parametric distribution or under the boundary of the interval. The energy-saving potential has structural properties, understood as the presence of levels of Z1 and Z2 potential, which have time-stable boundaries and are determined by the probabilistic patterns existing in the system and generated by the complex process of the influence of the technocenosis on units and each of the units on the technocenosis as a whole, according to the expression (2.2). Therefore, in the process of calculating the Z1 and Z2 energy-saving potentials of the technocenosis, expressions similar to (2.1) are used [24,28,29,37,38]:

$$\begin{cases} \Delta W_1 = \int_0^\infty W(r)dr - \int_0^\infty W_1(r)dr; \\ \Delta W_2 = \int_0^\infty W(r)dr - \int_0^\infty W_2(r)dr, \end{cases} \tag{2.3}$$

where ΔW_1 is Z1—energy-saving potential of units of mining enterprise; ΔW_2 is Z2—energy-saving potential of units of mining enterprise; $W(r)$ is approximation curve obtained for empirical values of power consumption of units; r is the rank of mining enterprise unit; $W_1(r)$ is ranked parametric distribution corresponding to the lower boundary of the variable confidence interval obtained in the interval estimation procedure based on the matrix of previous data; $W_2(r)$ is the ranked parametric distribution corresponding to the lower boundary of the variable confidence interval obtained in the interval estimation procedure after ZP-normalization.

Let us consider the standardization procedure, the essence of which is to recalculate the energy consumption of units within the functional groups of the mining enterprise units, based on real-life electricity consumption schedules and the best intragroup electricity consumption indicators, which, in turn, allows one to calculate a new variable confidence interval, where the lower limit will be used in the assessment of Z2-potential.

4 Solution method

Energy consumption of a mining enterprise is influenced by a large number of geological, technological, climatic, and organizational factors.

Strip mining technology has adopted a transportation-based system of development using large-capacity mechanical excavating machinery and dragline strippers with loading in heavy-duty 120–218 t dump trucks.

To obtain preliminary information, the analysis of Y data: 2018–19 power consumption of a mining enterprise was carried out, with consideration of the influence of the following factors:

1. x_1 – emergency shutdowns of 35 kV power lines;
2. x_2 – precipitation rate;
3. x_3 – average outdoor temperature; and
4. x_4 – idle time without 35 kV line voltage.

Factor analysis is presented in Table 2.1.

Pictures of Fig. 2.1 show charts of power consumption and the main technological indicators of the mining enterprise.

The dependence of electricity consumption Y on a number of technological indicators of the mining enterprise's production and the degree of dependence are presented in Table 2.2, namely: (1) X_1 — production volume, m^3; (2) X_2 — strip mining volume, m^3; (3) X_3 — bulk, m^3; (4) X_4 — re-excavation, m^3; (5) $X_5 = X_1 + X_2 + X_3 + X_4$ — run of mine, m^3.

Table 2.1 Matrix of paired coefficients of factor correlation.

	Y	x_1	x_2	x_3	x_4
Y	1	0.24	−0.27	−0.33	0.094
x_1	0.24	1	0.107	−0.32	−0.247
x_2	−0.027	0.107	1	0.205	0.107
x_3	−0.33	−0.32	0.205	1	0.332
x_4	0.094	−0.247	0.107	0.332	1

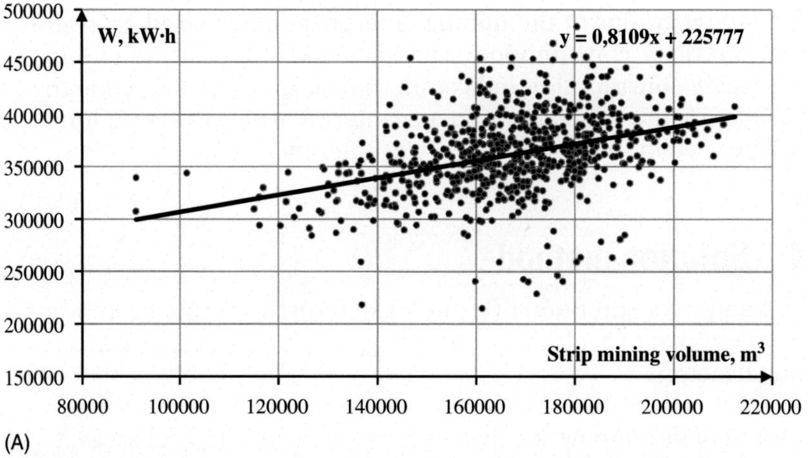

(A)

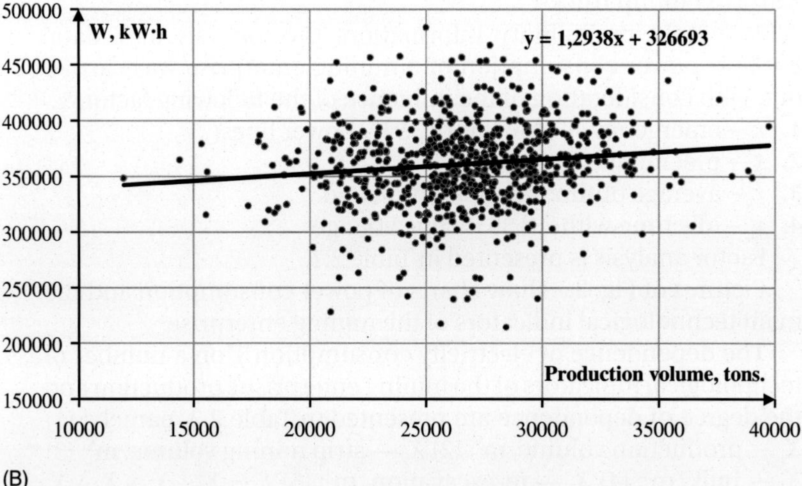

(B)

Figure 2.1. Dependence of power consumption on the technological indicators of a mining enterprise. (A) Volumes of strip mining, m³, (B) Coal production, tons.

Table 2.2 Correlation coefficients of technological indicators.

	Y	X_1	X_2	X_3	X_4
Y	1	0.13	0.35	0.35	0.03
X_1	0.13	1	0.08	0.02	−0.05
X_2	0.35	0.08	1	0.14	−0.09
X_3	0.35	0.02	0.14	1	0.06
X_4	0.03	−0.05	−0.09	0.06	1

Table 2.3 Result of the calculation of regression equations.

Description	Type of equation $Wa = f(X)$	Determination coefficient (R^2)
$W = f(X_1)$	$W = 1.293X_1 + 32{,}669$	0.016
$W = f(X_2)$	$W = 0.81X_2 + 22{,}577$	0.139
$W = f(X_5)$	$W = 0.872X_5 + 19{,}058$	0.18

The results of calculation of paired dependencies between the considered technological indicators are given in Table 2.3.

Modeling based on regression equations can be reduced to construction of a single equation with the inclusion of a large number of indicators (factors) or to construction of a system of equations. Based on the foregoing, we obtain a generalized model (2.4):

$$Y = 189540{,}4 + 1{,}4X_1 + 0{,}7X_2 + 4{,}2X_3 + 0{,}6X_4 \tag{2.4}$$

The value of the result (of power consumption) is influenced by all technological indicators (the more volumes, the greater the power consumption of the mining enterprise). If we consider the coefficient of multiple determination, it is important to note that $R^2 = 0.22$. This indicator is almost the lowest. This indicator is used to determine the number of unaccounted factors, which is 80%.

Analysis of these factors shows that they are all difficult to predict and virtually uncontrollable. In such cases, special methods should be used to determine the power consumption parameters, taking into account the fuzziness and ambiguity of the source data.

Fundamentally new idea is that all the units of the mining enterprise are included in one higher-level system and are subordinate to the Management Office, which undoubtedly has a systemic effect on all of its units through planning and limiting system, as well as administration system. Therefore, the totality of units (facilities) is a community of loosely linked and weakly interacting units of the mining enterprise (technocenosis). The connections are weak, but, nevertheless, they exist.

The mining enterprise units subordinate to the Management Office and are engaged in various types of activities aimed at single final result: coal mining and beneficiation. A high concordance coefficient (equal or greater than 0.8) for the aggregate data on the power consumption of mining enterprise units indicates the stability of the ranking surface and colligation as a whole. In addition, these units are limited in space and time; therefore, a mining enterprise can be classified as a technocenosis.

To identify trends in electricity consumption, it is necessary to analyze the power consumption structure of the mining enterprise. A rank analysis of the power consumption dynamics over time involves the construction of a mathematical model. Rank analysis allows one to take into account the particular features of a mining enterprise with its individual technological properties as an object of cenological research. Since the power consumption parameters of the mining enterprise are constantly changing under the influence of many different factors, the H-model allows one to adequately describe the power consumption of the studied units [3,4].

The staff of the departments "Power supply systems of enterprises" at Novosibirsk State Technical University and "Electrical complexes and systems" at the Siberian Federal University conducted an analysis of the dynamics of power consumption of units of a large mining enterprise (Fig. 2.2). The power consumption

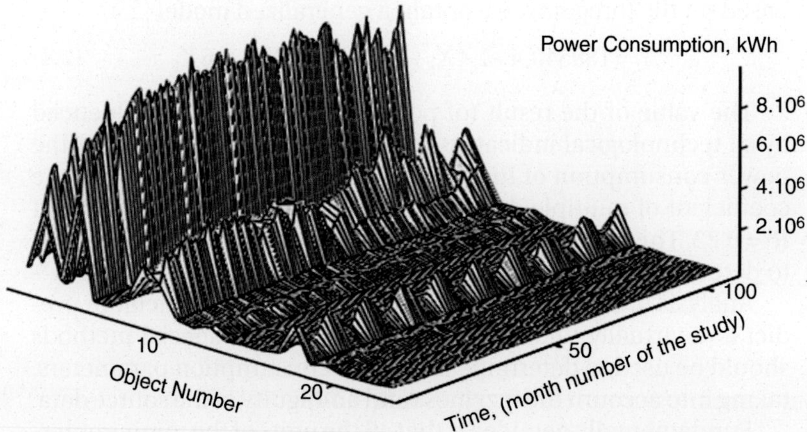

Figure 2.2. Dynamics of power consumption of mining enterprise objects.

statistics for the period from 2011 to 2019 with monthly payments to the electricity supplier were used. Analyzing and taking into account the influence of all direct and indirect factors that can affect the level of power consumption of an individual unit of a mining enterprise is virtually impossible due to the lack of numerical values for most of them. All of this overly complicates the model.

For subsequent statistical processing of data, approximation of power consumption dynamics is of great importance. It will be used to select the analytical dependence that best describes the set of points. The approximation can be carried out by various methods; in this case, the approximation is performed by the least squares method using three functions:

Exponential

$$y = b_0 x^{b_1}; \tag{2.5}$$

Logarithmic

$$y = b_1 \ln(x) + b_0; \tag{2.6}$$

Polynomial of the sixth degree

$$y = b_6 x^6 + b_5 x^5 + b_4 x^4 + b_3 x^3 + b_2 x^2 + b_1 x^1 + b_0. \tag{2.7}$$

The calculation results are presented in Table 2.4.

The accuracy of the mathematical model can be estimated by the coefficient of determination (the value of the reliability of the approximation):

$$R = 1 - \frac{\sum_{i=1}^{N} \left(W_i - W_{\phi i} \right)^2}{\sum_{i=1}^{N} \left(W_{\phi i} - W_0 \right)^2}. \tag{2.8}$$

Table 2.4 Approximation results of power consumption dynamics (December 2018).

Function	Equation coefficients	Approximation reliability
Exponential	$b_0 = 4 \cdot 10^6$, $b_1 = -1.44$	0.277
Logarithmic	$b_0 = 4 \cdot 10^6$, $b_1 = -1$	0.251
Polynomial	$b_0 = -2 \cdot 10^6$, $b_1 = 6 \cdot 10^6$, $b_2 = -2 \cdot 10^6$, $b_3 = 25109$, $b_4 = -16537$, $b_5 = 532{,}7$, $b_6 = -6{,}709$	0.352

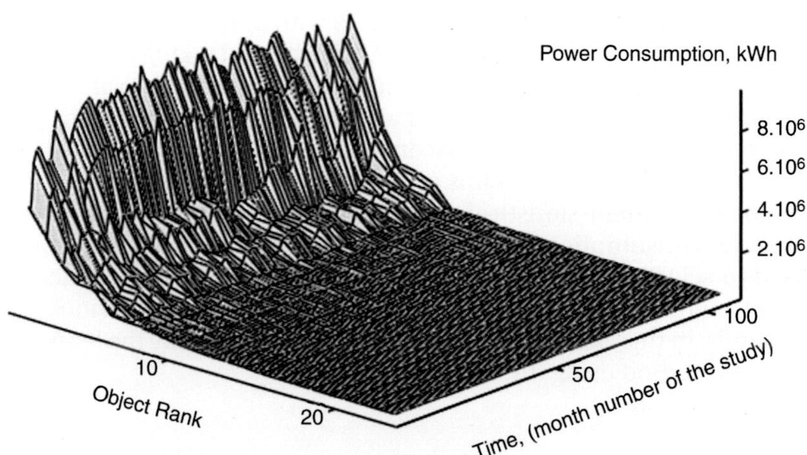

Figure 2.3. Three-dimensional ranking surface of power consumption of mining enterprise objects.

The closer R is to 1, the greater part of the initial change in power consumption this model is able to describe.

Analyzing the values presented in Table 2.4, we can conclude that the reliability of the approximation is low, which prohibits us from using the obtained mathematical models to analyze the dynamics of power consumption with acceptable accuracy.

At the same time, if we rank the mining enterprise units by the values of power consumption in descending order (Fig. 2.3) and apply the same mathematical apparatus to analyze the resulting ranked distribution, then the quality of the regression models increases significantly and the accuracy of the approximation tends to one.

Table 2.5 shows the results of approximating the ranked distribution of objects according to the values of electricity consumption (December 2018).

Table 2.5 Result of the approximation of ranking distribution of objects.

Function	Equation coefficients	Approximation reliability
Exponential	$b_0 = 5 \cdot 10^7,\ b_1 = -2.61$	0.903
Logarithmic	$b_0 = 7 \cdot 10^6,\ b_1 = -2$	0.813
Polynomial	$b_0 = 1 \cdot 10^7,\ b_1 = -5 \cdot 10^6$ $b_2 = 73084,\ b_3 = -57737,$ $b_4 = 2438,\ b_5 = -51{,}5,$ $b_6 = 0.417$	0.987

According to the established tradition, researchers engaged in rank analysis choose the two-parameter hyperbolic mathematical model (2.9) as the standard, described in [3,4,38]:

$$W = \frac{W_1}{r^{\beta}}, \tag{2.9}$$

where W_1 is the maximum power consumption; β is the rank coefficient.

Of course, this form is far from perfect, but its indisputable advantage is the determination of only two parameters in approximation: W_1 and β.

For the mining enterprise units, the regression coefficients of all distributions for the considered time period were calculated. An analysis of the parameters of the rank distribution and the nature of their changes are the criteria for stability of the infrastructure of a mining enterprise as a technocenosis-type system. By a change in β, one can estimate the state of technocenosis. The analysis of the time series β(t) (Fig. 2.4) makes it possible to determine whether the selected population is a technocenosis developing in a balanced and predictable way.

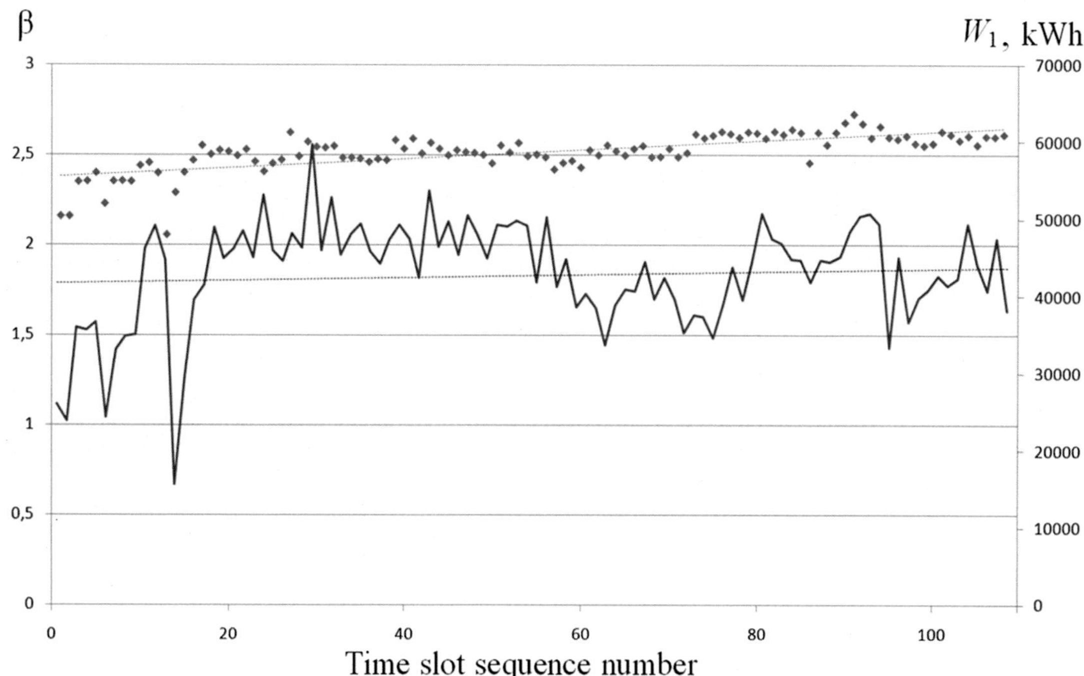

Figure 2.4. Dynamics of changes in ranking distribution parameters.

The trend of the time series of the rank coefficient β reflects the tendency in energy consumption:

1. an increase in the coefficient β allows one to conclude about a growing gap between large and small (in terms of energy consumption) units of the mining enterprise;
2. the constancy of the coefficient β means maintaining the ratio of large and small units of the mining enterprise; and
3. a decrease in the coefficient β characterizes a reduction in the gap between large and small units of the mining enterprise.

In the first and third cases, the relationship between large and small elements of the technocenosis is violated, which indicates a possible violation of its stability.

For more reasonable decision-making regarding the technical system under study as a whole, it is necessary, together with the trends in the time series of rank coefficients β, to study the trends in the time series of the so-called first points (in this case, units with maximum power consumption W_1).

For the mining enterprise under consideration, analysis of the time series of the rank distribution parameters shows that the values of the first point and the rank coefficient, although insignificantly, are growing (Fig. 2.4); the gap in the level of energy consumption between large and small units of the mining enterprise is growing. This means that large units have a tendency to increase energy consumption, and they need to be paid attention to, for a more detailed inspection.

One of the analytical procedures for rank analysis is the interval estimation of the parametric distribution [3,4,38]. It allows one to determine which of the units abnormally consume the resource. With regard to the power consumption of the mining enterprise: if the point on the rank distribution falls within the confidence interval, then, within the Gaussian spread of the parameters, it can be argued that this unit consumes electricity normally; if the point is below the confidence interval, then this, as a rule, indicates a violation of the normal technological process of power consumption at the given unit (frequent power outages, non-payments, excessive savings, etc.); if the point is above the confidence interval, then an abnormally high energy consumption takes place at the corresponding unit (Fig. 2.5). An in-depth energy inspection should focus on precisely these units. The consistent implementation of this methodology over a number of years ensures focused action on the most "weak" units. At the same time, funds provided for conducting energy surveys will be spent most efficiently, and the total energy

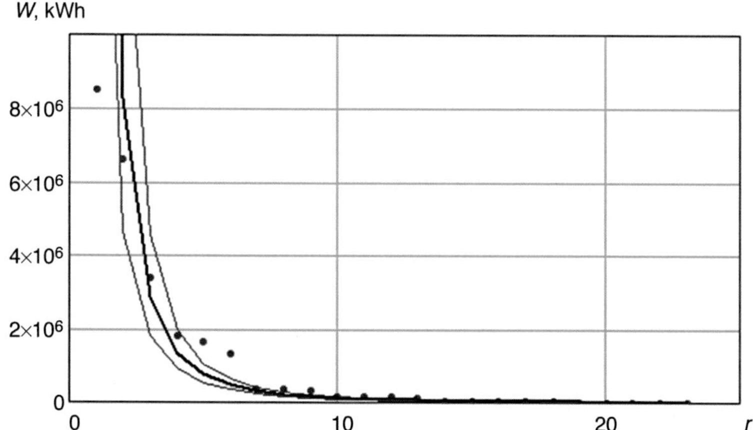

Figure 2.5. Confidence interval for the entire ranked parametric distribution. •••• – empirical data; ——– approximation curve; ——– upper and lower confidence limits

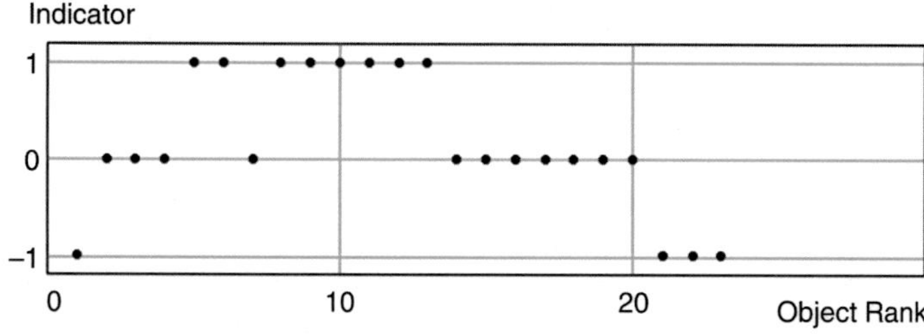

Figure 2.6. Diagram illustration of points falling into the confidence interval.

consumption of the mining enterprise will be constantly edging down [38,39].

Let us determine the number of points located above and below the confidence interval, as well as those that fall within the interval, using an indicator that takes the value of 0, 1 or –1, if the points are respectively inside, above or below the confidence interval (Fig. 2.6). It can be seen that 4 points lie below the confidence interval, 11 points fall within the confidence interval and 8 points lie above the confidence interval.

To get a list of audit priority for the mining enterprise units, the relative deviation value of points from the confidence interval boundaries is determined. Then the units were ranked by this parameter and the unit numbers were obtained in the initial database on power consumption (Table 2.6).

Table 2.6 List of audit priorities for mining enterprise units.

Queue no.	Unit numbers in the initial database	Facility
1	1	Coal intake
2	9	Strip mining unit
3	12	Explosives production unit
4	11	Road unit
5	20	Shop floor communications office
6	13	Specialized installation and commissioning site
7	10	Drilling site
8	19	Logistics management office
9	4	Drying and heating shop
10	7	Production unit
11	5	Administrative-household complex of the plant
12	16	Administrative-household complex of the surface mine

5 Discussion of the results

The interval estimation procedure allowed us to identify and grade the units of the mining enterprise, revealing those with abnormal electricity consumption. If their count is relatively large, in our case more than half, then additional research is required to determine the units where the audit is a priority for the infrastructure as a whole.

One of the key procedures for managing a technocenosis [3,4,38] is the timely identification of priority units for an in-depth energy inspection. For this, the so-called assessment of healthy energy consumption of the units is carried out [37,38,40–42]. In addition, this procedure allows to implement an additional study of the mining enterprise units, reaching conclusions regarding the dynamics of their overall infrastructure development [37].

When deciding an in-depth energy survey of the mining enterprise units as a key macro-indicator, it is necessary to take into account their health index [37,43]. This is represented by the angle between the X-axis and the trend line of the time series of the relative power consumption of the unit, which is defined as the ratio of the power consumption of this unit to the total power consumption of the technocenosis. As shown in [37,43], the value and sign of the health index allows one to draw conclusions regarding the dynamics of the development (or stagnation) of a unit, as well as to determine its number in the list of priority units for an in-depth energy inspection.

The power consumption ratios of each of the units to the total power consumption of the mining enterprise are calculated. The calculation results in Fig. 2.7 are displayed as a percentage. It should be noted that here, as an illustration, charts are given only for units with numbers 1, 9, 12, and 19, with the topmost (red) graph corresponding to No. 9, then No. 1, then No. 12, and then No. 19, the lowest.

The trend of time series of relative power consumption for each of the mining enterprise units is calculated. As has been noted, the angle of the trend to the X-axis reflects the dynamics of changes in the share of power consumption of each unit in the total power consumption of the mining enterprise, thus serving as a macro-indicator of the health of the unit in terms of power consumption. The approximating straight line $Y = ax + b$ determined by the least squares method is used as a trend. The a coefficient in the equation of the approximating trend line equals to the tangent of the slope of this line to the X-axis. Thus, to find the macro-indicator of the health of a unit, it is necessary to use the formula $\Theta = arctg(a)$. The calculation results are demonstrated on the example of the unit No. 9 in Fig. 2.8, amounting to Θ = –6.324. For clarity of illustration of the angle, a horizontal *red line* conjugated with the trend *(black line)* is added. As a result, the vector of the health index of each mining enterprise unit and the total health index of all units are determined.

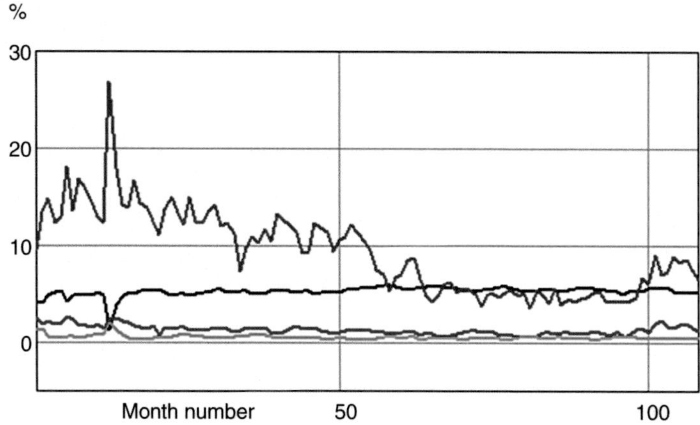

Figure 2.7. Time series of relative power consumption of mining enterprise units.

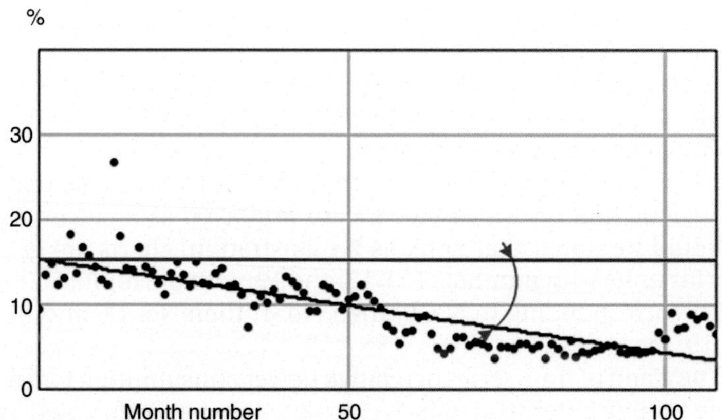

Figure 2.8. Macro-indicator reflecting the dynamics of power consumption of unit No. 9 (Θ = –6.324).

6 Conclusion

The conditions for the economic motivation to use resources in a cost-effective manner are a prime requirement if an organization needs to boost the efficiency of energy-saving technologies implementation, regardless of its purpose and type of activity. Energy-saving approach must be applied both at the technical level, and at the system level. The basis of energy saving is the systematic implementation of a set of technical and technological measures, which should be preceded by the optimization of organizations' energy consumption at the system level. The goal, which is to save money earmarked for paying for consumed energy resources, is obtained through organizational measures, as well as by creating scientifically sound preconditions for conducting targeted energy inspections, with the subsequent implementation of technical measures for energy conservation.

A study of the energy consumption for infrastructure of a mining enterprise begins with the collection of statistical information about the enterprise's facilities. By itself, this operation is extremely time-consuming and requires many years of systematic effort by the whole team. However, after completing the collection of information, there comes an equally difficult stage of deep statistical processing of the obtained data. Modern computer software ensures fast, efficient, and correct processing of data of any volume and complexity.

At the system level, improving the energy efficiency of mining enterprise units is carried out using the optimal energy management methodology and includes procedures for creating a database, interval estimation, forecasting, and rationing of energy consumption. The basis of the methodology is a dynamic model that allows one to reflect the process of power consumption of the coal production units for a time interval of up to 5 years or more.

Let us consider the prospects for the development of the theory of optimal energy management of a mining enterprise. This theory was developed in four stages, the first of which created a static model of optimal energy management. The standard procedures of the model were the interval estimation, forecasting, rationing, and potentiation, which, for clarification purposes, at the second stage were supplemented by fine-tuning procedures of difflex analysis (at the stage of estimating the interval), GZ analysis (at the forecasting stage), ASR analysis (at the stage rationing) and ZP analysis (at the stage of potentiation). The static model ensured a rather good handling of the problems of managing the energy consumption of a mining enterprise at small forecasting horizons, but it had a number of drawbacks, the main ones being the lack of internal control feedbacks and the underdevelopment of the criteria-based efficiency assessment system.

At the third stage of the theory development, the static model was supplemented with a data stream for feedback and transformed into a dynamic adaptive model, where the energy consumption process was reflected using transformation functions built on the basis of the Weibull-Gnedenko distribution law and the normal distribution law. As a performance criterion, the target functionality was used, based on the ratio of relative integral indicators of quality and the costs, as well as on the system of restrictions that are a consequence of the law of the optimal construction of technocenoses. This ensured significant expansion of the forecasting time frame and paved the scientific foundations for the implementation of internal control feedbacks. However, there was still a drawback in the dynamic adaptive model, namely, that it did not contain a methodology for implementing the actions of external control; this required further development of our theory at the next, fourth stage.

Thus, urgent scientific problem of the development of a dynamic adaptive model of energy consumption by adding the actions of external control has generated a new theoretical direction, focused mainly on study of the processes of technocenoses energy consumption at the so-called bifurcation stages. The introduction and comprehensive verification of a fundamentally new concept of bifurcation of energy consumption made it possible to supplement the theory of optimal management of energy consumption with potentiation procedures based on the Z3 potential for energy conservation, potentiation factored on existing constraints, MC prediction, regime-based rationing and DC analysis, in parallel adding a new concept of transformed ranked parametric distribution to the mathematical apparatus of rank analysis. In applied terms, a new approach to the methodology of optimal management of energy consumption opens very broad prospects for the creation of specialized software and hardware systems (up to situational centers), which could be used in regional energy grid, energy service, and energy retail companies, as well as in electrical profiles of regions, cities, municipalities, enterprises, and organizations. It is these aspects that are considered as key areas of development for the foreseeable future.

References

[1] A. Yuzhannikov, Yu. Sizganova, Yu E., V.V. Shevchenko, Cenoses and Fibonacci numbers in power supply systems, Bull. Assoc. Graduat. KSTU 16 (2008) 74–81.
[2] A. Yu Yuzhannikov, D.V. Antonenkov, Cenological parameters of power consumption of a mining enterprise, Izv. Universit. Mountain J. 6 (2009) 19–22.
[3] B.I. Kudrin, Introduction to technology: monograph, Tomsk (1993) 552.
[4] B.I. Kudrin, Coenological studies: Series. - Abakan: Systems Center. Through thorns to general and applied cenology. Basics of Cenology, Technology, Electricians. Technique, Moscow (1996) 549.
[5] V.V. Druzhinin, D.S. Kontorov, Problems of systemology: Problems of the theory of complex systems, Sov. Radio. (1976) 295.
[6] C.B. Williams, Patterns in the Balance of Nature and Related Problems in Quantitative Ecology, Acad Press, London; New York, (1964) 324.
[7] C. Darwin, Origin of Species. M.-L. OGIZ-Selkhozgiz (1937).
[8] R.A. Fisher, A.S. Corbet, C.B. Williams, The relation between the number of species and the number of individuals in a random sample of an animal population, Anim. Ecol. (1943).
[9] F.W. Preston, The canonical distribution of commonness and rarity, Ecology (1962) 43.
[10] R.W. MacArtur, On the relative abundance of bird species, Proc. Nat. Acad. of Sci. 43 (3) (1957).
[11] R.W. MacArtur, Note on Mr's Pielou Comments, Ecology (1966) 47.
[12] G.A. Watterson, Models for the logarithmic species abundance distributions, Theor. Popul. Biol. (1974) 6.
[13] M.V. Arapov, E.N. Efimova, Yu.A. Schreider, On the meaning of rank distributions, Sci. Tech. Inf. Ser. 2 (1) (1975) 9–20.
[14] L.S. Kozachkov, Systems of flows of scientific information, AN USSR, Institute of Cybernetics, Kiev, Naukova Dumka, (1973).
[15] Yu. K. Orlov, The generalized Zipf – Mandelbrot law and frequency structures of information units of various levels. M.: Nauka (1976).
[16] L.N. Zasorina, V.A. Agraev, V.V. Borodin, L.N. Zasorina, Frequency Dictionary of the Russian Language. M. Russian language (1977).
[17] A.I. Yablonsky, Stochastic Models of Scientific Activity. System Research. Yearbook. M.: Nauka (1975).
[18] G.K. Zipf, Human Behaviour and the Principle of Least Effort, Addison Wesley Press, Massachusetts, (1949).

[19] B. Mandelbrot, Information Theory and Psycholinguistics: Theory of Word Frequencies. Mathematical Methods in the Social Sciences. Sat articles. M.: Progress (1973).
[20] N.B. Andreev, T.S. Bakirov, N.N. Zavalishin, V.M. Efimov, Distribution of stable biocenosis organisms by energy flux values, Zoological problems of Siberia. Novosibirsk, "Science" 72 (1972) 6–8.
[21] G.E. Hutcliinson, The Niche: an Abstractly Inhabited Hypervolume, The Ecological Theater and the Evolutionary Play, Yale University Press, New Haven (Conn.), (1965).
[22] O.A. Jorge, El Modelo de Nicho Fundamental: Su Applicacion en la Investigacion Ecologica, Cie. e Invest. 26 (7) (1970).
[23] V.I. Gnatyuk, On the strategy for power sector development in the Kaliningrad region, I. Kant State University Magazine "Baltic Region" 1 (3) (2010) 67–77.
[24] V.I. Gnatyuk, D.V. Lutsenko, Systemic methods of energy saving management in housing stock: analytical review, Kaliningrad: Kaliningrad Region Government (2007).
[25] V.I. Gnatyuk, V.P. Zhdanov, Analysis of the Development of the Main Indicators of the Fuel and Energy Complex of the Kaliningrad Region: A Scenario of a Transition Period Associated with the Commissioning of Large Power Capacities (CHPP-2, NPP) and the Planned Separation of the Power Systems of the Baltic Countries from the Unified Energy System of Russia, RSU Press, Kaliningrad, (2009).
[26] B.I. Kudrin, The fundamentals of cenosises controling consumer electronics, Industr. Energet. 9 (2015) 41–52.
[27] A.L. Naumov, Energy audit - A tool of energy saving, Energy Saving 4 (2000) 12–13.
[28] V.I. Gnatyuk, G.V. Kretinin, O.R. Kivchun, D.V. Lutsenko, Potential of energy saving as a tool for increasing the stability, Int. J. Energy Econ. Policy 8 (1) (2018) 137–143.
[29] V.I. Gnatyuk, M.A. Nikitin, D.V. Lutsenko, O.R. Kivchun, Models and methods for predicting power consumption in managing objects of the regional electro technical complex, Mathematic. Model. 29 (5) (2017) 109–121.
[30] Y.V. Kosharnaya, Optimization of the structure of energy consumption of metallurgical enterprises for the purpose of assessing the potential for energy saving, Industr. Energet. 10 (2016) 22–29.
[31] D.V. Lutsenko, O.R. Kivchun, V.N. Vasiliev, V.I. Gnatyuk, Method for monitoring the electrical consumption of the electrical complex of the Kaliningrad region, Industr. Energet. 3 (2015) 26–35.
[32] V.N. Vasilev, O.R. Kivchun, Management of electricity consumption of the municipal entity on the basis of new information technologies. Proceedings of the Tenth Anniversary Conference Innovations in Science and Education. KSTU, Kaliningrad (2012) 455–458.
[33] B.I. Kudrin, Electricity today and problematic power supply, Industr. Energet. 10 (2016) 5–9.
[34] D.V. Lutsenko, O.R. Kivchun, L.V. Primak, V.I. Gnatyuk, Potenshirovanie in a technique of management of a power consumption of a technocenosis, Mechan. Construct. 8 (2014) 19–27.
[35] Y.V. Kosharnaya, Development of a system for standardizing energy consumption and estimating the amount of energy savings by the example of a metallurgical enterprise, Industr. Energet. 8 (2015) 13–22.
[36] D.V. Antonenkov, D.B. Solovev, Mathematic simulation of mining company's power demand forecast (by example of "Neryungri" coal strip mine), IOP Publishing IOP Conf. Series: Earth and Environmental Science 87 (2017) 032003, doi: 10.1088/1755-1315/87/3/032003.

[37] V.I. Gnatyuk, Rank analysis of technocenoses, Electrics 8 (2001) 14–22.
[38] V.I. Gnatyuk, The law of the optimal construction of technocenoses. Monograph. Zenological studies. M.: TSU Publishing House - Center for System Research (2005).
[39] B.I. Kudrin, B.V. Zhilin, O.E. Lagutkin, M.G. Oshurkov, Cenological Determination of Power Consumption Parameters of Multinomenclature Productions, Prince Publishing House, Tula: Priok, (1994).
[40] E.Y. Sizganova, D.V. Antonenkov, D.B. Solovev, Energy-saving efficiency and potential in educational establishments of Neryungrinsky Region, Int. J. Energy Technol. Policy 15 (2/3) (2019) 180, doi: 10.1504/IJETP. 2019.098963.
[41] D.V. Antonenkov, E. Yu. Sizganova, A. Yu. Yuzhnnikov, Energy saving of educational institutions: on the example of the Neryungri district of the Republic of Sakha (Yakutia) [Text]: monograph. M-in education and science of the Russian Federation, Siberian Federal University, North-Eastern Federal University. SFU, M.K. Ammosova, Krasnoyarsk (2015).
[42] V.V. Fufaev, Cenological determination of the parameters of power consumption, reliability, installation and repair of electrical equipment of the region's enterprises Monograph. M: System Center. Issled (2000).
[43] A.I. Saprykin, E.Y. Sizganova, Energy economy potential estimation of establishment needs objects of power distribution company, IOP Conference Series Materials Science and Engineering 537 (2019) 062064, doi: 10.1088/1757-899X/537/6/062064.

3

The contribution of energy crops to biomass production

S.V. Leontopoulos[a], **G. Arabatzis**[b]

[a]*General Department, University of Thessaly, Larissa, Greece;* [b]*Democritus University of Thrace, Department of Forestry and Management of the Environment and Natural Resources, Pantazidou, Orestiada, Greece*

Chapter outline

1 Introduction

1.1 General

The increase of human population results the intensification of food and energy production in order to cover the rising demands of standard of living needs, producing also, large volumes of residues and wastes. Usually, by-products and residues of agri-food processing have high management costs and may contaminate the environment when disposed out without any treatment affecting the food chain [1–3]. According to many studies, efforts by using appropriate methods of decontamination and extraction have been completed the last few decades [4,5]. These residues and wastes can be also used as raw materials for energy production or for high value-added products since they contain rich amounts of natural antioxidant, antimicrobial and bioactive substances that can benefit food-producers, consumers and environment [6–22] Potential use of bioenergy crops is depending by agronomic, socio-economic, and environmental factors such as agricultural

Low Carbon Energy Technologies in Sustainable Energy Systems. http://dx.doi.org/10.1016/B978-0-12-822897-5.00003-1

technology, food and energy demands, political stability, biodiversity conservation, investment security, etc. [23–27] . In their study, Erb et al. [23] estimated how the above-mentioned factors can affect primary bioenergy potentials by 2050 in 11 regions worldwide. From their findings it was mentioned that worldwide bioenergy potential would range from 26 to 141 EJ/year with the trend scenario results to be calculated at 77 EJ/year.

The aforementioned approach is also a European Union (EU) guideline under the term of bio-refinery which means the gradual full utilization of solid waste for energy production or for the production of a range of valuable materials. This approach leads to a reduction in the carbon footprint of solid waste due to its full utilization [28,29]. Thus, one of the main issues of the EU is to develop agriculture without the burden of the environment. Unfortunately, in many European countries, rural development is not yet fully harmonized with the promotion of energy crops to the desired extent. Many of the residual by-products of existing crops, such as cereal straw, corn, cotton residues, and olive oil by-products could become, after appropriate treatment, a new financial resource while obeying environmental protection rules. The use of these raw materials, residues, wastes or by-products, which are usually low or even negative costs, is believed to be able to reduce production costs per unit creating also new job positions. These conclusions are agreed with research completed by Ericsson et al. [30] where production costs of short rotation coppices (SRC) crops, perennial grasses, and annual straw crops are calculated to be about 4–5 €/GJ, 6–7 €/GJ, and 6–8 €/GJ, respectively. These costs are believed to be reduced to 3–4 €/GJ and 5–6 €/GJ in future when production conditions will be improved.

1.2 Energy production and demands

Energy resources are essential for human well-being, and improving quality of life [31–34]. Compared to 2008, this need is expected to be double to 32,922 TWh in 2035 [35]. According to Kaygusuz [36] energy demand increased from 9,208 million tonnes of oil equivalent (Mtoe) in 1980 to 14,475 Mtoe in 2008 and is projected to rise to 16,785 Mtoe in 2020 (Table 3.1). Although the use of RES share energy production between 1980 and 2008 was increased in absolute terms from 909 Mtoe to 1,552 Mtoe (≈71% increase) while its percentage of total energy production increased only from 10% to 11%, respectively.

Panwar et al. [37]. Represent a scenario of global RES development by 2040 (Table 3.2). In more detail, in 2010, the share of RES in final energy consumption was 16.6% while by 2040 approximately

Table 3.1 Global energy demands.

Source of energy	Global energy demands (Mtoe) 1980	2000	2008	2020
Carbon	1792	2292	3286	4124
Diesel	3107	3655	4320	4654
Natural gas	1234	2085	2586	3046
Nuclear	186	676	723	920
Water energy	148	225	276	389
Biomass and other wastes	749	1031	1194	1436
Other forms	12	55	82	196
Total	**9208**	**12019**	**14475**	**16785**

Source: From Ref. [36].

Table 3.2 Global scenario of RERs development by 2040.

Year	2001	2010	2020	2030	2040
Total energy consumption (Mtoe)	**10040**	**10550**	**11430**	**12350**	**13310**
Biomass	1080	1313	1791	2483	3271
Large scale hydroelectric	22.7	266	309	341	358
Geothermal	43.2	86	186	333	493
Small scale hydroelectric	9.5	19	49	106	189
Wind	4.7	44	266	542	688
Solar	4.1	15	66	244	480
Photovoltaic	0.1	2	24	221	784
Solar-thermal energy	0.1	0.4	3	16	68
Waves	0.1	0.1	0.4	3	20
Total RERs	**1365.5**	**1745.5**	**2964.4**	**4289**	**6351**
Percentage of RERs (%)	**13.6**	**16.6**	**23.6**	**34.7**	**47.7**

Source: From Ref. [37].

half of the energy production is projected to be produced from RES (47.7%). The participation of biomass as RES is expected to triple the energy produced from 1313 Mtoe in 2010 to 3271 Mtoe in 2040.

The use of traditional forms of fuel for energy production has caused significant environmental problems over the last 100 years mainly due to the use of fossil natural resources, which in most countries are responsible for the higher CO_2 emissions compared to other forms of power generation [38]. The energy sector is the primary polluter, as almost 95% of air pollution is due to the production, transformation, and use of conventional fuels. Furthermore, prices of fossil fuels, such as oil have risen in recent years depending on unstable political situations in producer's countries [39,40]. Thus, the search for new, more environmentally friendly, modes of energy resources and production methods is considered by many scientists to be the optimal solution to this situation [41].

Human energy needs has always been covered by the use of RES such as wind, water power, wood burn and various other forms of biomass and have been dominated by their abundance and ease of use [42]. The same authors have concluded that dependence or detachment from one form of energy to another can be motivated by several factors such as the depletion of the natural energy resources resulting increase of cost use, the exploitation, and use of more economical energy resources due to the technological advancements and innovations that were not available until then, the environmental degradation and in particular air and water pollution which can be so severe that they can have adverse effects on human health [32,33 43–47]. According to Kaygusuz [36] sustainable energy production and CO_2 reduction should increase the share of RERs in new power plants and generally promote RERs as a new alternative way of covering the increasing energy demands. Furthermore, renewable energy resources can make a powerful contribution to modern energy issues, as it is believed that they are the most efficient and clean energy technologies contributing to sustainability, energy self-sufficiency, energy mix diversification, and environmental protection providing additional economic benefit in both developed and developing countries [48–57] although RES consumption is higher in countries with a higher GDP than with those of lower GDP [58]. This is supported by a study completed by Ntanos et al. [59] where consumers had the willingness to pay about 26.5 € per quarterly electricity bill for a wider penetration of RES into the electricity mix in a developed country reach in RES like Greece. According to Li et al. [60] and Yan and Chen [61] it is believed that worldwide bioenergy production forecasts for 2050 amount to 100–450 EJ/year, with 64.78 EJ of them being produced in Latin America, 43.6 EJ in Africa, 23.55 EJ in Southeast Asia, 26.3 EJ in the EU, 15.89 EJ in the United States, 2.1 EJ in the UK, and 0.93 to 4.56 EJ in India.

Despite the above advantages of RES, existing technologies still remain commercially competitive compared to conventional energy production methods. This is due to the cost issues, efficiency, predictability, availability, stability, and the use of large areas of land for the production of RERs and especially biofuels. Although RERs is an alternative proposal, offering environmental, energy, economic and social benefits, this should not lead to unconditional acceptance of their technical exploitation projects. In conclusion, with the existing technology the construction of energy plants and sometimes the use of RES, raise environmental, economic, and social issues [62].

1.3 Biomass production

Etymologically, biomass comes from the word «βιοζ» (life) and the word «μάζα» mass. Biomass includes any type of residue or by living plant tissue that can be used to produce solid, liquid, and/or gaseous fuels. There are two types of biomass, residual forms produced by energy crops [63]. Residual forms of biomass divided into three main categories: Residues such as residues of cereal straw, cotton strips, twigs, etc. that are remaining in the field or forest after harvest of the main crop, residues of agricultural and forestry industries, such as olive kernels, ginning residues, sawdust, etc. and wastes such as the organic parts of industrial and municipal wastes.

Global biomass produced each year is estimated to amount to 172 billion tonnes of dry matter, with energy content 10 times the energy consumed worldwide at the same time. These enormous energy potential remains largely untapped, as recent estimates suggest that only 1/7 of global energy consumption is covered by biomass, mainly related to its traditional uses (firewood, etc.). According to Sinyak and Kolpakov [44] the increase in energy consumption is a result of scientific and technological progress coupled with an increase in population and an improvement in people's standard of living.

According to Eurobserver, [64] and 2012 primary energy production from solid biomass in the EU during 2019 was 99.3 Mtoe increased by 11.3% since 2012 production (82.3 Mtoe), although this production was still below the 100 Mtoe threshold. However, there was a reduction of −0.3% in solid biomass consumption in the EU between 2017 and 2018 due to a milder winter while there was a rise in electricity production. Among European countries the biggest drops were recorded for Italy (−502 ktoe), Germany (−310 ktoe), Austria (−265 ktoe), Hungary (−211 ktoe), and Sweden (−209 ktoe) while the highest consumption was recorded in the United Kingdom (885 ktoe), Bulgaria (339 ktoe), and Finland (273 ktoe). It was also noted that in 2018, 99.5 TWh of electricity was produced from solid biomass and 78.9 Mtoe of heat was consumed, compared to 79.5 TWh and 68 Mtoe in 2012, respectively. Finally, according to Eurostat in 2017 final energy consumption of charcoal in the EU was 196.6 ktoe. The report also stated that the EU consumes about 26.1 Mtoe of wood pellets increased by 2.1 Mtoe (8% growth) compared to 2018 and 16.6 Mtoe since 2011. However, the production of solid biomass sourced primary energy slipped by 0.4% compared to 2017 to 94.3 Mtoe in 2018 due to wood pellet imports mainly from the United States, Canada, and Russia and minus stock variations in EU. According to Eurobserver [65] about 30% of EU pellet consumption also came from imported sources with the United States participating in imports of about 1.764.000 Mtoe of wood pellets and Canada with 1.764.000 Mtoe. Among wood pellet use, 55% of wood pellet consumption was used by the residential and commercial sectors (56% in 2017), thus the remaining 45% was used by industry. Among European countries in 2019, United Kingdom was the EU's biggest wood pellet user with 8.5 Mtoe followed by Italy (3.3 Mtoe), Denmark (3.1 Mtoe), Germany (2.2 Mtoe), Sweden (1.8 Mtoe), France (1.6 Mtoe), Belgium (1.5 Mtoe), Austria (0.96 Mtoe), the Netherlands (0.57 Mtoe), Spain (0.57 Mtoe), Poland (0.45 Mtoe), and Finland (0.44 Mtoe). The major European operators of biomass plants for year 2018 was "DraxGroup" in United Kingdom with operational capacity at 2600 Mwe, "Orsted" in Denmark (1200 MWe and 1900 MWth), "Pohjolan Voima" and "Fortum Varme" in Finalnd (620 MWe–623 MWth and 667 MWe–622 MWth, respectively), "RWE" in Netherlands (660 MWe–350 MWth), "E.on" and "Zellstoff Stendal" in Germany, "Vattenfall" in Sweden (236 MWe) and "Engle" in France with operational capacity 285 MWe.

Germany is in the top of the list of the EU member states in terms of solid biomass production and consumption in 2017 and 2018 producing approximately 11.9 and 11.7 Mtoe and consuming 12.4 and 12.1 Mtoe of solid biomass, respectively. In France, an estimated 10.4 Mtoe of solid biomass were produced and consumed in 2017 while in 2018 solid biomass production and consumption were slightly reduced (10.2–10.3 Mtoe). In 2017, Sweden produced and consumed approximately 9.4 Mtoe of solid biomass while in 2018; the solid biomass production and consumption were slightly lower than the previous year. However, solid biomass production and consumption was slightly increased in 2018 compared to 2017 reaching 8.9 Mtoe. Italy produced 7.1 Mtoe of solid biomass while it consumed 8.5 Mtoe.

However, production-consumption gap was greater in United Kingdom, which for 2018 produced 4.5 Mtoe while it consumed 7.3 Mtoe. Estonia produces in 2018 more solid biomass that its consumption (1.6 instead of 1.0 Mtoe). Furthermore, no solid biomass production or consumption was recorded for Cyprus and Malta.

Regarding the Gross heat production from soiled biomass, Sweden is in the top of the list of EU member states in terms of total heat generation from solid biomass. The country produced 2484 Mtoe of thermal energy from solid biomass in 2018, including 0.685 Mtoe of combined heat only plants and 1799 Mtoe in CHP plants. Finland and Denmark were next in line for the top three biomass producers for thermal energy in 2017 and 2018. United Kingdom finally, was the largest producer of solid biomass electricity with 23.532 TWh total electricity, followed by Finland and Germany, where 11.8212 and 10.827 TWh of energy was produced.

Due to the growing needs of EU member states for biofuel energy, the EU plans to double the use of crops and woods for biofuels. However, it is calculated that in order to achieve this target the land required is equivalent to 3 times the size of the United Kingdom. The EU has a target of 20% of energy consumption coming from renewable energy sources by 2030. Research carried out by the University of Vienna has found that doubling the use of bioenergy in order to meet the EU's goals, more than 70 million hectares of land is needed.

Although the EU's global energy footprint for 2010 is about the same size of land as Sweden's, researchers commented that achieving its goals would require energy crops to grow by a third, which is almost equivalent to an area in the size of Poland. According to this study, the EU is expected to use 40 million hectares of land for biomass production by 2030, 11 million hectares for biofuels and 19 million hectares for energy crops. In a study completed by [66] scenarios were examined that freed-up surplus agricultural area in the EU ranges from 0 to 6.5 Mha (0–15.0 Mha, if fallow land is included) until 2050. These studies reported that the EU has a high biomass potential from freed-up surplus land (7–48 Mha) mainly due to crop yield improvements.

Thus, it is strongly believed that increasing demand for bioenergy could lead to an indirect change in land use to support domestic production of biomass from plants. This is expected to have potential environmental impacts, including soil degradation, nitrate fertilizer pollution, deterioration of the greenhouse effect, as well as the reduction of land used in production of food crops and rising prices [67]. For example, the proliferation of energy crops and the expansion of maize crop production for energy purposes could lead to serious practical, ethical, nutritional, and economic problems as it would lead to a reduction in the use of maize for animal feed, thereby increasing the cost of livestock production and thus increase in prices for meat, dairy products, and poultry would ultimately lead to reduced consumption of these products [68].

In addition, the intensification of the cultivation of energy crops could also have an impact on the soil structure as a result of the erosion and compression caused by the use of agricultural machinery. For example, soil erosion in sugarcane production is considered high, mainly because of the time the land remains bare between harvest and the new growing season. In addition, the cultivation of energy plants requires the use of inputs such as fertilizers and other plant protection products resulting in water and environmental pollution. Their combustion causes air, water, and soil pollution. According to [69] during the production of corn bio-ethanol, large amounts of CO_2 released, soil erosion and the use of more nitrogen fertilizer than any other crop are environmental constraints. Concerns have also been raised about deforestation, especially in developing countries. However, some studies concluded that sugarcane production and extension cannot be directly linked to deforestation while others are directly linked deforestation through sugarcane production [70].

Moreover, the acceptance of energy produced by farmers is also a key factor in the availability of the product [71], which of course depends on factors such as its technological utilization by the domestic industries their acceptance by consumers and biomass supply chain networks since biomass supply chain studies are still limited due to complexity and the several parameters that must have been taken under consideration [72,73]. In consumer surveys regarding the acceptance of bioethanol use, it has been found that the majority of consumers realize that the greatest potential benefit of using an ethanol-gasoline blend compared to gasoline is to reduce their external dependence on gasoline oil with consequent environmental and economic benefits For this reason, many of them stated that they would be willing to buy ethanol-gasoline mixture if available [74]. Similar conclusions were drawn from a study carried out in rural areas which concerned, among other things, the prospects for job positions and the existence of subsidies [75,76].

Furthermore, Steininger and Voraberger [77] commented that the use of biomass energy could also have a negative impact on GDP in cases where the public subsidies required to start using biomass are very high and this is why the costs of energy services provided by the use of biomass are determined by both the cost of biomass products and the cost of technology utilization [78].

Nowadays, with the use of energy crops, developed countries try to reduce, in addition to their environmental and energy issues, the problem of agricultural crop over production. As is well known, in the EU countries agricultural over production inevitably lead to a decline in agricultural land and agricultural production. At the same time in United States, energy crops compete for land with food crops and forest with energy crops to be suitable for growing in regions where they have a comparative advantage relative to other crops [79]. In EU countries, it is estimated that by 2030 a reduction of 2–5 or 13–15 million hectares of cropland for other crops including energy crops, depending on the scenario could be faced [80]. That is why today, biomass is perhaps the only RERs that faces reactions not only locally but also globally. This criticism even comes from ecological organizations and non-profit organizations (NGOs) and focuses on the fact that large crops are committed to the production-conversion of fuels to food, which could lead to malnutrition [81]. Furthermore, limitation factors are the difficulty in collecting, processing, transporting, and storing it as compared to fossil fuels, the costly installations, and equipment required for biomass utilization in relation to conventional energy sources and the large dispersion and seasonal production.

Due to the above disadvantages, the cost of biomass remains high compared to other fossils and especially oil. However, there are already applications in which the utilization of biomass has economic benefits. In addition, this problem is gradually being eliminated because of the rising oil prices, and more importantly, due to the improvement and development of its exploitation technologies. Finally, environmental benefits must be taken into account at all times. It is believed that using biomass for power generation can prevent the greenhouse effect since energy crops and biomass produced by them does not contribute to the concentration of this pollutant in the atmosphere because during its production and through photosynthesis, significant amounts of CO_2 are re-bound.

1.4 Biofuel

In addition to producing heat and electricity, biomass can be used for the production of liquid fuels (biofuels). Biochemical production of liquid fuels mainly focuses on the production of bioethanol (CH_3CH_2OH) by fermenting sugars, starch, cellulose, and semi-cellulose derived from various types of biomass, such as maize, sorghum, wood, algae, etc. [82–88]. In addition, a biofuel from tobacco plant is being developed by American aircraft maker Boeing, in partnership with South African Airlines, in an effort to reduce CO_2 emissions and promote green energy in Africa's most advanced economy. The fuel will be manufactured on the basis of a hybrid tobacco plant known as Solaris, produced by SkyNRG. The plant contains a large amount of oilseeds that turn into fuel. Test crops are already taking place in South Africa. Biofuels are recently used in transport in many European countries, such as France, Germany, Sweden, Austria, Italy, Denmark, Brazil, United States, and some Asian countries. Biofuel consumption for transport use was increased by 10.1% between 2017 and 2018 in EU, reached to 17.0 Mtoe.

Concerning the effects of biofuel production on the environment, economy, and society, it has been observed that plant productivity varies, resulting in their inaccurate identification of energy production and their dependence on various key climate factors. So, while Brazilian bioethanol produces 8 times more energy than what is used for its production, American produces only 1,1–1,7 times more energy than what is spent on its production. In addition, it has been found that 29% more energy is needed to produce bioethanol from corn than energy produced by bioethanol itself. It is also believed that biofuels are more efficient when used for heat or energy production than for transport [89–91].

EU countries are among the largest biodiesel producers [92] with France, Germany, Spain, and United Kingdom, to be on the top four producers in Europe. According to Palmieri et al. [93] the biodiesel plants produced approximately 2.3 Tg/year. Analysis carried out by the European Council for Renewable Energy, it is commented that by 2050, biomass may provide about 36% of total European primary energy consumption [94]. However, the industrial viability of fermentation requires the resolution of some technological difficulties [95]. In terms of world bioethanol production, the top five world producers for 2012 were United States, Brazil, China, and Canada with the United States and Brazil together accounting for 87% (61% and 26%, respectively) of world production (Table 3.3).

1.5 Energy crops distribution, contribution, and utilization

Conventional fossil fuels, such as coal, oil, and gas, which have been a major source of energy since the industrial revolution, face the risk of depletion. The development of RERs, such as biomass energy, biogas, and biodiesel is imperative and timely [96].

Table 3.3 Biofuels consumption for transport in the EU-28 in 2018 (in ktoe).

Country	Bioethanol Biodiesel	Biogas	Total consumption	
France	586.0	2812.0	0.0	3398.0
Germany	756.0	1929.0	34.0	2719.0
Spain	160.0	1568.0	0.0	1728.0
Sweden	96.9	1342.6	118.0	1557.2
UK	376.8	897.1	0.4	1274.3
Italy	32.6	1217.0	0.1	1249.7
Poland	173.0	770.0	0.0	943.0
Netherlands	170.7	330.5	7.2	508.4
Austria	57.9	423.1	0.3	481.3
Belgium	93.6	381.0	0.0	474.6
Finland	80.7	315.6	0.3	396.7
Czech Republic	61.3	247.4	0.0	308.7
Romania	91.1	206.2	0.0	297.2
Denmark	0.0	286.0	0.3	286.3
Portugal	7.6	272.3	0.0	279.9
Greece	0.0	169.0	0.0	169.0
Bulgaria	26.7	139.6	0.0	166.2
Ireland	27.3	127.0	0.0	154.2
Slovakia	19.6	129.9	0.0	149.5
Hungary	40.0	108.4	0.0	148.4
Luxemburg	10.1	109.5	0.0	119.6
Lithuania	8.0	69.8	0.0	77.8
Slovenia	8.6	34.7	0.0	43.3
Latvia	7.9	1.4	0.0	9.3
Cyprus	0.0	9.0	0.0	9.0
Malta	0.0	9.0	0.0	9.0
Estonia	1.0	0.0	0.0	1.0
Croatia	0.2	0.3	0.0	0.5
Total EU 28	2892.9	13905.6	160.6	16959.1

Source: From Ref. [96a].

Several cultivated plant species produce not only edible products but also residues both at the harvest and cultivation stages, which could be used to produce biomass energy [97]. As energy crops have been proposed a number of plant species that are annual or perennial, herbaceous or woody, producing solid biomass or raw material for liquid or gaseous biofuels. Energy crops, which include both certain crops and wild plants, are intended to produce biomass, which can then be used for various energy purposes. Energy products such as electricity, heat, and biofuels (bioethanol, biomethanol, biodiesel, biogas, biodimethyl ether, Bio-E TBE -ethyl tert-butyl-, Bio-M TBE -methyl tert-butyl ether-, synthetic biofuels, and pure vegetable oils produced from oily plants) are produced from biomass. Among energy crops, perennial species in tropical and sub-tropical regions yield the highest by providing greater amounts of energy. For example, in terms of clean energy production and energy efficiency, crops such as sugarcane, and palm tree offer the best and most promising opportunities for sustainable biofuel production [98].

From studies carried out in Europe over the last 20 years have been recorded, suitable plant species for energy production purposes in the different climatic conditions [99,100]. Plants such as wild artichoke, eucalyptus, sweet sorghum, or cannabis are only grown in the Mediterranean region, while others such as willow and rapeseed are more adapted to the colder climates of Central and Northern Europe. Plant species such as mishanthus can under certain conditions to be cultivated throughout Europe, from Sicily to Denmark and Ireland [101]. Depending on the area of cultivation, there has been a great variation in the productivity and hence in the economic benefit of each crop. For example, high yields (3–4 tonnes/ha dry weight) have been recorded in sorghum, legume, and reed in southern Europe, but these refer to irrigated crops and small experimental plots, while large- scale willow crops in Sweden yielded only 0.8–1 tonnes of dry weight/acre.

According to Zegada-Lizarazu and Monti [102] depending on crop characteristics, economic importance, geographical distribution and climate adaptability, energy crops could be used in crop rotation systems to control pests, improve soil fertility, maintain long-term land productivity, and thereby increase yields and profitability of farms. For example in the southern Europe, the combination of conventional crops such as wheat, pulses, corn, sunflower with energy crops such as sweet and fibrous sorghum, kenaff, etc., applied to a crop rotation system could optimize the utilization of soil resources. For northern European countries where the climatic conditions are different, species such as rapeseed, cereals (e.g., wheat, barley, oats), flax, and various other legumes would be a better choice. Stolarski et al. [103] the potential use of perennial industrial crops as important sources of lignocellulosic biomass after six successive rotations. More specific, during their experiments the highest energy yield value was obtained from the Ekotur willow variety (1857 GJ/ha), and it was followed by *Miscanthus sinensis, Spartina pectinata, Sida hermaphrodita*, and *Helianthus salicifolius*

Furthermore, according to Shortall [104] some energy crops due to their resistance to adverse climatic and soil conditions can be used in order to provide agricultural income in remote areas with non-fertile soils. The idea of using less productive or "marginal land" for cultivation of energy crops is being promoted. It is argued that the use of this type of land will not compete with food production, it is widely available and soil is suffocated to the least harmful environmental impacts due to its desertification [105].

European countries such as Ukraine, Romania, Poland, and France could be the main energy producing countries from energy crops producing up to 50% of Europe's total annual biomass resource potential [106].

1.5.1 Energy crops in Sweden

According to Wright [107] Sweden is seen as one of the examples of successful development and use of energy crops for energy production and the use of willow as biofuel in district heating were widespread even before 1990 [108]. Sweden's target is that by 2020, 49% of the total energy consumed it must be produced from renewable energy which is the highest proportion of any of the 28 EU Member States [109]. Almost about 2% of Sweden's arable land is used for energy production [109] with 13 energy crops available to Swedish farmers. Among them, cereal residues cover the largest area, which is approximately 30,000 hectares. For this reason, there are many scattered heating installations in the country that operate on the burning of straw raw materials. Oilseed crops also cover an area of about 25,000 hectares, mainly in southern Sweden. Other energy crops cultivated are, oats, poplars, *Sativa cannabis*, grass, sugar beet, cane, *Phalaris arundinacea*, willow, *Populus tremula*, and *P. tremuloide.*

According to Paulrud and Laitila [110] in 2006, the use of solid biofuels from biomass of agricultural origin in Sweden was about 1 TWh, while the total energy supply from biomass fuels, including peat was 112 TWh. According to Lonnqvist et al. [111] in 2012, it was estimated that the Swedish potential of energy resources for biogas production from residues and energy crops, amounted to 8.86 TWh, which was equivalent to approximately 9% of current domestic consumption energy in transportation sector. In more details Gissen et al. [112] analyzed six crops (hemp, sugar beet, maize, triticale, grass/clover, wheat) in southern Sweden for biogas production. From his findings, it was suggested that the higher biomass yields and biomass production were observed in sugar beet cultivation. The crops with the least negative environmental impact were those of hemp and clover while Triticale cultivation also presented a low risk of adverse environmental impacts, yielding energy similar to that of cereals and maize.

Furthermore, cultivation of cereals for energy use is associated with various advantages stemming from various factors stimulating their use by farmers [113,114]. According to the Swedish Government, these factors include esthetic assessments of the rural landscape, commercial financial agreements, opportunities for land leasing, know-how (regarding cultivation methods), use of existing machinery and better management of environmental hazards from agricultural residues. Regarding the cultivation of cannabis for energy use, this is considered to be quite recent and is believed that cause less undesirable interference in the rural landscape than the cultivation of cereals [115]. However, limiting factors such as legal restrictions, lack of know-how of cultivation methods, costs of changing the production system, limited market opportunities, failure to secure substantial financial gain, etc., are serious obstacles to the adoption of the cultivation [116].

Regarding hybrid poplar, it is noted that it is a perennial plant introduced into the Swedish energy production system quite recently and there are currently no studies available on the factors that stimulate or inhibit the propagation of cultivation in Sweden since studies usually focus on a single crop [117,118] on a specific aspect of energy agriculture [109,119].

In conclusion, despite Sweden's small contribution to world energy production from energy crops, there are requirements for the agricultural sector to change. The production of more energy-related products or energy crops tends to take the place of traditional agriculture [115]. In order to develop strategies and guidelines for involving more farmers in energy crop, policy makers should have information on how farmers evaluate the characteristics of energy crops [110].

1.5.2 Energy crops in Austria

In Austria, biomass energy production is driven by the country's energy policy, which is in line with EU objectives for CO_2 mitigation impacts [120]. Especially rural areas, which are not financially well developed can benefit from the extra income that can be generated Stocker et al. [121]. For example, the future use of land cultivated with rapidly growing tree species, such as poplars and willows for biomass energy production could be of great importance for the Eastern Sytria region. However, biomass thermal power plants are distributed all over the country. According to [122] *Sida hermaphrodita* a versatile, perennial crop (15–20 years) has been in the center of researcher's interest due his potential use as energy crop, which produces solid fuel for thermal utilization. Due to different experimental setups, a broad range of dry matter yields are mentioned in the literature, from 2.0 up to 25.0 t/ha. However, in Austrian conditions the dry matter production at the end of the first growing season ranged between 1.2 and 2.4 t/ha when harvested at the end of October (biogas utilization) and between 1.0 and 2.1 t/ha when harvested at the beginning of February (thermal utilization). Changes in harvest strategy onwards the 2nd year of cultivation had a result the increase of dry matter production for biogas utilization to 6.6 and 9.5 t/ha on 2nd year of installation and between 7.3 and 8.6 t/ha in the following 3rd year. Regarding the cost contribution of energy crops a study completed by Trink et al. [78] showed that the lowest cost per hectare and per year of cultivation in Austria results from the cultivation of perennial crops, such as willow, poplar, mishanthus, and alfalfa, is depending mainly by the installation costs which is spread through the years. Regarding the cost of transporting biomass, this depends on many factors such as the energy density of the material being transported, the transport distance, the level of automatism, and the mode of transportation. For example, lower energy costs can be achieved in mishanthus cultivation; however low energy density, results high and specific transportation and storage costs. Cultivation of cereals and maize as energy crops in Austria has the highest energy costs in cents/kWh. In addition to agricultural crops, forestry products such as logs, shavings, etc., also plays an important role in generating energy and heat. As most forests in eastern Austria are privately owned, cultivation costs are relatively small. High investment, consumer interest, and low operating costs indicate that biomass technologies in Austria can become profitable.

1.5.3 Energy crops in Germany and Denmark

Germany is considered to be one of the largest energy crop and biomass producer [92,123] within EU countries, with land occupied for energy crop's production to be estimated in 2011, at approximately 16% of which 40% was attributed to biogas production [124]. Thus, biomass energy plants are distributed all over the country producing 6,944 MW from a total energy production of 191,153 MW. Regarding the potential effects of energy crops Bunzel et al. [123] commented that their adoptability will depend rather on the future design of the agricultural systems. A good example is Bavaria area where short rotation coppice covers about the 15% of total land area cultivated with SRC in Germany. Among tree species cultivated in the Bavaria, 98% of the cases were poplar [125]. However, monoculture of energy crops is not the most appropriate production system due to the possible risks associated with the increased cultivation of pesticide-intensive energy crops. Instead, energy crops should be integrated into the existing food production systems. Regarding acceptance of energy crops by farmers and energy consumers it is suggested that further education is needed in order to encourage the use of sustainable crop rotations, innovative cropping systems and perennial energy crops. However, according to an evaluation of 238 datasets Beer and Theuvsen [126] commented that German farmers' attitudes are generally negative thus they participate relatively little in agricultural wood growing. Concluding Bunzel et al. [123] suggested that optimized cultivation systems with diverse crop rotations could help to improve monotonous agricultural landscapes, increase biodiversity, and minimize pesticide exposure in German agriculture.

Similarly, Denmark is considered to be one of the leaders in RES worldwide. By achieving 2020 Denmark's target of 30% of energy consumption for electricity and heat needs from renewable sources since 2016, Denmark faces new goals in the use of RES. Nowadays, biomass such as wood, straw, and renewable waste in the form of wood pellets, is the largest source of renewable energy covering the 67% of total renewable energy consumption([127]; Energy Statistics 2016). So far, energy crops have not reached yet the target set by the Danish government of 100,000 ha and are not currently widely used in commercial biofuels production.

1.5.4 Energy crops in United Kingdom

Although biomass production is often portrayed as a new activity, historically there has been a tradition of utilizing forest land. According to Booth et al. [128] and the European Environment Agency [129], it has been foreseen that 0.8 hectares of land in the United Kingdom could be allocated for biomass cultivation by 2010, while this could be as high as 1.1 million hectares in 2020 and 1.6 million hectares in 2030. In order to achieve the necessary energy potential by using energy crops by 2020 it is required about 350,000 hectares of land [130]. According to the RCEP (2004) the land required for energy crops will increase from 1 million hectares in 2020 to 5.500.000 hectares in 2050. Thus, biomass energy plants are spread all over the countryside. According to Convery et al. [131] in a study conducted in the Cumbria area, it was observed that there was an awareness of new renewable energy markets, along with the increased volatility of traditional crops in the area. On the basis of this observation, it was concluded that short-term opportunities for increasing biomass production can be achieved faster by intensifying the management of existing forest areas. In a medium term, as biomass production becomes more and more reliable, marginal agricultural land will be used for biomass production. However, profit and demographic factors may affect the acceptance of energy crop cultivation by farmers [131,132]. According to Defra [130], in order to increase the available biomass to more than 1.000.000 tones of dry weight of wood per year, wood pulp from British forests, use of manure and significant increase in intake of perennial energy crops should be further promoted. In Scotland, the area planted or approved for energy crop planting by the end of 2006 was 300 hectares. With the application of modern harvesting techniques and the dissemination of energy crops, these areas increased to about 600 hectares in 2007 and 2008 [133]. In Northern Ireland, 400 hectares have been planted or approved for planting while an additional of 410 hectares have been approved for planting in 2007 [134]. In Wales in 2007, 40 hectares of reed (SRC) were cultivated [135]. In 2007 the total area of energy crops in the United Kingdom was 15,546 hectares for willow and mulch SRCs [130]. Concerning the subsidy to farmers in England, the Energy Crop Scheme (ECS) under the English Rural Development Program (ERDP) 2000–06 provided a subsidy of 1000 £/ha for willow SRC and 920 £/ha for grassland while in Wales, the subsidy for the installation of willow SRC was only 600 £/ha. Finally, in Northern Ireland, the average aid rate was 1920 £/ha [136].

1.5.5 Energy crops in Poland

In Poland, RES in 2007 represented 4.8% (200 PJ) of final energy consumption; 15% of the target under Directive 2009/28/EC for RES in 2020. Among RES, solid biomass was the most important accounting for 93% of RES energy production [137]. Although limited use of forest biomass is allowed for energy production, energy crops appear to be the most important source of biomass to be used in order to achieve the 2020 targets in Poland, both in the sectors of energy production and transportation. According to Ignaciuk and Dellink [138] biomass in Poland produced from crop residues of traditional agriculture, forestry, and energy crops. According to Borkowska et al. [139] the most valuable non-food perennial species of agricultural plants in Poland are willow (*Salix* sp.), giant miscanthus (*Miscanthus gigantheus*) Amur silvergrass (*Miscanthus sacchariflorus*), Virginia fanpetals (*Sida hermaphrodita*), Jerusalem artichoke (*Helianthus tuberosus*), prairie cordgrass (*Spartina pectinata*), switchgrass (*Panicum virgatum*), and big bluestem (*Andropogon gerardii*).

Regarding biomass utilization Iglinski et al. [140], reported that plants using RES in Poland were local required no central technological infrastructure while there were over 100 energy plantations of at least 5 hectares producing about 0.5 MW of energy. Among them, at least 40 plantations used for biomass production for energy-heat purposes and at least 39 farmers produce biofuel of 1 million dm^3/year. Of the total number of energy plantations, 80 were near municipal waste collection sites where they were involved in combustion in energy generation and urban heating, 56 use biomass produced for biogas production near sewage treatment plants, 8 energy plantations were located in agricultural biogas plants and 46 were near medical waste incineration facilities.

Among the energy crops that can be cultivated in Polish soil conditions is willow (*Salix viminalis* L.) which in 2007 was the most common plant [141], and cereals [140], from which residues could be harvested for energy purposes since straw production is estimated at 25 million Mg per year. *Sida hermaphrodita* also covered an area of 750 hectares with yield per hectare ranging from 12 to 17 Mg dry mass. The use of mallow plants as fresh biomass which can be used as silage feed or for energy production during the process of fermentation of methane, was also being investigated [140]. Other plant species such as *Helianthus tuberosus*, miscanthus (*Miscanthus sacchariflorus*), and *Andropogon gerardi* were investigated for potential use as energy crops. Furthermore, Jankowski et al. [142] in a long-term field experiment conducted in northeastern Poland compared the biomass yield and energy efficiency in low and high inputs of Virginia fanpetals (*Sida hermaphrodita*) a highly promising perennial plant for the production of bioenergy and bioproducts [143–145]. Jankowski et al. [142] commented that the high-input technology characterized by higher above-ground biomass yield compared to low-input technology (5.1 vs. 4.4 Mg/ha/year DM). However, in studies completed by Borkowska and Molas [146]; Nabel et al. [147]; and Nahm and Morhart [148] the dry matter yield ranged from 10 to 20–25 Mg/ha depending on factors such as the propagation method (seeds or root cuttings), the useful lifespan of the plantation, harvest date, and environmental conditions.

As an agricultural country, Poland needs to develop technologies based on biomass, biofuels and biogas. However, future investors are discouraged by the high cost of investing in RES technologies, the high cost of preparing the investment over operating costs, the lack of a defined economic and tax policy by the Polish State and the differences in productivity of agricultural land.

According to Chodkowska-Miszczuk and Szymanska [149] apart from the aforementioned limitations socio-demographic factors and diversification of economic activities in rural areas can affect the development of energy crops. For example, according to Krasuska and Rosenqvist [137] willow-derived biomass values make the crop profitable, whereas those of triticale make the crop unprofitable. In Poland, the cultivation of willow and juniper is lower than triticale. In general, perennial crops are considered to be more favorable than annual crops due to the relatively high yields of cereals per unit area and also due to the lower environmental impacts [150,151]. In conclusion, although energy crops in 2007 covered 8700 hectares with periodic logging and perennial grasses [152], the area of these crops was estimated to be 1.6 million hectares. However, bio-energy farmers are planning to increase their activity in the future [153].

1.5.6 Energy crops in Czech Republic

In Czech Republic, half of the agricultural land is in non-fertile areas while one-third of this is forest. According to Havlıckova and Suchy [154], has been recorded that since 1995 the area of agricultural crops has decreased by 15,000 hectares, while the area of forests has increased by approximately 16,000 hectares and permanent grassland has increased by about 71,000 hectares. It is suggested that non-fertile land can be used for energy crops cultivation instead of food crops. In terms of its energy policy, the Czech Republic considers that biomass production is the dominant source of renewable energy. However, various obstacles limit current growth. One of the most important barriers to biomass production is the lack of consistent mapping of the potential of individual biomass forms. By choosing the most appropriate energy crop plant species will allow to create a State strategy for energy crop species suitable to be cultivated for biomass production [155]. In terms of technical operation, biomass is the form of energy with the best prospects from all RERs for electricity and heat generation in the Czech Republic. For this reason, electricity production from biomass increased from 731 GWh in 2006 to 968 GWh in 2007. This increase is continued over the next years. This increase is partly due to the increase in the number of young farmers using cogenerated biomass. More specifically, in 2007 a total of 665,000 tones of biomass were used to generate electricity, significantly more than the 512,000 tons in 2006. An example of biomass production is the research completed by Brant et al. [156] who investigated the potential use of several catch crops such as *Brassica napus, Lolium multiflorum, L. perenne, Phacelia tanacetifolia, Sinapis alba, Trifolium incarnatum, Raphanus sativus* var. *oleiformis*, and *T. subterraneum* for biomass production. From their results, it was demonstrated that the highest biomass production and the highest cover of catch crops were observed in treatments with *S. alba* (1382.0 kg/ha). Moreover, they observed that the average biomass production (sum of catch crops, volunteers and weeds) was highest in treatments with *S. alba, R. sativus*, and *P. tanacetifolia* and lowest in treatments with *B. napus, L. multiflorum*, and *L. perenne* demonstrating that an increase in the percentage share of volunteers caused a decrease in the biomass production of catch crops. According to a research by Havlıckova and Suchy [154] on modeling the future use of energy crops in the Czech Republic, mishanthus crop was not included in the model because there was not sufficient planting material in the country and planting was relatively expensive.

1.5.7 Energy crops in Ukraine

Agricultural residues such as straw, corn, sunflower, etc. can be used for energy production in Ukraine. However, cereal straw for energy production is limited due to its use as a feed in animal production. According to [157] in 2008 the total allocation of forest firewood biomass amounted to 14.1 million tones with a total volume of 1.8 billions of m^3. According to Raslavicius et al. [158] the low contribution of bioenergy to the country's rural development was limited due to the energy policy that prevents the supply and use of modern energy sources in rural areas of Ukraine. While Ukraine has the highest potential for energy production by renewable sources in Europe, it may have the smallest share of bioenergy use compared to any other European country (Table 3.4)

However, according to Dolinsky [159], bioenergy technologies are widely introduced using modern boilers for heat production by the combustion of wood, straw, and peat waste. The feasibility study of the application of this technology have shown that biomass electricity production is difficult to sustain due to the low cost of electricity in the local market, but biomass heat generation is competitive and has accepted commercialization prospects. The situation is still unstable due to the recent geopolitical argument and the cold war confrontation with Russia, which is also the main provider of cheap gas. It seems that Ukraine should rely on its own energy sources such as bioenergy and not those from gas imports from Russia. It is generally considered that the development of RERs can boost Ukraine's agriculture and rural areas, create new markets for agricultural products offering decentralized power supply solutions for rural communities. Biomass heat

Table 3.4 Ukraine grain crop production and global production rank in 2015–16.

Grain crop	Production in million tones	Rank worldwide
Corn	39.0	6
Wheat	27.3	7
Sunflower	11	1
Sugar beet	11	7
Barley	9.5	7
Soya	4	8
Rapeseed	2.2	9

Source: UN FAOSTAT and worldatlas.com.

generation is feasible in Ukraine and the investments have a payback period of approximately 2 years. According to Raslavicius et al. [158] there are 3 large biogas plants in operation in the country, and another 12 plants under construction/design. However, opportunities created by RERs should not be used as a new excuse for the old, ineffective agricultural subsidy policy [158]. Farms that have failed to restructure to improve management techniques and increasing their efficiency in food production is almost certain to fail in RERs production as well [160].

1.5.8 Energy crops in Lithuania

Tilvikiene et al. [161] compared *Artemisia dubia*, *Miscanthus giganteus*, and *Sida hermaphrodita* with the traditional *Festuca arundinacea* grass in order to identify their cultivation as potential energy crops under the northern European-Lithuanian-weather conditions. From their results it was clear demonstrated that the productivity of crops varied between the species. The highest dry matter production was achieved by *Miscanthus giganteus* (21.54 t of dry matter/ha) while *M. giganteus* productivity varied from 9.70 to 21.54 t/ha, followed by *A. dubia*. However, after the third year of cultivation, the biomass yield from *A. dubia* was decreased while *M. giganteus* resulted the highest yield. *M. giganteus* is a quite promising energy crop cultivated worldwide with yield productions varied among different countries. For example, the crop in Italy has been recorded to produce 28–48 t/ha [162], in Denmark 22–29 t/ha [163], in Germany biomass productivity varied from 17 to 25 t/ha [164,165] and in Poland was about 13.7 t/ha [166]. As dry matter yield of *M. giganteus* is not dramatically lower compared to the countries it can be assumed that is a promising crop not only for warm climate regions, but also for temperate areas.

1.5.9 Energy crops in Italy

Piscioneri et al. [167] studied the use of Cardoon (*Cynara cardunculus*), a perennial herbaceous plant that has been cultivated for many years as a traditional food source in Italy as potential energy crop. From their findings it was commented that the dry matter content of capitula varies for all the genotypes with a maximum of 431.99 g/m^2 in the genotype Cy10 and a minimum of 257.06 g/m^2 in the genotype Cy8. In general, the good dry biomass yields ranging from 10 to 15 t/ha obtained during the third year. Furthermore, Manzone and Calvo [168] studied the biomass production of two poplar species in a 6 and 3 years rotation cycle with maize. In their study they conclude that the average biomass production for poplar was 13.9 Mg DM/ha per year and 19.2 Mg DM/ha for maize while the energy consumption for the poplar cultivations was about 15 GJ/ha per year, represented only the 6% of the energy biomass production (about 257 GJ/ha per year). Moreover, Mantineo et al. [169] studied the biomass yield and energy balance of *Arundo donax, Miscanthus x giganteus* [170] and *Cynara cardunculus* var. *altilis* for energy use in different irrigation and nitrogen fertilization levels in the semi-arid Mediterranean environment. In their findings it was concluded that on the third year, above ground dry matter yield increased over all studied factors, in *A. donax* from 6.1 to 38.8 t/ha, in *M. x giganteus* from 2.5 to 26.9 t/ha while in *C. cardunculus* biomass production was decreased from 24.7 t/ha to 8.0 t/ha. In the fourth and fifth years, above ground dry matter yields of all crops decreased, but *A. donax* and *M. x giganteus* still maintained high productivity levels in both years. Furthermore, energy inputs of *A. donax* and *M. x giganteus* were higher in the year of establishment than that of *C. cardunculus* (34 GJ/ha for *A. donax* and *M. x giganteus* and 12 GJ/ha for *C. cardunculus*), mainly due to irrigation. In particular, *A. donax* appeared the highest biomass production compared to the other two crops. *C. cardunculus* as an Mediterranean autumn-winter crop, could be suited to this environment because it can grow without irrigation treatment while *M. x giganteus* exhibited the highest nitrogen use efficiency and energy nitrogen use efficiency. Moreover, from findings of Solinas et al. [171] it is suggested that crops without irrigation generate lower impacts on the environment than rainfed species and that annual crops are more damaging than perennial crops. In a study conducted by Angelini et al. [172] biomass yield and energy balance of giant reed (*Arundo donax* L.) cropped in central Italy was about 17 MJ/kg dry matter and it was not affected by fertilization, or by plant density or harvest time. Fertilization enhanced crop biomass yield from 23 to 27 dry t/ha (years 1–6 mean value). Maximum energy yield output was 496 GJ/ha, obtained with 20,000 plants per ha and fertilization. In the mature crop the energy efficiency was highest without fertilization both with 20,000 (131 GJ/ha) and 40,000 plants per ha (119 GJ/ha). Finally, according to Carneiro and Ferreira [173] the results of the promotion of energy crops in Italy shown that the value of the supposed feed-in tariff (FIT) may not be enough to attract the interest of private investors for the projects. The creation of specific FIT for this type of biomass has been highlighted and can be justified by both the project risk perception and the expected strategic and environmental value of these investments. Research carried out has shown that this sector is innovative; investment can bring significant strategic advantages and can benefit the population financially and socially. However, the inexperience of energy crops regarding the development and the use of technologies required for this effort may delay the effective implementation of these projects.

1.5.10 Energy crops in Spain

Like other Mediterranean countries, energy crop species suitable for Spain agricultural land conditions are more less the same. Among them, *Eucalyptus* spp. Cultivated in about 20 million hectare worldwide are the main source of lignocellulosic biomass used by commercial plantations. Thus, Fernández et al. [174] studied biomass production of *Eucalyptus x urograndis* for energy purposes under different irrigation and fertilization conditions. From their findings it was suggested that irrigation and fertilization significantly increased aboveground biomass production, averaging 20.6–55.4 t/ha per year while commercial eucalyptus plantations in the region usually produce 3–25 t/ha per year of dry woody biomass with a fertilization range of 0–100 kg/ha of N per year. In order to estimate residual biomass from tree plantations, several new alternative techniques using crop growth models to forecast biomass [175], macroscopic and real-time RS information [176] or by using Terrestrial Laser Scanning (TLS), and Airborne Laser Scanning (ALS) instead the classical use of dendometric parameters such as tree height, crown height and crown diameter were investigated [177,178]. Furthermore, beside crop growth models, artificial neural networks (ANN) and models such as autoregressive integrated moving average (ARIMA are used to predict the future selling prices of the fuelwood [41]. Another important energy crop studied in Spain and general in Mediterranean climate conditions of low rainfall and hot dry summers, is the perennial plant *Cynara cardunculus* suitable for set-aside lands Gominho et al.

[179] with its productivity ranged to 10–20 Mg/ha/year [180,181]; while Lag-Brotons et al. [182] reported that its biomass yield production was ranged from 4.9 Mg/ha [167] to 27.6 Mg/ha [183].

1.5.11 Energy crops in Greece

The reduction of the imported energy consumption, the efficient use of RES in energy production system, the significant reduction in carbon dioxide (CO_2) emissions and the consumer protection are some of the main factors of the National energy policy in Greece. The share of RES on total primary energy supply in Greece has been increased significantly in recent years, reaching at 12.5% in 2016. Biofuels (including small amounts of waste) are the largest source of RES, representing approximately half of the share of RES in the total supply of primary energy. More specific, it has been calculated that the supply of biofuels amounted to 1.4 million tonnes of equivalent oil in 2016, representing a 36% increase since 2006. Solid biofuels accounted for three-quarters of all biofuels and waste on total primary energy supply, and are mainly used for heating in domestic boilers.

According to Panoutsou [184], energy crops can be attractive alternatives if properly integrated into existing agricultural activities. In terms of Greek production conditions, sweet sorghum, rapeseed, and sunflower are the main energy crops. Wild artichoke as a perennial and winter crop outperforms annual sorghum and sunflower crops because it produces a large amount of dry biomass of high-energy value with the lowest inputs. Among the other two crops, sorghum (for bioethanol and solid fuel production) has a large production potential in Greece, advantageous in many places over many annual summer crops, such as corn, but disadvantaged in underdeveloped industrial support, as well as the way and timing of harvesting.

In Greece, the agricultural energy crops grown based on data from the Ministry of Rural Development and Food (OPEKEPE) are wild artichoke, rapeseed, sunflower, and soybeans. Among energy crops, in recent years, sunflower is the most distributed energy crop in Greek agriculture cultivated at over 60,000 hectares. The second place in agricultural land coverage is occupied by the cultivation of oilseed rape, followed by the cultivation of soybeans and wild artichokes [185].

According to Mondol and Koumpetsos [186] the use of biomass in Greece plays an important role in energy production and about 450,000 tones of firewood were used to generate heat, cooling, and electricity.

However, consumption for house heating and cooking is depending on environmental issues and household sociological and economical characteristics [187–189]. Furthermore, another factor affecting fuel wood consumption is fuel wood prices observed among different forest species [190]. It has been found that the available agricultural and forestry residues annually equate to 3–4 million tons of oil, while the potential of energy crops, such as sunflower, rapeseed, and sweet sorghum can exceed agricultural and forestry residues and forest plantations that have been established covering an area of approximately 33,500 ha [41].

Compared to agricultural and forestry residues, these crops have the advantage of higher yield per unit area and easier harvesting process since it has been estimated that the amount of biomass that can be produced per acre is 3–4 tones of dry matter, equivalent to 1–1.6 TYP (Tons of Oil Equivalent) while in the case of dry crops it may reach 2–3 tones of dry matter, that is, 0.7–1.2 TYP. It was also estimated that the biomass availability in Greece consists of approximately 7.500.000 tones of agricultural crop residues (cereals, maize, cotton, tobacco, sunflower, twigs, sprouts, kernels, etc.) and 2.700.000 tones of forest logging residues (branches, bark, etc.). More specific, in the region of Thessaly, where large amounts of primary agricultural biomass is available, it has been estimated that about 707.164 tons producing 619 GWh of electricity and 894 GWh of thermal energy on annual basis, could be used for energy production purposes [191]. However, their economic viability is insufficient due to the limited evidence from the Greek crop to substantiate crop yields on biomass. In addition, research carried out by the National Kapodestrian University of Athens regarding the production of biodiesel from seed oils, have shown that the cost of biodiesel was estimated at 0.70 €/L, of which 0.15 €/L is the cost of industrialization and 0.55 €/L was the cost of seed oil [192]. According to the same author, the most efficient energy crop is sweet sorghum, which is clearly profitable and can therefore replace unsustainable sugar beet, maize and cotton cultivation as long as market prices for biofuels is acceptable. However, the reluctance of Greek farmers, however, seems to be reversing, and in recent years there has been an increase in areas where energy crops are grown. This increase reflects, on the choice of producers to switch to alternative crops due to shrinkage of subsidies and yields for traditional crops [71]. In a study completed by Panoutsou [193], it was found that cotton farmers are more reluctant to replace cotton crops with energy crops than cereal farmers, and this is because the variability of market prices of cereals significantly affects the competitiveness of crops. Thus, alternative forms of motivation should be provided to the farmers beyond the financial ones, in order the extensive use of energy crops to be achieved [194]. Concluding, Greece, with significant and valuable potential in more than one RES, can make the switch to energy use, although the financial situation of the country over the last few years has significantly limited any efforts in this area [195–197].

1.5.12 Energy crops in Turkey

During 2013, the biomass establishment capacity for electrical energy based on biomass productions was calculated to 237MW. Nazli et al. [198] studied the potential use of miscanthus (*Miscanthus x giganteus*), switchgrass (*Panicum virgatum*), giant reed (*Arundo donax*), and bulbous canary grass (*Phalaris aquatica*) as energy crop in the Mediterranean environment of Turkey. From his results it was concluded that giant reed produced the highest average biomass yield (12.86–36.78 t/ha) over 3 years, followed by miscanthus (12.75–23.54 t/ha), switchgrass (11.88–18.91 t/ha), and bulbous canary grass (5.21–10.83 t/ha). These results suggested that satisfactory biomass production could be achievable from miscanthus, switchgrass, and giant reed in the semi-arid Mediterranean environment under adequate moist conditions. However, irrigation requirement increases the energy cost, thus decreasing the energy ratio. Among them, bulbous canary grass may be evaluated as an alternative bioenergy crop in the dry marginal lands of Mediterranean for sustainable biomass production [199,200]. However, it may be suggested that all four crops have a remarkable potential as dedicated bioenergy crops in the Mediterranean environment due to their adaptation capabilities and satisfactory biomass yields although biomass yields of miscanthus, switchgrass, and giant reed reduced significantly when the harvest was delayed from autumn to winter. However, for bulbous canary grass an autumn harvest should primarily be preferred. Agricultural residues can be also used as biomass for energy production. Total amounts of biomass residues obtained from arable field crops and horticultural crops in Turkey were estimated at 59.432 kilotonne and 15.652 kilotonne, respectively with a theoretical energy potential calculated to 908.119 TJ and 90.354 TJ, respectively.

1.5.13 Energy crops in United States of America and Canada

According to Villamil et al. [201], RES in 2009 share in the US energy market reached 8% with biomass being the largest domestic energy source accounting for almost 4% of the total energy consumption. The interest in biofuels in the United States is increasing since biomass cropping systems have also the potential to alter the ecosystem services provided by agricultural landscape by increasing pest suppression and pollination, decreasing greenhouse gas emissions, and enhance grassland bird communities [202]. For these reasons, Federal and State regulations require the increase in the minimum amount of energy produced from RES. For example, the State of Illinois supports that by 2025 at least 25% of electricity should be generated from RES including potential miscanthus' adoption [203,204]. Rockwood and Dippon [205] studied the potentials use of *Eucalyptus grandis* and slash pine as biomass energy crops for southern and northern Florida. From their findings it was suggested that *E. grandis* progenies may produce up to 70 dry Mg/ha in 24 months while Slash pine's maximum productivity, can be reached to 15.2 dry Mg/ha after 8 years of under the intensive culture practices, when approximately 1518 trees/ha superior progenies were planted in the field. Furthermore, Paine et al. [206] commented that the short-rotation woody crop *Populus* spp. L is considered to be the best candidate for biofuel plantations. Furthermore, regarding land use it was mentioned that energy crop species could be cultivated in highly erodible land; wetlands presently converted to agricultural uses; and marginal agricultural land in selected regions. In Tennessee region a short rotation (6–10 years) woody crop (SRWC) was investigated for potential biomass production. Projected SRWC yield varied from 5.4 to 9.7 dry Mg/ha compared to switchgrass yields, ranged from 9.7 to 19.8 dry Mg/ha. In the same study, it was concluded that these two energy crop species could potentially support 18 GW (assuming 10,000 t, Btu of wood/kWh) if wood were the only feedstock and 30 GW if switchgrass was the only fuel [207]. In the same study Zamora et al. [208] evaluates the biomass production, chemical composition, and energy content of selected hybrid poplar commercial clones such as *Populus nigra, P. maximowiczii, P. deltoid* established in polyculture systems. After 7 years of growth, biomass produced the highest amount of biomass (7.9 Mg/ha) and largest theoretical ethanol yield (425 L/Mg of dry biomass). After 13 years of growth the biomass production reached to 11 Mg/ha producing approximately 3.85–4.92 m^3/ha of ethanol.

Moreover, study completed by Surendra et al. [209] examined for 3 years the composition of Napier grass (*Pennisetum purpureum*) and energy cane (*Saccharum hybrid*). From their findings, it was concluded that fiber content was higher in Napier grass than energy cane. In more details, significantly higher fiber content was found in the leaves of energy cane than stems, while fiber content was significantly higher in the stems than the leaves of Napier grass. In a study conducted by Pedroso et al. [210], it was evaluated the biomass potential production of several plant species such as *Miscanthus* × *giganteus*, switchgrass (*Panicum virgatum*), big-bluestem (*Andropogon gerardii*), bermudagrass (*Cynodon dactylon*), and elephant grass (*Cenchrus purpureus*), tall wheat grass (*Thinopyrum ponticum*), and tall fescue (*Festuca arundinacea*) in Davis-California area where climate conditions are similar to those in Mediterranean region. Their results suggest that *Miscanthus* and switchgrass have greatest potential as bioenergy crops yielding about 10.9 and 8.4 Mg/ha respectively. In order to maximize efficiency of energy crops, biomass production is suggested to be pelletized. According to Uslu et al. [211] this process can increase energy density, can be implemented to reduce transportation costs since the high energy density means that less feedstock needs to be transported. Moreover, Shahrukh et al., [212] developed a process model to conduct a comparative energy analysis for steam-pretreated pellet production of agricultural residues (wheat straw) and energy crops (switchgrass). The results of the analysis have shown that the heating value of the fuel can be improved by the steam pretreatment process. At the same time, the steam pretreatment process results in an increase in the process energy requirement for drying and steam pretreatment while the drying energy requirement is higher for switchgrass pellets than for wheat straw pellets.

1.5.14 Energy crops in Mexico

Energy production in Mexico is highly dependent on fossil fuels. Between 2007–2017 an average of 8935 PJ per year have been produced of which 88.2% consisted of fossil fuels, 7.2% of renewable sources, 3.4% of coal, 1.2% of nuclear while biomass comprises only the 4% of the total energy production. It is also mentioned that 56% of agricultural land is cultivated by feed crops such as maize, beans, wheat, barley, rice, sorghum, soybeans, sesame, and sunflower; while sugarcane, cotton, coffee, and agave are cultivated in the 8.7% of the area. Out of these crops, only maize can be considered at the National level, because it is grown practically in every part of Mexico [213]. The harvest and processing of these crops generates large amounts of residues. According to Lozano-García et al. [216] the potential electricity generation by gasification and combustion, based on crop residues in Mexico is estimated to be around 71 GWh or 79.3 GWh using hydrogen for fuel cells. Thus, Mexico has shown an increasing interest in producing energy from biomass and other renewable energy sources. For this reason, policies, laws, financial transition, and national strategies regarding energy sector have been issued last year by the Mexican government. According to Honorato-Salazar, [214] energy production generated from residues of agricultural crops is estimated to 87.94 megatons of dry matter per year (Mt DM/year) with total energy potential estimated at 670.3 PJ/year, of which 542.5 PJ/year from primary residues and 127.8 PJ/year from secondary residues. Among cultivated species, maize, sorghum, and sugarcane offer about 43.3%, 25.5%, and 18.1% of the energy production, followed by wheat (6.3%), barley (1.6%), and beans (1.0%). However, sugar cane, sweet sorghum, and beetroot are envisaged as the main crops for bioethanol production, while Jatropha, castor oil, and oil palm are the crops for biodiesel production.

Furthermore, forest species such as Pine (*Pinus* spp.), fir (*Abies* spp.), other conifers, oak, other temperate hardwoods, valuable tropical hardwoods, and common tropical hardwoods are also used for energy production accounting for about 1.42 Mt DM/year with an energy potential of 30.72 PJ/year. Firewood and sugarcane bagasse are still the main biomass energy sources in Mexico, supplying 5.2% and 0.9% of the National primary energy demand, respectively.

1.5.15 Energy crops in Brazil

Brazil is one of the leaders in biofuel production worldwide with sugarcane sector to play an important role in the National economy, considering energy as a substitute for conventional gasoline and sugar production [215] since 1970. At an International level it is agreed that the Brazilian ethanol program is regarded as an important initiative toward fuel diversification and a bridge between conventional fossil fuels and experimental energy sources [216]. Thus, several studies have indicated that Brazil's experience in the sugar-energy sector could make it a model for other countries [217,218]. Sugarcane cultivations are mainly focused on the State of Sao Paolo, where 4.5 million hectares are cultivated, in Western-Center region where the sugarcane area is covered 1.8 million hectares and in North-East region where 0.84 million hectares are cultivated [219]. Sugarcane production during 2014–15 was approximately 634 megatons (Mt) of harvested cane and almost 29 billion liters of ethanol [220] cultivated in approximately 9 million hectares.

However, climate change does not leave Brazilian sugarcane production unaffected. It has been reported by de Carvalho et al. [221] that sugarcane yield could be affected by possible future reductions in rainfall levels especially in the Northeastern region of Brazil. Regarding sugarcane production for energy purposes Barros et al. [222] calculated ethanol and gasoline consumption along with the price ratio series. In their study examined the dynamics of ethanol consumption in Brazil by taking into account two important features of the data: its degree of persistence and the seasonality. From their results it concluded that strong policy measures must be adopted on prices in the event of shocks since they do not recover by themselves in the long run. de Freitas and Kaneko [213] also analyzed the characteristics of ethanol demand at the regional level taking into account the peculiarities of the developed Center-South and the developing North-Northeast regions and found that demand and prices for ethanol in Brazil differs between regions. While in the Center-South region the price elasticity for both ethanol and alternative fuels is high, consumption in the North-Northeast is more sensitive to changes in the stock of the ethanol-powered fleet and income. Another factor affecting Brazilian sugarcane energy sector is that beside the important productive investments that they were made by oil, energy, and chemical companies in the first decade of the 2000s, since then, Brazilian energy biofuel sector may not be prepared to face the technological and market challenges emerging the present day in the biofuels domain worldwide [223]. This is also supported from the average productivity, which is far from what it could be and varied over 35% within the same region although, during the previous decade, Leite et al. [224] commented the possibility of Brazil supplying ethanol to replace 5% of the global demand for gasoline by 2025. However, Brazil among United States (with double ethanol production than Brazil) are the main bioethanol producers accounted for 85% ethanol production worldwide [225]. Finally, another important source of biomass is land forest planted with Eucalyptus species.

1.5.16 Energy crops in China

China's primary energy source is consumed mainly by coal (66% of China's domestic energy consumption), crude oil and natural gas (19% and 6%, respectively) [226]. In order to reduce over-reliance on coal for power and heat generation and to mitigate associated environmental and health problems, China has implemented a series of regulatory policies with main purpose to cover 11% of its primary energy consumption from non-fossil fuel sources by 2015 and 15% by 2020. Thus, the potential of bioenergy production from cultivated plant species has aroused the interest of scientists, political leadership and consumer farmers in China and the various aspects of their production and their impact have been studied. As the major obstacle to the development of bioenergy in China is large-scale, steady supply of raw materials has been the focus of attention in recent years with regard to biomass resources. Biomass resources include energy crops, forest residues, wood fuels, manure, and more. Li et al. [227], estimated the theoretical amount of biomass production to 2.61–3.51 TIA (tones of oil equivalent) while the available raw material is approximately 440–640 millions of TIA. However, only 1.5%–2.5% was used in 2009 as bioenergy. Plant species suitable for non-fertile soils are Salix, Tamarix, Caragana, and *Prunus* sp. The estimated amount of biomass in these soils is expected to reach to 100 million tonnes/year in 2020, and 200 million tones/year in 2050. It was reported that crop residues in 2003 amounted to 308–360 million TIA, with the available raw material reaching from 150 to 216 million TIA. Similarly, wood and forest residues reached 1.242.000.000 TIA, with available reserves amounting from 166 to 300 TIA. In addition Chen [228] estimated potential biomass supply from crop residues in China using a mathematical programming model. From his analysis, it was concluded that China can potentially produce about 174.4–248.6 million dry metric tons of crop residues per year when biomass prices are larger than $100 per metric ton. As mentioned earlier, China also produces large quantities of biomass as crop residues such as cereal straw, and other plants such as oilseeds, cotton, potatoes, and sugar beets. More specifically, it has been found that approximately 687.4 million tones of cereal straw was produced as a residue per year, of which more than 400 million tonnes could be used for bioenergy production. The main areas for cereal straw production are located in Northwest and South China in the provinces of Heilongjiang, Hebei, Henan, Shandong, Jiangsu, and Sichuan [229]. In addition, more than 1000 species of wild plants recorded in the Chinese territory are suitable for biodiesel production. Number of studies have been conducted on the control and breeding species of these plants [230–232]. For example, the cultivation of *Jatropha curcas* has been studied with emphasis on cultivation techniques and the extraction of biodiesel [233]. However, only a limited number of plant species and varieties are used for biodiesel production. Other important plant species used as energy plants for biodiesel production in China are *J. curcas*, a shrub grown in southwestern China, *Vernicia fordii, Sindora glabra, Sapium sebiferum, Pistacia chinensis, Euphorbia tirucalli,* [230,232], and species of Salix (Salicaceae), Hippohae (Elaeagnaceae), Tamarix (Tamaricaceae), and Caragana (Leguminosae) [234]. The Chinese leadership believes that bio-energy will play an important role in future energy supply. However, the feasibility of developing bio-energy can vary widely across regions, depending on financial, environmental and technological factors.

1.5.17 Energy crops in Pakistan

Economic development, improved lifestyle and increasing population have lead increases on energy needs for Pakistan [235,236]. According to Pakistan economic survey [237], Pakistani agriculture contributes about 21% to Gross Domestic Product (GDP) and is the profession of over 45% of the population. It has calculated that contribution of biomass waste is approximately 802 Tera BTUs (TBTUs). Although energy crops cultivation is growing at a rate of up to 10% per year, only a small percentage of that is used for energy production purposes. With no petroleum resources and no significant funds, the country needs to shift its energy needs to other forms of energy such as biomass by exploiting the most barren areas. According to Khan and Dessouky [238] in 2009, ethanol production was at a high level as there were 76 sugar mills operating with a crushing capacity of

300.000 tones of cane per day. In Pakistan, sugar cane and molasses are the main by-products used for ethanol production. In addition, corn stalk, sugarcane trash, rice straw, wheat straw, and cotton stalks are the major crop residues having a production of 6.43, 8.94, 17.86, 35.6, and 50.6 Mt, respectively. For molasses process, there were 21 distilleries with a processing capacity of 2 million tons of molasses and the production of 400.000 tons of ethanol in 2009. In order to evaluate and forecast the current and future availability of selective crop residue to generate renewable energy Kashif et al. [239] developed a forecast model incorporating historical trends in the crop yield such as wheat straw, rice husk, rice straw, cotton straw, corn stover, and bagasse. Using this model, it was found that about 40 million tones of crop residue was available in Pakistan for power generation in the year 2018 considering the available residue removal factor of 50%. This translates to an estimated potential of about 11,000 MW of electricity generation capacity using crop residue derived biomass for 2018. This capacity is predicted to gradually increase up to 16,000 MW by the year 2035 based on the trends in the growth of crop production since 2001.

1.5.18 Energy crops in Iran

As Iran is one of the most important oil producer's worldwide, uses of alternative energy sources are still remaining limited. However, the growth rate in electricity generation from biomass reached about 495% (11.6 MW) from 2009 to 2015. This high percentage increase caused mainly due use of biogas and waste incineration plants. It is estimated that renewable energy production in Iran is 22 million barrels of oil equivalent (MBOE) mainly based on solid biomass including charcoal, wood, wood waste, municipal solid waste, animal manure and plant waste. According to the Agricultural Statistics of Iran, the crop cultivation area in 2017/18 was 11.77 million hectares and 20% of the total crop production is estimated to be wasted as residues [240] as cereals is one of the most cultivated plant species in the country. Thus, the use of the crop residuals should particularly be considered for energy production. According to Samadi et al. [241] total energy produced by gasification technology was 341,290 TJ. Among this amount electricity and heat obtained from this energy was 66,075 and 3 99,112 TJ, respectively.

1.5.19 Energy crops in India

India produces 686 MT gross crop residue biomass on annual basis, of which 234 MT (34% of gross) are estimated as surplus for bio-energy generation. Amongst all crops, sugarcane produces the highest amount of surplus residue followed by rice. The estimated annual bio-energy potential from the surplus crop residue biomass is 4.15 EJ, equivalent to 17% of India's total primary energy consumption [242]. Thus, technological developments and government policy help to promote the production of electricity from forest biomass through gasification technology. Although, forest biomass to generate electricity through the process of gasification technology is expected to bring economic and environmental benefits to India, attention has to be paid to the negative environmental aspects of promoting plantation as monoculture. According to Dwivedi and Alavalapati [243] in India, about 5 million MWh of electricity can be produced annually resulting in a reduction of CO_2 emissions by 4 million tons per year. The same study found that the cost of generating an electricity unit was inversely proportional to the scale of production. Regarding production costs, it was found that the cost per unit of electricity used as fuel for *Acacia nilotica* was the lowest among all selected species. More specifically, the total annual dry biomass available after 3 years of energy plantation amounts to approximately 2.7 million tones. In 2012 it was estimated that this amount of biomass could generate 5.2M MWh of electricity for home use, accounting for about 67% of total electricity needs. Furthermore, cultivation of *Jatropha curcas* an adaptable plant on a wide variety of soils and climatic zones including in nutrient depleted poor soils, is being undertaken mainly for fuel oil production, but it also results in biomass residues such as fruit husks, seed shell, seed cake, or kernel meal. The energy density of jatropha residues resulting an estimated total marketable energy yield of 90 GJ/year/ha of a 5-year-old jatropha plantation compared to a total yield of 30 GJ/year when jatropha oil alone is marketed. However, several by-products of jatropha are considered to be not commercially marketed due to relatively low energy density. Furthermore, according to Patel et al., [244] varieties of bamboo (*Bambusa balcooa*), sorghum, and pearl millet (*Pennisetum glaucum*) were examined as energy crops resulting a positive net energy balance of 602.49 GJ/ha/year, 454.29 GJ/ha/crop cycle, and 278.20 GJ/ha/crop cycle, respectively.

1.5.20 Energy crops in Thailand

Production of biodiesel from palm oil began in Thailand on a small scale 4 decades ago, and recently there is an interest in further expansion. According to Siriwardhana et al. [245] the Thailand government's strategy for biodiesel development was to replace 10% of oil use in the transport sector by 2012. The plan was to increase the use of biodiesel by 365 million liters in 2007 to 3.100 billion liters by 2012. The biodiesel development strategy proposed to increase the demand for biodiesel from 0.03 million liters/day in 2005 to 8.5 million liters/day in 2012. This goal has been achieved by using palm oil cultivation. In 2009 there was about 800 petrol stations in the country supplying B5 (a mixture of 5% biodiesel and 95% diesel) for trucks, with total sales of 548 million liters/year. Since 2007 biodiesel consumption increased from 3.43 million liters to 65.13 million liters in 2009. Due to Thailand government subsidies, in 2009 price of B5 biodiesel was lower than the diesel price by about 2.1 cents per US dollar (0.70 Bahtper liter). To meet this achievement in 2008, there was about 72 local small-scale plants that produced 100 L of biodiesel per day. These plants were used palm oil, Jatropha, and cooking oil residues as raw material for biodiesel production [246]. According to [247] 1 hectare of palm oil crop could produce 5 tons of crude palm oil per year, which was 5 times the amount produced from rapeseed. Compared to soybean yields in the United States, biodiesel production per hectare in Thailand was in 2009, up to 10 times higher. Finally, it was recorded that Jatropha cultivation could replace palm oil and could be the first choice for biodiesel production. In 2007, 16.000 hectares of Jatropha were cultivated in Thailand, but only 3200 of them were commercially separable.

1.5.21 Energy crops in Malaysia

As one of the main palm oil producer's worldwide, residues of biomass in Malaysia are estimated to reach on 2020 to 100 million tons (dry weight) compared to 83 million tons (dry weight) in 2012. However, according to Chin et al. [248] the development of biofuels in Malaysia is still essential not only in terms of energy security but also in achieving the goal of reducing carbon emissions by 2020. The support system for financial contracts between farmers and biodiesel industries are important for the development of biofuels, as high production costs are still an important obstacle to their development. Finally, the same study argues that the development of biofuels will face problems if the issues of social acceptance are neglected and the social acceptance of energy crops and biofuels by the inhabitants of the country is not advanced. Thus, the exploitation of energy crops and crop residues as alternative energy sources has been investigated by several authors the last few years. Sohni et al. [249] studied the physicochemical characterization of Malaysian crop and agro-industrial biomass residues such as oil palm frond, oil palm trunk, empty fruit bunches, palm kernel shell, rice husk, rice straw, and kenaf biomass as renewable energy biomass resources (Table 3.5).

Table 3.5 Relationship between climatic conditions and soil types with energy crop species.

Energy Crop	Country	Climate conditions (temperature, climatic zones, humidity)	Soil type (including nutrients' composition in N,K,P)
Willows *Salix alba*, *S.viminalis*	Sweden, Austria, United Kingdom, Poland, Czech Republic, United States, China	Thrives in a wide range of moisture conditions. It requires full sun and partial shade are best for this tree. It prefers a minimum of four hours of direct, unfiltered sunlight daily. The plant species are found primarily in cold and temperate regions of the Northern Hemisphere while there are some species that are low-growing or creeping shrubs suitable for the arctic and alpine conditions [250]	Grows well in acidic, alkaline, loamy, moist, rich, sandy, well-drained, and clay soils. It grows well near water but has some drought tolerance.
Poplars-Aspen *Populus tremula* and *P. tremuloide*	Sweden, Austria, Denmark, Germany, Poland, Italy, United States	*Populus* sp., *is* the most widely distributed tree in North America. Climatic conditions vary greatly from −57°C to 41°C in summer) over the range of the species, especially winter minimum temperatures and annual precipitation. Aspen is a thermophilic species that is expected to benefit from rising temperatures in boreal forests [251].	It grows on many soil types, especially sandy and gravelly slopes, and it is quick to pioneer disturbed sites where there is bare soil. It prefers good aspen soils witch are usually well drained, loamy, and high in organic matter, calcium, magnesium, potassium, and nitrogen. Nevertheless, growth on sandy soils is often poor because of low levels of moisture and nutrients. Fertilizer interacted strongly with irrigation increasing mean height about 19% and stem volume 92% after three growing seasons. Foliage analysis indicated that the main growth response was to the P supplied by the fertilizer. For seedlings growth it is suggested 25 g N, 25 g P, 15 g K, and 0.32 g B per tree [252].
Grass and switchgrass *Panicum virgatum*	Sweden, Poland, Turkey, United States	Is native to North America in Canada southwards into the United States and Mexico. Is a versatile and adaptable plant growing in many weather conditions, lengths of growing seasons, soil types, and land conditions. It appears widespread adaptability in temperate climates.	Is a native perennial warm season grass with the ability to produce moderate to high yields on marginal farmlands. Crop appears drought and flooding tolerance, relatively low herbicide and fertilizer input requirements, and hardiness in poor soil conditions.
Grass *Festuca arundinacea*	Lithuania, United States	Is native to Europe especially in cool-season areas. However, it is adaptable to climate conditions similar to those in Mediterranean region since is the most heat tolerant of the major cool season grasses.	Increases soil organic carbon and soil fertility, reduce erosion [253].

(*Continued*)

Table 3.5 Relationship between climatic conditions and soil types with energy crop species. (*Cont.*)

Energy Crop	Country	Climate conditions (temperature, climatic zones, humidity)	Soil type (including nutrients' composition in N,K,P)
Napier grass, Elephant grass *Pennisetum purpureum*	United States	It is one of the highest yielding tropical grasses. It can be grown under a wide range of climate conditions and systems such as dry or wet conditions. It is mainly found from 10°N to 20°S where temperatures range from 25°C to 40°C [254] and annual rainfall over 1500 mm. Is sensitive to frost.	Is tolerant of drought and will grow in areas where the rainfall range is 200–4000 mm. However, is not tolerant of flooding and prefers well-drained soils. It grows better in rich, deep soils, such as friable loams, but can also grow on poorly drained clays, with a fairly heavy texture, or excessively drained sandy soils with a pH ranging from 4.5 to 8.2 [254–256].
Bermuda grass *Cynodon dactylon*	United States	Is native to most of the eastern hemisphere. It is widely cultivated in warm climates all over the world between 30°S and 30°N latitude, when rainfall varies between 625 and 1.750 mm. It prefers climate conditions similar to those in Mediterranean region.	Soil acidity seems to be the main factor determining the distribution of the species cytotypes [257].
Elephant grass *Cenchrus purpureus*	United States	It needs good moisture for production. Does not tolerate prolonged flooding or water logging. Best growth achieved between 25 and 40°C, and little growth below about 15°C. It appears moderate shade tolerance	It grows best in deep, well-drained friable loams with pH varied between 4.5 and 8.2. It grows on a wide range of soil types where provided fertility is adequate. However, it should be planted into fertile soil. Once established, requires, 150–300 kg/ha/year N, together with other nutrients as indicated by soil tests. Responses at much higher levels of applied N have been obtained. Yields decline rapidly if fertility is not maintained [258,259].
Tall wheat grass *Thinopyrum ponticum*	United States	It requires a minimum annual rainfall about 425 mm, although it is generally regarded as a pasture for higher rainfall zones, up to as much as 800mm. It is frost tolerant and recovers well after frost damage.	It is widely planted in Western Australia on saline land in the southern agricultural areas, on land that shows signs of salt and is not permanently waterlogged. Nitrogen fertiliser will greatly increase production. It is tolerant of acid and alkaline soils [260,261].
Sugar beet *Beta vulgaris*	Sweden, Ukraine, United States, Germany, United Kingdom, Poland, France	It can be grown in a wide range of climatic conditions and are noted for tolerance to soil salinity [262]. Humid conditions. Resistant in low temperatures. Most varieties of beets are frost hardy, but they do not tolerate extreme heat. It requires regular, deep watering and the soil should never be allowed to dry out [263].	Well-drained soil types, with a pH between 6.5 and 7. Poor fertilizer limits high yield and quality of beets. An adequate supply of nitrogen is essential for optimum yield. However, excessive nitrogen decreases sugar content and increases impurities (Na, K, and amino-N) [264,265].

Cane *Arundo donax*	Sweden, Italy, Turkey	It has been widely planted and naturalised in the mild temperate, subtropical and tropical regions of both hemispheres [266–268].	It grows well in damp soils, either fresh or moderately saline. It appears resistance in wet fields. Pre-plant fertilizer is distributed according to the initial soil fertility, but usually P application (80–100) and N fertilization might be reduced to intermediate levels (60 kg N/ha) to achieve high agronomic use efficiency [269]
Sugar cane *Saccharum afficinarum, S. hybrid*	Mexico, Brazil, Pakistan, India, United States	It needs a precipitation between 1500 and 2000 mm/year rainfall and a well defined dry season of about 4–5 months. Nevertheless, is susceptible to drought. Some cane varieties are well adapted to semi-arid conditions. Optimal growth conditions are achieved with 1800 hours sunshine/year [270].	It can adapt to a large variety of soil types (70% clay–75% sand) but prefers medium-heavy texture soil with neutral-slight acid pH. It usually requires 85–100 kg N, 60 kg P_2O_5, and 180–250 kg K_2O [270].
Phalaris arundinacea, P. aquatic, P. canriensis	Sweden, Turkey, United States	European strains may survive colder temperatures than some North American strains [271].	It grows well in clay/loam soil and in sand (if the water content is high enough) but do not in peaty soils..It is suitable for dry marginal lands. It is moderately tolerant of drought and saline or alkaline soils [272]
Maize *Zea mays*	Sweden, Austria, Ukraine, Italy, Greece, Mexico, Pakistan	Humid conditions at the beginning of the cultivation and dry conditions for the seed maturation. Exposition to rainfall variability and climate change may affect resource use efficiency and appear a negative impact in crop yield [273,274].	N application rates between 50 and 100 kg/ha depending soil types.
Cereals (wheat, triticale, barley)	Sweden, Austria, Poland, Ukraine, Italy, Greece, Mexico, Pakistan	Humid conditions during winter and dry conditions during seed maturation. Resistant to low temperatures when plants are in early stage development.	In wheat, 7–14 kg N per and 15 kg P/acre should be applied in the fall either before or at the time of seeding. The remainder of the amount of fertilizer (25 kg) N needed should be top dressed early in the following spring.
Oat *Avena sativa*	Sweden, United Kingdom, Germany, United States, Canada, Poland, China	It is grown at any moderately fertile soil in full sun and tolerates cool moist and mild frost conditions.	It prefers, dry wasteland, cultivated ground and meadows, especially on heavier soils. High N content in soil is not a desirable condition may lead to lodging.
Pearl millet *Pennisetum glaucum*	India and many African countries	It is well adapted to growing areas characterized by drought, low soil fertility, and high temperature. Climate change may result yield reduction and/or total crop failure caused by erratic seasonal rainfalls and floods [275].	It performs well in soils with high salinity or low pH. Suitable for areas where wheat and maize would not survive.

(*Continued*)

Table 3.5 Relationship between climatic conditions and soil types with energy crop species. (*Cont.*)

Energy Crop	Country	Climate conditions (temperature, climatic zones, humidity)	Soil type (including nutrients' composition in N,K,P)
Sweet sorghum *Sorghum bicolor*	Mexico, India, Nigeria, United States, Greece,	It is well adapted to warm and dry conditions. Fiber sorghum can successfully grow in temperate climates.	It tolerates salts and alkaline soils and it can grow in a wide range of pH (5–8). Recommended fertilization rates are 80–100 kg N/ha, 70 kg P_2O_5/ha, and 140 kg K_2O/ha [276].
Rice *Oryza sativa*	Mexico, Pakistan, India, Malaysia	Tropical wet conditions. The optimum temperature for rice cultivation is between 25°C and 35°C [277]. A further temperature warming by 1–2°C will have negative impact on rice productivity [278].	It prefers wetlands. For 1 tn/acre rice production is required/acre 16–20 kg N, 4–8 kg P_2O_5, 6–10 kg K_2O, and 2–3 kg SO_3.
Clover *Trifolium repens, T. pretense*	Sweden	Some species are more tolerant of certain climatic and soil conditions.	It does not require N fertilization since is a N-fixing species.
Virginia fanpetals *Sida hermaphrodita*	United States, Austria, Poland, Lithuania	Climate and soil requirements are moderate [279].	In Poland is cultivated on soils with medium and low production capacities.
Miscanthus giganteus	Austria, Poland, Lithuania, Turkey, United States	Climate conditions similar to those in Mediterranean region. Good survival of established *M.* x *giganteus* with winter air temperatures dropping as low as −29°C [280] has been reported, other results indicate a major risk to viability when soil temperatures drop below −3°C at the 5-cm soil level [281].	It grows well in marginal land and needs low fertilizer inputs. It appears high nitrogen use efficiency. N fertilization of 180 kg N/ha^{-1} has little or no effect on the yields. Conversely, in other studies it was observed significant increases in yield [282].
Amur silvergrass *Miscanthus sacchariflorus*	Poland	Can overwinter in very low temperatures (−21°C) [276].	
Alfalfa *Medicago sativa*	Austria	Alfalfa is a widespread leguminous fodder grasses cultivated for feeding livestock animals. Alfalfa seeds have been germinated when the soil temperature is more than 5°C.	Soil planted with *M. sativa* L. is high in fertility and productivity. It is advised to cultivate in poor saline soil and its seeds are poorly developed in high salinity and acid soils [283].
Rapeseed *Brassica napus*	United States, Germany, Czech Republic, Ukraine, Lithuania, Greece	It is used as a cover crop in the US during the winter as it prevents soil erosion, producing large amounts of biomass. Optimal temperature for seed germination is 10–20 °C. It can overwinter even at −15°C.	It can grow in several soil types with pH ranges from 5,5 to 8,5. Optimal soil pH for plant growth is 6–7,5. For better yields it requires 25 kg/acre N early in the Spring combined with S.

Jerusalem artichoke *Helianthus tuberosus*	Poland	The plant is a suitable crop in any soil and climate where *Zea mays* will grow. Plants are sensitive to day-length hours, requiring longer periods of light from seedling to maturation of plant, and shorter periods for tuber formation. Climate conditions similar to those in Mediterranean region are ideal for cultivation.	It is suitable for sandy, loamy and heavy clay soils with a preference to well-drained soil. It can also grow in nutritionally poor soil. However, heavy soils produce the highest yields. Acid, neutral, and basic (alkaline) soils are suitable for cultivation.
Prairie cordgrass *Spartina pectinata*	United States, Canada, Poland	It grows up to 61°N in its native range in North America. *S. pectinata* rhizomes shown no signs of winter injury and they are very tolerant in cold conditions [284].	It is suitable for sandy, loamy and lay soils. Suitable pH: acid, neutral and basic (alkaline) soils and can grow in saline soils. It can grow in semi-shade (light woodland) or no shade. It prefers moist or wet soil.
Big bluestem *Andropogon gerardii*	Poland, United States	It responds differently to the combined effects of warming and reduced water availability [285].	It requires light porous sandy soil. Plants tolerate clay, alkaline pH, drought, moderate salt levels, and seasonal flooding. Fertilizer applications, excess shade, or abundant water can cause plants to develop weak stems that sprawl or collapse.
Lolium multiflorum, L. perenne	Czech Republic	It is best suited for growing in temperate climates where optimum growth temperature 18–20°C occurs [286]. Minimum annual rainfall requirement is 457–635 mm [287].	It grows best on fertile, well-drained soils but has a wide range of soil adaptability, and tolerates both acidic and alkaline soils [288]. It responds well to applications of N and P and is moderately tolerant of acid soils although there is a sensitivity to aluminum concentration when soil pH is low ($pH_{Ca} < 4.4$) [289]. Nitrogen application has been proposed as a measure to improve persistence of perennial ryegrass in intensively grazed dairy pastures [290]
Phacelia tanacetifolia	Czech Republic	The plant is being used in a wide range of climatic regions. Best-planted soil temperature is between 2.7 and 20°C. It is highly attractive to honeybees. Its use as a fall/winter cover crop may be appropriate when it will be followed by a vigorous cash crop (e.g., potatoes) in early spring [291].	It is adaptable in many soil types.
White mustard *Sinapis alba*	Czech Republic	*S. alba* is among the widest developed wild crucifers in the Mediterranean areas [292]. It grows best where the annual precipitation varies from 35 to 179 cm, annual temperature from 5.6 to 24.9°C and pH from 4.5 to 8.2.	It can grow in sandy, loamy and heavy clay well-drained moist soils with suitable pH acid, neutral and alkaline. For best production, it requires high nutrient soils with a high level of nitrogen

(*Continued*)

Table 3.5 Relationship between climatic conditions and soil types with energy crop species. (*Cont.*)

Energy Crop	Country	Climate conditions (temperature, climatic zones, humidity)	Soil type (including nutrients' composition in N,K,P)
Trifolium incarnatum	Czech Republic	Similar to alfalfa	
Fobber radish *Raphanus sativus* var. *oleiformis*	Czech Republic	*R. sativus* is a temperate crop, usually heat-sensitive [293].	The crop requires well-drained, light, sandy, deep, soils, with pH 6–6.5.
Sunflower *Helianthus* sp.	Ukraine, Greece, Mexico	Sunflower's native environment is dry, requiring 6 or more hours of direct sunshine per day. While, it can survive extreme heat, however, optimal temperatures for growing sunflowers are between 21 and 26°C. It prefers wet, humid weather, although well-draining soil or containers are essential in such climates.	It grows at any soil conditions, except those that are waterlogged. However, ideal soil environment is a well-draining sand or loam mixture, with a pH range of 6.0–7.5. Basic fertilization with 25–30 kg of fertilizers 20-10-10, 25-5-5, 20-7-12, 15-15-15, 15-9-15, etc. Spring fertilization with 5 units of N corresponding to approximately 11 kg of urea (46% N), 13 kg of urotheic (40% N), 16 kg of ammonia nitrate (33.5%), and 20 kg of nitro-sulfur ammonia (26%). However, Increasing fertilizer rates appear a negative influence on sunflower rooting [294].
Artemisia dubia	Lithuania	Is a promising energy crop in the northern part of the temperate climate zone [295].	Nitrogen fertilization at 90 and 170 kg/ha N rates slightly increased plant height and accumulation of biomass [296].
Cardoon *Cynara cardunculus*	Italy, Spain, Greece	It grows in dry conditions with severe summer droughts. Conditions those are similar to the Mediterranean climate [297].	For better growth It requires fertilizer with 50 kg/ha N, 90 kg/ha P_2O_5, and 40 kg/ha K_2O [298].
Eucalyptus sp.	Australia, Spain, United States, Brazil	Eucalypts grow well in a wide range of climate conditions. However, most trees are not tolerant of severe cold [299,300].	Fertilization range of 0–100 kg/ha of N per year.
Cotton *Gossypium* sp.	Greece, Mexico, Pakistan	Climate conditions similar to those in South Mediterranean region and East Asia.	

Jatropha	Mexico, China, India, Thailand	It is mainly cultivated in tropical savannah and monsoon climates and in temperate climates without dry season and with hot summer, while very few varieties were found in semi-arid and none in arid climates. Plantations in arid and semi-arid areas hold the risk of low productivity or irrigation requirement. Plantations in regions with frost risk hold the risk of damage due to frost [301].	Adaptable plant on a wide variety of soils and climatic zones including in nutrient depleted poor soils.
Pine *Pinus* sp.	Mexico	Pine tree is the world's most widespread conifer as it can be found throughout Eurasia from the western Mediterranean to the Russia [302].	Black charcoal, phosphate and microorganisms can be applied in *Pinus* sp., as fertilizer and as root promoted microorganisms (AMF fungi) in a sand dune environment [303].
Fir *Abies* sp.	Mexico	It can be grown on Atlantic, Mediterranean, warm, continental, and cold climates. However, they are highly vulnerable to any climate change, which with the increase in temperature would place them in a critical situation [304].	N fertilization affects Abies seedlings growth having a greater root and cone growth potential 305,306].
Tamarix chinensis	China	*T. chinensis* is a fast growing and adaptable species in extreme environments. Is unable to tolerate winter temperatures below −20°C. It appears good adaptability to dry atmospheric conditions, high temperatures in the dry season and low temperatures during winter.	It grows on a variety of soils, tolerating dry, waterlogged and saline-alkaline soils in coastal land, effectively improve saline-alkali soils. It can grow well even when partially covered by sand [307–310,342]. However, plant growth and photosynthetic efficiency is significantly increased under fresh water conditions compared with saline water and salt–water conditions [311].
Siberian pea shrub *Caragana arborescens*	China	It is an important species due to its pharmacological properties [312,313]. It is cold tolerant even in temperatures at −30°C. It prefers a continental climate with long hot summers and cold fairly dry winters. It does not grow so well in areas that do not have very cold winters or they have wet conditions. Suitable for dry conditions.	Suitable for poor soils. It succeeds in most well-drained soils, preferring full sun and light sandy dry or well-drained conditions. It tolerates very alkaline soils and will also do well in very poor conditions and on marginal land

(*Continued*)

Table 3.5 Relationship between climatic conditions and soil types with energy crop species. (*Cont.*)

Energy Crop	Country	Climate conditions (temperature, climatic zones, humidity)	Soil type (including nutrients' composition in N,K,P)
Vernicia fordii	China	Its contents of FAME and other physicochemical properties produced from the seed oils for biodiesel production makes it acceptable and very promising for the limit of European biodiesel qualities for BD100 [314,315]. It is widely introduced in tropical and subtropical regions where mean annual temperature ranging from 18 to 28°C and mean annual rainfall from 650 to 3500 mm [316,317].	It prefers moist and well-drained soils. It has the capability to grow on acidic and alkaline soils (pH 4.5–7.5).
Diesel trees *Sindora glabra*	China	Suitable for cultivation in non-agricultural coastal land, used by the local as kerosene due to its flow out yellowish or amber liquid when wounded [318].	
Chinese tallow tree, soap tree, popcorn tree, pau de sebo (Brazil), tarcharbi or pahari shishum (India), arbor de la cera (Cuba) *Sapium sebiferum*	China	It is native to semitropical areas of central South China from higher latitudes to coastal areas. It is tolerate frosts and, in the dormant state, will tolerate brief lows of −10°C. However, there is considerable variability in cold tolerance among strains. It tolerates in a wide variation in quantities of annual rain (including flooding) [319].	It possesses a remarkable capacity for thriving in in drought, frost and arid soils [320]. It tolerates a wide range of soil conditions, including poorly drained and saline soils. It appears strong positive response to fertilization and unusually positive interactions with the soil biota in fertilized conditions (72.3% increase in biomass in field, fertilized soils) [321].
Pistacia chinensis	China	It tolerates temperatures down to between −5 and −10°C. [322]. It also tolerates drought resistance.	It succeeds in an ordinary loamy soil and in dry soils. It prefers well-drained soil, but it will grow also in poor, saline, acid or alkaline soils [323]. A slow-release 10-10-10 fertilizer is applied during spring.
Indian tree spurge *Euphorbia tirucalli*	China, India	It produces biomass even under very marginal soil and extreme climatic conditions [324]. It can grows in low annual precipitation (250–500 mm) and temperature ranging from 21 to 28°C [256].	It is well adapted to drought, salinity and infertile soil, but it does not tolerate frost [325]. It grows at any pH ranging from 6 to 8.5

Gum arabic tree *Acacia nilotica*	India	It should not be introduced into humid and subhumid areas, or into dry areas. It prefers dry conditions, with an annual rainfall of 250–1500 mm. Average annual temperatures commonly vary from 15 to 28°C, although it can withstand daily maximum temperatures of 50°C, and is frost tender when young (CABI website).	It will grow on saline, alkaline soils, and on those with calcareous pans, but requires sufficient moisture in the soil or subsoil. It will grow but remain stunted on shallow soils with underlying bedrock or beds of nodular kankar and also on poor argillaceous soils [326,327].
Bamboo *Bambusa balcooa*	India, China	The high annual carbon accumulation rates suggest the potential for successful carbon farming using bamboo [328]. Bamboo trees grow best in areas where temperatures are within 22–28°C, but can tolerate from 9 to 35°C. It prefers tropical, wet conditions with a mean annual rainfall in the range 2,300–3,000mm, but tolerates 700–4,500mm.	It succeeds in any type of soil but it prefers heavy textured soils with good drainage and pH of about 5.5 tolerating from pH values from 4.5 to 7.5 (Useful tropical plants website). In vivo propagation of planting materials through culm cutting, is optimized using coarse sand over soil, soil plus sand (1:1;v/v) or vermiculite [329].
Palm tree	Thailand, Malaysia	Tropical, wet conditions	
Sativa cannabis	Sweden	Optimal day temperature range from 24 to 30°C. Ideal humidity for optimal growth are 40%–60% RH. For optimal plant growth it is required a period of light and a period of dark Flowering is induced by providing at least 12 hours per day of complete darkness.	Plant needs more N than P and K during all life phases. The presence of secondary nutrients (calcium, magnesium, sulfur) is recommended. Soil should be allowed to dry adequately before re-watering.
Kenaf *Hibiscus cannabinus*	Malaysia	It prefers tropical, wet conditions with annual temperature ranges from 11.1 to 27.5°C. However, water stress had no effects on biomass allocation [330]. Kenaf is frost sensitive and damaged by heavy rains and strong winds [331]. However, there are limitations emerged due low chilling tolerance, photoperiod sensitivity, and low genetic variability [332].	It prefers a well-drained humus rich fertile soil in full sun. However, it tolerates most soils but prefers a light sandy soil. It grows at a wide range pH values between 4.3 and 8.2 (though it prefers neutral to slightly acid (Plants for a future website).

2 Biomass conversion to biomass production

Since the prehistoric ages, biomass utility for human beings was in great importance. Even before the Industrial Revolution and the invention of the electricity, biomass use was in great interest since it was the main energy-heat provider for several human activities. Use of biomass residues is still an important energy provider especially for the cover of daily needs of energy consumption in rural areas in developing countries. Beside this usage, biomass residues it is believed that can be also used for the production of synthetic fuels replacing and limiting the use and dependence from fossil fuels. As a renewable energy source, bioenergy has been calculated that contest of about 14%–18% of the global renewable energy production supplying with about 62.5 exajoules (EJs) of energy production [333–335]. Thus, it is believed that biomass can potentially be utilized for the production of sustainable heat and power production [336] appearing strongly competitiveness to other fossil fuels in EU, countries by 2050 [337]. The use of different biomass plant residues as raw materials, affects biomass conversion. In a recent study Goffe and Ferrasse [338] concluded that lignocellulosic biomass provide usually lower conversion efficiency than microalgae biomass. Another important factor is that biomass production is occurs usually in short cycle. It is well known that fossil fuels are formed through complex cycles that last millions to billions of years within earth core and plaques and when extracted they require technical complex procedures for their final transformation. In contrary, biomass production cycle it is completed in a shorter period lasting usually a year.

Regarding the use of biomass conversion to biofuel for transport use, it is commented that the low price of traditional oil and gas fuels may cause an adverse effect on the competitiveness of biofuels used for this purpose. In order to improve biofuel efficiency properties several research studies have been conducted. Butera et al. [339] analyzed and modeled several operating modes using biomass gasification as route to produce methanol. Biomass gasification and electrolysis for biofuels production are combined in order to achieve fuel requirements by increasing hydrogen content on the syngas. So far, different biomass gasifiers have been developed. For example, the TwoStage Gasifier uses two different reactors where pyrolysis and char gasification occurs producing a tar-free syngas with high cold gas efficiency [340,341].

According to Goffe and Ferrasse [338] the increase of energy density and usefulness of biomass source is carried out through biomass conversion. Transesterification and fermentation was the main conversion processes for production of first generation biofuels like biodiesel and ethanol, which are used as mixtures with conventional fuels in existing engines respectively. The main biomass sources of the first generation biofuels was oil extracted from energy crops such as rapeseed, sunflowers, palm tree, soya, etc. However, thermo-chemical and biochemical processes such as fermentation, anaerobic digestion, incineration or combustion, oxidation as HydroThermal oxidation (SHTO) and Wet Air Oxidation (WAO), torrefaction, pyrolysis, gasification (better known as supercritical water gasification-SCWG), liquefaction (hydrothermal liquefaction-HTL), and hydrothermal upgrading (HTU) are some of the processes uses non-food sources of biomass obtained from agricultural residues, forest waste, and non-edible plants converting these residues into valuable products, such as oil and gas. The above-mentioned processes are useful for steam production, used to initiate turbines for electricity production purposes.

More specific, combustion oxidation produce incombustible gases like CO_2 and H_2O used for heat and electricity purposes. Additionally, pyrolysis, gasification, liquefaction, SHTO, WAO, and SCWG produces incombustible gases like CO_2 and H_2O, gases (H_2, CO) and bio-oil used for heat, transportation, electricity, and industry purposes [342–344]. Digestion mainly produces methane used for heat, transport, and electricity purposes, while fermentation produces alcohols mainly used in industry and in heat, transport, and electricity production. Finally, transesterification mainly produces esters used in industry while it also produces biofuel-biogas used to initiate engines for heat, transport, and electricity production. For the quantification of the biomass source, several analytical techniques, such as thermogravimetric analyzer (TG), Fourier Transform Infrared Spectroscopic (FTIR), thermogravimetric analysis (TGA), nuclear magnetic resonance (NMR) spectroscopy, scanning electron microscope (SEM), X-ray diffraction (XRD), mass spectrometry (MS), gas chromatography (GC), and pyrolysis–gas chromatography/mass spectrometry (Py-GC/MS) are used in order to determine the combining techniques, and to analyze the main parameters of biomass improving the energy crop growing techniques [345].

Regarding gasification process they are used biomass sources like spiroulina, poplar wood or almond shell and lignocellulose wood of several species and energy crops, producing mainly methanol, syngas, hydrogen, dimethyl ester, and FT diesel [346–350]. In fermentation process they are used biomass sources, such as corn, wood containing lignocelluloses, glucose, and ley crops, producing mainly ethanol, cellulosic, butanol, ester-diesel, methane, and ester microdiesel [351,352]. Finally, regarding hydrogen production from biomass Li et al. [353] commented that several factors such as feedstock, temperature and steam/biomass ratio, and different kind of catalysts may affect its productivity. In general, thermochemical conversion methods include some of the most common used methods and can be either applied on high value biomass fuel, such as wood pellets, or even at the lowest value biomass fuels, such as manures, and the agricultural by-product of wheat straw [354].

3 Conclusions

The energy sector is one of the main sector affecting overall economic growths. However, energy production plans and decisions must take into account all dimensions of energy use and production (social, economic, environmental), as climate change raises the urgent need to adapt a low-carbon energy production model. Decisions on climate change prevention and mitigation affect the global energy market in the long and medium term, as well as the quality of human life. In recent years, issues related to the secure energy supply of states have been taken into a new dimension globally. These issues are implementing policies regarding the maximum use of domestic energy sources and the energy savings.

Nowadays, the improvident use of natural resources and the pollution of the environment have been taken into serious account the necessity to limit the use and the gradual replacement of fossil fuels with renewable energy sources, where the cultivation of energy plants is of particular importance. In addition, the limitation of the world's oil and gas reserves and the agricultural overproduction in developed countries has accelerated the installation of energy plants. The change to biofuels production is now a reality.

Many studies have highlighted the important role of uncertainty in the application of innovative agricultural technologies and techniques. This is why a significant part of the acceptance and promotion of energy crops is also financed by the State (in the form of subsidies). Many studies have also shown that the liquidity constraint of farms, their size, the type of crop, its profitability and productivity, the time of establishment, the level of debt of the farms are some of the important and crucial element in adopting a technology. Thus, besides the risk of financial constraint, a number of other factors can influence the decision making of energy crops such as cane and mishanthus. Indeed, these crops require a longer set-up period before production starts, so that farmers invest without a direct result. For example, cane and *Miscanthus* cultivation in France, Italy, and in Ireland, it is less profitable for a farmer than a traditional crop rotation system. Despite the fact that crop cultivation is highly competitive with winter cereals, the price of the product produced was particularly low during the last few years. It is believed that with notary farming the cultivation of such plants may become more attractive and may be a prerequisite for long-term crop development. Possible solution to the above situation is the increase in yield production of energy crops. In an input-output analysis, the higher the amount of biomass used, the higher the impact on the value added of the product. However, positive effects on value added and employment do not necessarily lead to a positive net financial contribution.

Energy crops lead to changes in forms of agricultural activity and contribute to the financial diversification of the countryside. Land use for energy crops should be considered as a possible and reliable solution to problems such as land abandonment, increased energy demand, and growth opportunities in rural areas. This will strengthen the declining agricultural economy and lead to the development of the local agricultural industry as well as the creation of agro-industrial clusters. In particular, the development of biofuel production is expected to offer new opportunities for income diversification and employment in rural areas. Furthermore, the use of crop residues for energy purposes offsets the increase in energy demand by enhancing clean energy production.

On the contrary, the economical exploitation of residual biomass was favorable only in the case of high yields of crop residues. However, extensive and intensive cultivation of energy crops leads to monoculture, degradation of land uses and significant impacts on biodiversity, water supply due to increased demands on irrigation of energy crops, and effects on soil quality, while the use of fertilizers and pesticides increases the cost production as well as the acidity of the soil and water, creating at the same time additional economic and environmental problems.

In particular, EU should invest in sustainable alternative fuels, including biofuels and bioeconomics in general. This will not only reduce energy dependence on fossil fuel imports, diversify energy supplies, maintain technological and industrial leadership, but also achieve climate change goals.

Deploying a literature review upon contribution of energy crops in biomass production and balancing out the strengths and the weaknesses of each energy crop in different countries and climate conditions, our scope is that this study will benefit farmers, decision-makers, and researchers who are engaged at any way on the exploration and development of alternative and renewable energy sources.

Websites

https://www.cabi.org/isc/datasheet/17463
Plants for a future website: https://pfaf.org/
Useful tropical plants website: http://tropical.theferns.info/viewtropical.php?id=Bambusa+balcooa

References

[1] A. Fiorentino, A. Gentili, M. Isidori, P. Monaco, A. Nardelli, A. Parrella, Environmental effects caused by olive mill wastewaters: toxicity comparison of low-molecular-weight phenol components, J. Agric. Food Chem. 51 (2003) 1005–1009.

[2] C. Makridis, C. Svarnas, N. Rigas, N. Gougoulias, L. Rokam, S.V. Leontopoulosm, Transfer of Heavy Metal Contaminants from Animal Feed to Animal Products, J. Agr. Sci. Tech. USA 2 (1A) (2012) 149–154.

[3] N. Gougoulias, S. Leontopoulos, Ch. Makridis, Influence of food allowance in heavy metal's concentration in raw milk production of several feed animals, Emirates J. Food and Agric. 26 (9) (2014) 828–834.

[4] K.B. Petrotos, M.I. Kokkora, P.E. Gkoutsidis, S. Leontopoulos, A comprehensive study on the kinetics of Olive Mill wastewater (OMWW) polyphenols absorption on macroporous resins. Part II. The case of Amperlite FPX66 commercial resin, Desalin. Water Treat. (2016a) 1–8.

[5] K. Petrotos, D. Lampakis, G. Pilidis, S.V. Leontopoulos, Production and encapsulation of polyphenols derived from clarifies waste by using a combination of macroporous resins and spray drying, International Journal of Food and Biosystems Engineering 1 (1) (2016) 40–50.

[6] N.H. Aziz, S.E. Farag, L.A. Mousa, M.A. Abo-Zaid, Comparative antibacterial and antifungal effects of some phenolic compounds, Microbiosystems 93 (1998) 43–54.

[7] C. Ethaliotis, K. Papadopoulou, M. Kotsou, I. Mari, C. Balis, Adaptation and population dynamics of Azotobacter vinelandii during aerobic biological treatment of olive mill wastewater, FEMS Microbiology Ecology 30 (1999) 301–311.

[8] C. Paredes, J. Cegarra, A. Roig, M.A. Sfinchez-Monedero, M.P. Bernal, Characterization of olive mill wastewater (alpechin) and its sludge for agricultural purposes, Bioresour. Technol. 67 (1999) 111–115.

[9] H.K. Obied, M.S. Allen, D.R. Bedgood, P.D. Prenzler, K. Robards, R. Stockmann, Bioactivity and analysis of biophenols recovered from olive mill waste, J. Agric. Food Chem. 53 (2005) 823–827.

[10] V. Ani, M.C. Varadaraj, A.K. Naidu, Antioxidant and antibacterial activities of polyphenolic compounds from bitter cumin (*Cuminum nigrum* L).., Eur. Food Resid. Technol. 224 (2006) 109–115.

[11] N.G. Baydar, O. Sagdic, G. Ozkan, S. Cetin, Determination of antibacterial effects and total phenolic contents of grape (*Vitis vinifera*) seed extracts, Int. J. Food Sci. 41 (2006) 799–804.

[12] G. Bonanomi, V. Giorgi, G. Del Sorbo, D. Neri, F. Scala, Olive mill residues affect saprophytic growth and disease incidence of foliar and soilborne plant fungal pathogens, Agricultural Ecosystems Environment 115 (2006) 194–200.

[13] J.H. Chavez, P.C. Leal, R.A. Yunes, R.J. Nunes, C.R. Barardi, A.R. Pinto, et al., Evaluation of antiviral activity of phenolic compounds and derivatives against rabies virus, Veter. Microbiol. 116 (2006) 53–59.

[14] E. Xia, G. Deng, Y. Guo, H. Li, Biological activities of polyphenols from grapes, Int. J. Mol. Sci. 11 (2010) 622–646.

[15] M. Kokkora, P. Vyrlas, Ch. Papaioannou, K. Petrotos, P. Gkoutsidis, S. Leontopoulos, et al., Agricultural use of microfiltered olive mill wastewater: effects on maize production and soil properties, Agric. Agric. Sci. Proc. 4 (2015) 416–424.

[16] M.I. Kokkora, K.B. Petrotos, Ch. Papaioannou, P.E. Gkoutsidis, S.V. Leontopoulos, P. Vyrlas, Agronomic and economic implications of using treated olive mill wastewater in maize production, Desalination and Water Treatment (2016) 1–7.

[17] P. Skenderidis, C. Mitsagga, I. Giavasis, C. Hadjichristodoulou, S. Leontopoulos, K. Petrotos, et al., Assessment of antimicrobial properties of water and methanol UAE extracts of goji berry fruit and pomegranate fruit peels in vitro. Third I.C., FaBE 2017, Rhodes Island 01-04/06-17. FaBE Proceedings, (2017) 541–549.

[18] P. Skenderidis, C. Mitsagga, I. Giavasis, K. Petrotos, D. Lampakis, S. Leontopoulos, et al., The in vitro antimicrobial activity assessment of ultrasound assisted Lycium barbarum fruit extracts and pomegranate fruit peels, Journal of Food Measurement and Characterization (2019) 1–15.

[19] P. Skenderidis, D. Lampakis, I. Giavasis, S. Leontopoulos, K. Petrotos, C. Hadjichristodoulou, et al., Chemical properties, fatty-acid composition, and antioxidant activity of goji berry (Lycium barbarum L. and Lycium chinense Mill) fruits, Antioxidants 8 (60) (2019) 1–13.

[20] S.V. Leontopoulos, I. Giavasis, K. Petrotos, M. Kokkora, Ch. Makridis, Effect of different formulations of polyphenolic compounds obtained from OMWW on the growth of several fungal plant and food borne pathogens. Studies in vitro and in vivo, Agric. Agric. Sci. Proc. 4 (2015) 327–337.

[21] S.V. Leontopoulos, M.I. Kokkora, K.B. Petrotos, In vivo evaluation of liquid polyphenols obtained from OMWW as natural bio-chemicals against several fungal pathogens on tomato plants, Desalin. Water Treat. (2016) 1–15.

[22] S. Leontopoulos, P. Skenderidis, H. Kalorizou, K. Petrotos, Bioactivity potential of polyphenolic compounds in human health and their effectiveness against various food borne and plant pathogens A review, Int. J. Food .Biosyst. Eng. 7 (1) (2017) 1–19.

[23] K.H. Erb, H. Haberl, C. Plutzar, Dependency of global primary bioenergy crop potentials in 2050 on food systems, yields, biodiversity conservation and political stability, Energy Policy 47 (2012) 260–269.

[24] M. Marra, D.J. Pannell, A.K. Abadi Ghadim, The economics of risk, uncertainty and learning in the adoption of new agricultural technologies: where are we on the learning curve?, Agric. Syst. 75 (2–3) (2003) 215–234.

[25] O. Flaten, G. Lien, M. Koesling, P.S. Valle, M. Ebbesvik, Comparing risk perceptions and risk management in organic and conventional dairy fanning: empirical results from Norway, Livestock Product. Sci. 95 (2005) 11–25.

[26] T. Serra, D. Zilberman, J.M. Gil, Differential uncertainties and risk attitudes between conventional and organic producers: the case of Spanish arable crop farmers, Agric. Econ. 39 (2008) 219–229.

[27] R. Greiner, L. Patterson, O. Miller, Motivations, risk perceptions and adoption of conservation practices by farmers, Agric. Econ. 99 (2–3) (2009) 86–104.

[28] Council of the European Union. Brussels European Council 8/9 March (2007). Presidency conclusions. Brussels, 7224/1/07 REV 1; 2007.

[29] European Commission. European Union-Energy and Transport in Figures-2007.

[30] K. Ericsson, H. Rosenqvist, L.J. Nilsson, Energy crop production costs in the EU, Biomass and Bioenergy 33 (2009) 1577–1586.

[31] I. Yilmaz, M. Ilbas, An experimental study on hydrogen–methane mixtured fuels, Agric. Econ. 35 (2) (2008) 178–187.

[32] G.W. Crabtree, M.S. Dresselhaus, The hydrogen fuel alternative, Mrs Bull. 33 (4) (2008) 421–428.

[33] I. Dincer, M.A. Rosen, Sustainability aspects of hydrogen and fuel cell systems, Energy Sustain. Dev. 15 (2011) 137–146.

[34] G.L. Kyriakopoulos, G. Arabatzis, Electrical energy storage systems in electricity generation: Energy policies, innovative technologies, and regulatory regimes, Renew. Sustain. Energy Rev. 56 (2016) 1044–1067.

[35] N. Bagheri Moghaddam, S.M. Mousavi, M. Nasiri, E.A. Moallemi, H. Yousefdehi, Wind energy status of Iran: Evaluating Iran's technological capability in manufacturing wind turbines, Renew. Sustain. Energy Rev. 15 (8) (2011) 4200–4211.

[36] K. Kaygusuz, Energy for sustainable development: A case of developing countries, Renew. Sustain. Energy Rev. 16 (2) (2012) 1116–1126.

[37] N.L. Panwar, S.C. Kaushik, S. Kothari, Role of renewable energy sources in environmental protection: a review, Renew. Sustain. Energy Rev. 15 (3) (2011) 1513–1524.

[38] F. Dincer, The analysis on wind energy electricity generation status, potential and policies in the world, Renew. Sustain. Energy Rev. 15 (9) (2011) 5135–5142.

[39] M. Balat, H. Balat, Political, economic and environmental impacts of biomass-based hydrogen, Int. J. Hydrogen Energy 34 (2009) 3589–3603.

[40] J. Nowotny, N.T. Veziroglu, Impact of hydrogen on the environment, Int. J. Hydrogen Energy 36 (20) (2011) 13218–13224.

[41] T. Koutroumanidis, K. Ioannou, G. Arabatzis, Predicting fuel wood prices in Greece with the use of ARIMA models, artificial neural networks and a hybrid ARIMA–ANN model, Energy Policy 37 (2009) 3627–3634.

[42] B.D. Solomon, K. Krishna, The coming sustainable energy transition: History, strategies, and outlook, Energy Policy 39 (11) (2011) 7422–7431.

[43] B. Sorensen, A history of renewable energy technology, Energy Policy 19 (1) (1991) 8–12.

[44] Y.V. Sinyak, A.Y. Kolpakov, Economic efficiency of synthetic motor fuels from natural gas, Stud. Russian Econ. Dev. 23 (1) (2012) 27–36.

[45] J. Ebner, The Sino-European race for Africa's minerals: When two quarrel a third rejoices, Resourc. Policy 43 (2015) 112–120.

[46] A.A. El Anshasy, M.S. Katsaiti, Are natural resources bad for health?, Health Place 32 (2015) 29–42.

[47] K. Söderholm, P. Söderholm, H. Helenius, M. Pettersson, R. Viklund, V. Masloboev, T. Mingaleva, V. Petrov, Environmental regulation and competitiveness in the mining industry: Permitting processes with special focus on Finland Sweden and Russia, Resourc. Policy 43 (2015) 130–142.

[48] B. Moreno, A.J. Lopez, The effect of renewable energy on employment, Renew. Sustain. Energy Rev. 12 (3) (2008) 732–751.

[49] E.L. Sastresa, A.A. Uso, I.Z. Bribia, S. Scarpellini, Local impact of renewables on employment: Assessment methodology and case study, Renew. Sustain. Energy Rev. 14 (2010) 679–690.
[50] C. Tourkolias, S. Mirasgedis. Quantification and monetization of employment benefits associated with renewable energy technologies in Greece, 2011.
[51] J.A. Grahovac, J.M. Dodić, S.N. Dodić, S.D. Popov, D.G. Vučurović, A.I. Jokić, Future trends of bioethanol co-production in Serbian sugar plants, Renew. Sustain. Energy Rev. 16 (5) (2012) 3270–3274.
[52] T. Silalertruksa, S.H. Gheewala, K. Hünecke, U.R. Fritsche, Biofuels and employment effects: Implications for socioeconomic development in Thailand, Biomass Bioenergy 46 (2012) 409–418.
[53] E. Llera, S. Scarpellini, A. Aranda, I. Zabalza, Forecasting job creation from renewable energy deployment through a value-chain approach, Renew. Sustain. Energy Rev. 21 (2013) 262–271.
[54] R. Pollin, J. Heintz, H. Garrett-Peltier, The economic benefits of investing in clean energy, Center for American Progress and Political Economy Research Institute (2009).
[55] R. Prakash, I.K. Bhat, Energy, economics and environmental impacts of renewable energy systems, Renew. Sustain. Energy Rev. 13 (9) (2009) 2716–2721.
[56] M.K. Farooq, S. Kumar, R.M. Shrestha, Energy, environmental and economic effects of Renewable Portfolio Standards (RPS) in a Developing Country, Energy Policy 62 (2013) 989–1001.
[57] S. Ntanos, G. Kyriakopoulos, M. Chalikias, G. Arabatzis, M. Skordoulis, S. Galatsidas, D. Drosos, A social assessment of the usage of Renewable Energy Sources and its contribution to life quality: The case of an Attica urban area in Greece, Sustainability 10 (1414) (2018) 1–15.
[58] S. Ntanos, M. Skordoulis, G. Kyriakopoulos, G. Arabatzis, M. Chalikias, S. Galatsidas, A. Batzios, A. Katsarou, Renewable energy and economic growth: Evidence from European countries, Sustainability 10 (2626) (2018) 1–13.
[59] S. Ntanos, G. Kyriakopoulos, M. Chalikias, G. Arabatzis, M. Skordoulis, Public perceptions and willingness to pay for Renewable Energy: A case study from Greece, Sustainability 10 (687) (2018) 1–16.
[60] J.F. Li, L. Shi, L.Y. Ma, Review of the international renewable energy development, Int. Fossil Oil Econ. 2 (2006) 35–37.
[61] L.G. Yan, J.W. Chen, Key issues on energy sustainable development in China, Science Press, Beijing, 2007, pp. 354–378 2007.
[62] S. Tampakis, G. Tsantopoulos, G. Arabatzis, I. Rerras, Citizens' views on various forms of energy and their contribution to the environment, Renew. Sustain. Energy Rev. 20 (2013) 473–482.
[63] M. Christou, I. Eleftheriadis, C. Panoutsou, I. Papamichail, 2007. Current situation and future trends in biomass fuel trade in Europe. EUBIONET II, Greece /www.eubionet.netS.
[64] Eurobserver, Solid Biomass barometer, 2018.
[65] Eurobserver, Solid Biomass, barometer (2013).
[66] H.S. Choi, S.K. Entenmann, Land in the EU for perennial biomass crops from freed-up agricultural land: A sensitivity analysis considering yields, diet, market liberalization and world food prices, Land Use Policy 82 (2019) 292–306.
[67] A. Muscat, E.M. de Olde, I.J.M. de Boer, R. Ripoll-Bosch, The battle for biomass: A systematic review of food-feed-fuel competition, Global Food Secur. 25 (2020) 100330.
[68] R. Kikuchi, R. Gerardo, S.M. Santos, Energy lifecycle assessment and environmental impacts of ethanol biofuel, Int. J. Energy Res. 33 (2) (2009) 186–193.
[69] M Balat, Possible methods for hydrogen production Energy Sources, Part A: Recovery, Utilization, and Environmental Effects 31 (1) (2008) 39–50.

[70] H. Azadi, S. Jong, B. Derudder, P. Maeyer, F. Witlox, Bitter sweet: How sustainable is bioethanol production in Brazil?, Renew. Sust. Energ. Rev. 16 (6) (2012) 3599–3603.

[71] S. Leontopoulos, G. Arabatzis, S. Ntanos, Ch. Tsiantikoudis, Acceptance of energy crops by farmers in Larissa's regional unit. A first approach. 7th International Conference on Information and Communication Technologies in Agriculture, Food and Environment (HAICTA), Kavala, 17-20/09/15, CEUR Workshop Proc; 1498 (2015) 38–43.

[72] E. Grigoroudis, K. Petridis, G. Arabatzis, RDEA: A recursive DEA based algorithm for the optimal design of biomass supply chain networks, Renew. Energy 71 (2014) 113–122.

[73] K. Petridis, E. Grigoroudis, G. Arabatzis, A goal programming model for a sustainable biomass supply chain network, Int. J. Energy Sector Manage. 12 (1) (2018) 79–102.

[74] J.D Ulmer, R.L Huhnke, D.D Bellmer, D.D. Cartmell, Acceptance of ethanol-blended gasoline in Oklahoma, Biomass and Bioenergy 27 (5) (2004) 437–444.

[75] A.B. Delshad, L. Raymond, V. Sawicki, D.T. Wegener, Public attitudes toward political and technological options for biofuels, Energy Policy 38 (7) (2010) 3414–3425.

[76] M.A. Cacciatore, D.A. Scheufele, B.R. Shaw, Labeling renewable energies: How the language surrounding biofuels can influence its public acceptance, Energy Policy 51 (2012) 673–682.

[77] K. Steininger, H. Voraberger, Exploiting the medium term biomass energy potentials in Austria: a comparison of costs and macroeconomic impact, Environ. Resourc. Econ. 24 (2003) 359–377.

[78] Th. Trink, Ch. Schmid, Th. Schinko, K.W. Steininger, Th. Loibnegger, C. Kettner, A. Pack, Ch. Toglhofer, Regional economic impacts of biomass based energy service use: A comparison across crops and technologies for East Styria, Austria. Energy Policy 38 (10) (2010) 5912–5926.

[79] R.D. Sands, S.A. Malcolm, S.A. Suttles, E. Marshall, Dedicated energy crops and competition for agricultural land. United States Department of Agriculture (USDA), A report summary from the economic research service. (2017) Availabel at: www.ers.usda.gov.

[80] B. Allen, B. Kretschmer, D. Baldock, H. Menadue, S. Nanni, G. Tucker, Space for energy crops-assessing the potential contribution to Europe's energy future, Instit. Eur. Environ. Policy (2014) Available from www.birdlife.org.

[81] J. Popp, Z. Lakner, M. Harangi-Rákos, M. Fári, The effect of bioenergy expansion: Food, energy, and environment, Renew. Sustain. Energy Rev. 32 (2014) 559–578.

[82] A. Demirbas, Biomethanol production from organic waste materials, energy sources, Part A: Recovery, Util. Environ. Effects 30 (6) (2008) 565–572.

[83] G. Eggleston, Future sustainability of the sugar and sugar-ethanol industries, ACS Symp. Series 1058 (2010) 1–19.

[84] M. Dwidar, S. Lee, R.J. Mitchell, The production of biofuels from carbonated beverages, Appl. Energy 100 (2012) 47–51.

[85] C.W. He, The current situation of the development of biofuels and main technical problems, Adv. Mater. Res. 827 (2013) 244–249.

[86] M. Takaki, L. Tan, T. Murakami, Y.Q. Tang, Z.Y. Sun, S. Morimura, K. Kida, Production of biofuels from sweet sorghum juice via ethanol-methane two-stage fermentation, Industr. Crops Product. 63 (2015) 329–336.

[87] P. Fasahati, H.C. Woo, J.J. Liu, Industrial-scale bioethanol production from brown algae: Effects of pretreatment processes on plant economics, Appl. Energy 139 (2015) 175–187.

[88] H.M. Kim, S.G. Wi, S. Jung, Y. Song, H.-J. Bae, Efficient approach for bioethanol production from red seaweed Gelidium amansii, Bioresourc. Technol. 175 (2015) 128–134.
[89] D. Connolly, B.V. Mathiesen, A. Ridjan, Comparison between renewable transport fuels that can supplement or replace biofuels in a 100% renewable energy system, Energy 73 (2014) 110–125.
[90] D. Djurović, S. Nemoda, B. Repić, D. Dakić, M. Adzić, Influence of biomass furnace volume change on flue gases burn out process, Renew. Energy 76 (2015) 1–6.
[91] V.S. Yaliwal, N.R. Banapurmath, S.Y. Adaganti, K.M. Nataraja, P.G. Tewari, Effect of bioethanol and thermal barrier coating on the performance of dual fuel engine operating on Honge oil methyl ester (HOME) and producer gas induction, Int. J. Sustain. Eng. 9 (2015) 17.
[92] B. Flach, K. Bendz, R. Krautgartner, S. Lieberz, EU-27 biofuels annual report. USDA Foreign Agricultural Service GAIN, Report No: NL3034, 2013.
[93] N. Palmieri, M. Bonaventura-Forleo, A. Suardi, D. Coaloa, L. Pari, Rapeseed for energy production: Environmental impacts and cultivation methods, Biomass Bioenergy 69 (2014) 1–11.
[94] R. Picchio, R. Spina, A. Sirna, A.L. Monaco, V. Civitarese, A.D. Giudice, Characterization of woodchips for energy from forestry and agroforestry production, Energies 5 (2012) 3803–3806.
[95] A.G. Gayubo, A. Alonso, B. Valle, A.T. Aguayo, J. Bilbao, Selective production of olefins from bioethanol on HZSM-5 zeolite catalysts treated with NaOH, Appl. Catal. Environ. 97 (1) (2010) 299–306.
[96] D.Y. Leung, Y. Yang, Wind energy development and its environmental impact: A review, Renew. Sustain. Energy Rev. 16 (1) (2012) 1031–1039.
[96a] Biofuels barometer (2019). Eurobserver. Available from: https://www.eurobserv-er.org/biofuels-barometer-2019/.
[97] L.A. Ioannou, G.L. Puma, D. Fatta-Kassinos, Treatment of winery wastewater by physicochemical, biological and advanced processes: A review, J. Hazard. Mater. 286 (2015) 343–368.
[98] S. Fazio, L. Barbanti, Energy and economic assessments of bio-energy systems based on annual and perennial crops for temperate and tropical areas, Renew. Energy 69 (2014) 233–241.
[99] R. Venendaal, U. Jorgensen, C.A. Foster, European energy crops overview, Biomass and Bioenergy 13 (3) (1997) 147–185 Elsevier Science Ltd., Great Britain.
[100] G. Bocquehon, F. Jacquet, The adoption of switchgrass and miscanthus by farmers: Impact of liquidity constraints and risk preferences, Energy Policy 38 (2010) 2598–2607.
[101] D. Styles, F. Thorne, M.B. Jones, Energy crops in Ireland: an economic comparison of willow and Miscanthus production with conventional farming systems, Biomass Bioenergy 32 (2008) 407–421.
[102] W. Zegada-Lizarazu, A. Monti, Energy crops in rotation: A review, Biomass Bioenergy 35 (2011) 12–25.
[103] M.J.A Stolarski, M. Krzy aniak, K. Warmi ski, E. Olba-Zi ty, E. Penni, D. Bordiean, Energy efficiency indices for lignocellulosic biomass production: Short rotation coppices versus grasses and other herbaceous crops, Industrial Crops and Products 135 (2019) 10–20.
[104] O.K. Shortall, Marginal land for energy crops: Exploring definitions and embedded assumptions, Energy Policy 62 (2013) 19–27.
[105] R. Schubert, H.J. Schellnhuber, N. Buchmann, A. Epiney, R. Grieshammer, M. Kulessa, D. Messner, Future Bioenergy and Sustainable Land Use, London, (2008).
[106] De Wit M.P., Faaij A., Fischer G., Prieler F., Van Velthuizen H., (2007). The potential of European biomass resources and related costs in the EU-27 and

the Ukraine. In: 15th European Biomass Conference and Exhibition, Berlin, Germany, 2–11 May 2007.

[107] L. Wright, Worldwide commercial development of bioenergy with focus on energy crop-based projects, Biomass and Bioenergy 30 (2006) 706–714.

[108] B. Johansson, P. Borjesson, K. Ericsson, L.J. Nilsson, P. Svenningsson, The use of biomass for energy in Sweden—critical factors and lessons learned, IMES/EESS Report 35, Energy Environmental System Studies, Lund, Sweden, (2002).

[109] M. Ostwald, A. Jonsson, V. Wibeck, T. Asplund, Mapping energy crop cultivation and identifying motivational factors among Swedish farmers, Biomass Bioenergy 13 (2013) 25–34.

[110] S. Paulrud, T. Laitila, Farmers' attitudes about growing energy crops: A choice experiment approach, Biomass Bioenergy 34 (2010) 1770–1779.

[111] T. Lonnqvist, S. Silveira, A. Sanches-Pereira, Swedish resource potential from residues and energy crops to enhance biogas generation, Renew. Sustain. Energy Rev. 21 (2013) 298–314.

[112] C. Gissen, T. Prade, E. Kreuger, I. Nges, H. Rosenqvist, S. Svensson, M. Lantz, J.E. Mattsson, P. Borjesson, L. Bjornsson, Comparing energy crops for biogas production and yields, energy input and costs in cultivation using digestate and mineral fertilization, Biomass Bioenergy 64 (2014) 199–210.

[113] A. Roos, H. Rosenqvist, E. Ling, B. Hektor, Farm-related factors influencing the adoption of short-rotation willow coppice production among Swedish farmers, Acta Agric. Scand. 50 (2000) 28–34.

[114] K. Jensen, C.D. Clark, P. Ellis, B. English, J. Menard, M. Walsh, D. Ugarte, Farmer willingness to grow switchgrass for energy production, Biomass Bioenergy 31 (2007) 773–781.

[115] A.P.C. Faaij, Bio-energy in Europe: Changing technology choices, Energy Policy 34 (3) (2006) 322–342.

[116] Forsberg, M., Sundberg, M., & Westlin, H. (2006). Smaskalig brikettering av hampa-forstudie [Small scale briquette production from hemp a pre-study] 2006. Available at: http://www.jti.se/uploads/jti/R-351-MF, MS, HW. pdf.

[117] H. Rosenqvist, A. Roos, E. Ling, B. Hektor, Willow growers in Sweden, Biomass Bioenergy 18 (2) (2000) 137–145.

[118] B. Hillring, Rural development and bioenergy experiences from 20 years of development in Sweden, Biomass Bioenergy 23 (6) (2002) 443–451.

[119] E. Skarback, P. Becht, Landscape perspective on energy forests, Biomass Bioenergy 28 (2) (2005) 151–159.

[120] R. Madlener, M. Koller, Economic and CO_2 mitigation impacts of promoting biomass heating systems: an input–output study for Vorarlberg, Austria. Energy Policy 35 (2007) 6021–6035.

[121] Stocker, A., Großmann, A., Madlener, R., & Wolter, M.I. (2008). Renewable energy in Austria: Modeling possible development trends until 2020. Paper presented to the International Input Output Meeting on Managing the Environment, Seville, Spain, July 9-11, 2008.

[122] P. Von Gehren, M. Gansberger, W. Pichler, M. Weigl, S. Feldmeier, E. Wopienka, G. Bochmann, A practical field trial to assess the potential of Sida hermaphrodita as a versatile, perennial bioenergy crop for Central Europe, Biomass Bioenergy 122 (2019) 99–108.

[123] K. Bunzel, M. Kattwinkel, M. Schauf, D. Thran, Energy crops and pesticide contamination: Lessons learnt from the development of energy crop cultivation in Germany, Biomass Bioenergy 70 (2014) 416–428.

[124] FNR Fachagentur Nachwachsende Rohstoffe, 2012. Bundesministeriums fur Ernahrung, Landwirtschaft und Verbraucherschutz, available at: www.fnr.de; 2012.

[125] S. Hauk, T. Knoke, S. Wittkopf, Economic evaluation of short rotation coppice systems for energy from biomass-A review, Renew. Sustain. Energy Rev. 29 (2009) 435–448.
[126] L. Beer, L. Theuvsen, Conventional German farmers' attitudes towards agricultural wood and their willingness to plant an alley cropping system as an ecological focus area: A cluster analysis, Biomass Bioenergy 125 (2019) 63–69.
[127] D.E. Agency, Energy Statistics 2016, Copenhagen, (2018).
[128] E. Booth, R. Walker, J. Bell, D. Mc Cracken, J. Curry, An Assessment of the potential impact on UK agriculture and the environment of meeting renewable feedstock demands, Scottish Agricultural College, Aberdeen (2009).
[129] European Environment Agency, (2012). Available at: http://www.eea.europa.eu/data-and-maps/figures/land-available-for-biomass-producti on-for-energy.
[130] Defra, (2007a). UK Biomass Strategy. Department for Environment, Food and Rural Affairs, May 2007.
[131] I. Convery, D. Robson, A. Ottitsch, M. Long, The willingness of farmers to engage with bioenergy and woody biomass production: A regional case study from Cumbria, Energy Policy 40 (2012) 293–300.
[132] M.A. Robinson. To what extent should UK farmers be' entre preneurialised'? Ru Source Briefing No. 477, 2007. Available from: http://www.arthurrankcentre.org.uk/projects/rusource_briefings/rus07/477.pdfS
[133] SAC. Commercial viability of alternative non-food crops and biomass on Scottish Farms-a special study supported under SEERAD Advisory Activity 211, March 2007. Available from: http://www.sac.ac.uk/mainrep/pdfs/nonfoodbiomass.pdfS.
[134] DARDNI. Renewable Energy Action Plan. Department of Agriculture and Rural Development, Northern Ireland, January 2007. Available from: http://www.dardni.gov.uk/renewable-energy-action-plan-2.pdfS.
[135] Welsh Assembly Government. Personal communication from Vicky Davies, Rural Development Adviser, Technical Services Division, Welsh Assembly Government, Llandrindod Wells, 2007.
[136] C. Sherrington, J. Bartley, D. Moran, Farm-level constraints on the domestic supply of perennial energy crops in the UK, Energy Policy 36 (2008) 2504–2512.
[137] E. Krasuska, H. Rosenqvist, Economics of energy crops in Poland today and in the future, Biomass Bioenergy 38 (2012) 23–33.
[138] A.M. Ignaciuk, R.B. Dellink, Biomass and multi-product crops for agricultural and energy productionAn AGE analysis, Energy Econ. 28 (2006) 308–325.
[139] H. Borkowska, R. Molas, A. Kupczyk, Virginia fanpetals (*Sida hermaphrodita*) cultivated on light soil; height of yield and biomass productivity, Polish J. Environ. Studies 18 (2009) 563–568.
[140] B. Iglinski, A. Iglinska, W. Kujawski, R. Buczkowski, M. Cichosz, Bioenergy in Poland, Renew. Sustain. Energy Rev. 15 (2011) 2999–3007.
[141] E. Ganko. Technological potential of growing plants for energy purposes in Poland. In: The materials from XII Science Conference "Energy crop plantation and use of agricultural production area in Poland", 2008.
[142] K.J. Jankowski, B. Dubis, M.M. Sokolski, D. Załuski, P. Borawski, W. Szemplinski, Biomass yield and energy balance of *Virginia fanpetals* in different production technologies in north-eastern Poland, Energy 185 (2019) 612–623.
[143] H. Borkowska, R. Molas, Yield comparison of four lignocellulosic perennial energy crop species, Biomass Bioenergy 51 (2013) 145–153.
[144] K. Michalska, M. Bizukojc, S. Ledakowicz, Pretreatment of energy crops with sodium hydroxide and cellulolytic enzymes to increase biogas production, Biomass Bioenergy 80 (2015) 213–221.

[145] M. Nabel, V.M. Temperton, H. Poorter, A. Lücke, N.D. Jablonowski, Energizing marginal soilse. The establishment of the energy crop Sida hermaphrodita as dependent on digestate fertilization, NPK, and legume intercropping, Biomass Bioenergy 87 (2016) 9–16.

[146] H. Borkowska, R. Molas, Two extremely different crops, Salix and Sida, as sources of renewable bioenergy, Biomass Bioenergy 36 (2012) 234–240.

[147] M. Nabel, D.B. Barbosa, D. Horsch, N.D. Jablonowski, Energy crop (*Sida hermaphrodita*) fertilization using digestate under marginal soil conditions: A dose-response experiment, Energy Procedia 59 (2014) 127133.

[148] M. Nahm, C. Morhart, Virginia mallow (*Sida hermaphrodita*) as perennial multipurpose crop: Biomass yields, energetic valorisation, utilization potentials, and management perspectives, Global Change Biol. Bioenergy 10 (2018) 393–404.

[149] J. Chodkowska-Miszczuk, D. Szymanska, Update of the review: Cultivation of energy crops in Poland against socio-demographic factors, Renew. Sustain. Energy Rev. 15 (2011) 4242–4247.

[150] P. Brjesson, Energy analysis of biomass production and transportation, Biomass Bioenergy 11 (4) (1996) 305–318.

[151] European Environment Agency. Estimating the environmentally compatible bioenergy potential from agriculture. Copenhagen, p. 134. EEA Technical report No 12/2007, 2007.

[152] Agency for Restructuring and Modernizing of Agriculture. Information on Declared Energy Crop Area. Warsaw, 2007.

[153] A. Faber, R. Pudełko, Prognosis on biofuel and renewable energy demand versus land use in Poland till 2020. XIII scientific conference, 16-17 June 2009; the Institute of Soil Science and Plant Cultivation e National, Research Institute; Puławy (2009).

[154] K. Havlıckova, J. Suchy, Development model for energy crop plantations in the Czech Republic for the years 2008-2030, Renew. Sustain. Energy Rev. 14 (2010) 1925–1936.

[155] Havlıckova, K., Knapek, J., & Vasıcek, J. (2004). The economics of short rotation coppice. In: Proceedings of the 2nd World Conference Biomass for Energy, Industry and Climate Protection. Rome: ETA Florence and WIP-Munich; 561-564, ISBN 88-8940704-2.

[156] V. Brant, J. Pivec, P. Fuksa, K. Necka, D. Kocourkov, V. Venclova, Biomass and energy production of catch crops in areas with deficiency of precipitation during summer period in central Bohemia, Biomass Bioenergy 35 (2011) 1286–1294.

[157] ERA-ARD Ukraine, Bioenergy in Ukraine - possibilities of rural development and opportunities for local communities. Reporton ERA-ARD activity in Ukraine - Cognitive broshure. Taurapolis, Kaunas, 2009.

[158] L. Raslavicius, A. Grzybek, V. Dubrovin, Bioenergy in Ukraine - Possibilities of rural development and opportunities for local communities, Energy Policy 39 (6) (2011) 3370–3379.

[159] A. Dolinsky, Resolution. In: Fourth International Conference on Biomass for Energy, Kyiv, Ukraine, 22–24 September 2008.

[160] IERPC, Renewable energy policy in Ukraine. Working Paper no. V6. Institute for Economic Research and Policy Consulting, Kiev, 2006.

[161] V. Tilvikiene, Z. Kadziuliene, I.E Liaudanskiene, Z.Z. Cerniauskiene, A. Cipliene, A.J Raila, J. Baltrusaitis, The quality and energy potential of introduced energy crops in northern part of temperate climate zone, Renewable Energy 151 (2020) 887–895.

[162] L.G. Angelini, L. Ceccarini, N. Nassi, O. Di Nasso, E. Bonari, Comparison of *Arundo donax* L. and *Miscanthus x giganteus* in a long-term field experiment

in Central Italy: Analysis of productive characteristics and energy balance, Biomass Bioenergy 33 (2009) 635–643.

[163] R. Wahid, S.F. Nielsen, V.M. Hernandez, A.J. Ward, R. Gislum, U. Jørgensen, H.B. Møller, Methane production potential from *Miscanthus* sp.: effect of harvesting time, genotypes and plant fractions, Biosyst. Eng. 133 (2015) 71–80.

[164] Y. Iqbal, M. Gauder, W. Claupein, S. Graeff, H.V. Onninger, I. Lewandowski, Yield and quality development comparison between miscanthus and switchgrass over a period of 10 years, Energy 89 (2015) 268–276.

[165] J.C. Clifton-Brown, I. Lewandowski, Screening Miscanthus genotypes in field trials to optimise biomass yield and quality in Southern Germany, Eur. J. Agron. 16 (2016) 97–110.

[166] B. Kołodziej, J. Antonkiewicz, D. Sugier, *Miscanthus giganteus* as a biomass feedstock grown on municipal sewage sludge, Industr. Crops Product. 81 (2016) 72–82.

[167] I. Piscioneri, N. Sharma, G. Baviello, S. Orlandini, Promising industrial energy crop. Cynara cardunculus: A potential source for biomass production and alternative energy, Energy Convers. Manage. 41 (10) (2000) 1091–1105.

[168] M. Manzone, A. Calvo, Energy and CO_2 analysis of poplar and maize crops for biomass production in north Italy, Renew. Energy 86 (2016) 675–681.

[169] M. Mantineo, G.M. D'Agosta, V. Copani, C. Patane, S.L. Cosentino, Biomass yield and energy balance of three perennial crops for energy use in the semi-arid Mediterranean environment, Field Crops Res. 114 (2009) 204–213.

[170] A. Monti, S. Fazio, V. Lychnaras, P. Soldatos, G. Venturi, A full economic analysis of switchgrass under different scenarios in Italy estimated by BEE model, Biomass Bioenergy 31 (2007) 177–185.

[171] S. Solinas, P.A. Deligios, L. Sulas, G. Carboni, A. Virdis, L. Ledd, A land-based approach for the environmental assessment of Mediterranean annual and perennial energy crops, Eur. J. Agron. 103 (2019) 63–72.

[172] L.G. Angelini, L. Ceccarini, E. Bonari, Biomass yield and energy balance of giant reed (*Arundo donax* L.) cropped in central Italy as related to different management practices, Eur. J. Agron. 22 (2005) 375–389.

[173] P. Carneiro, P. Ferreira, The economic, environmental and strategic value of biomass, Renew. Energy 44 (2012) 17–22.

[174] M. Fernández, J. Alaejos, E. Andivia, J. Vázquez-Piqué, F. Ruiz, F. López, R. Tapias, *Eucalyptus x urograndis* biomass production for energy purposes exposed to a Mediterranean climate under different irrigation and fertilisation regimes, Biomass Bioenergy 111 (2018) 22–30.

[175] R. Jiang, T. Wang, Jin. Shao, S. Guo, Zhu, Wei, Y. Yu, S. Chen, R. Hatano, Modeling the biomass of energy crops: Descriptions, strengths and prospective, J. Integr. Agric. 16 (6) (2017) 1197–1210.

[176] Z. Chao, N. Liu, P. Zhang, Ti. Ying, K. Song, Estimation methods developing with remote sensing information for energy crop biomass: A comparative review, Biomass Bioenergy 122 (2019) 414–425.

[177] S. Srinivasan, S.C. Popescu, M. Eriksson, R.D. Sheridan, N.-W. Ku, Terrestrial laser scanning as an effective tool to retrieve tree level height, crown width, and stem diameter, Remote Sensitive 7 (2) (2015) 1877–1896.

[178] A. Fernandez-Sarria, I. Lopez-Cortes, J. Estornell, B. Velazquez-Marti, D. Salazar, Estimating residual biomass of olive tree crops using terrestrial laser scanning, Int. J. Appl. Earth Observ. Geoinf. 75 (2019) 163–170.

[179] J. Gominho, M.D. Curt, A. Lourenço, J. Fernández, H. Pereira, *Cynara cardunculus* L. as a biomass and multi-purpose crop: A review of 30 years of research, Biomass Bioenergy 109 (2018) 257–275.

[180] J. Fernandez, M.D. Curt, P.L. Aguado, Industrial applications of *Cynara cardunculus* L. for energy and other uses, Industrial Crop Production 24 (2006) 222–229.

[181] C. Bolohan, D.I. Marin, M. Mihalache, L. Ilie, A.C. Oprea, Total biomass and g rain production of *Cynara cardunculus* L. species grown under the conditions of southeastern Romania, AgroLife Sci. J. 3 (2014) 31–34.

[182] A. Lag-Brotons, I. Gómez, J. Navarro-Pedreño, A.M. Mayoral, M.D. Curt, A. Sewage, Sludge compost use in bioenergy production-a case study on the effects on *Cynara cardunculus* L energy crop, J. Clean Product. 79 (2014) 32–40.

[183] A. Ierna, R.P. Mauro, G. Mauromicale, Biomass, grain and energy yield in *Cynara cardunculus* L. as affected by fertilization, genotype and harvest time, Biomass Bioenergy 36 (2012) 404–410.

[184] C. Panoutsou, Biomass, grain and energy yield in *Cynara cardunculus* L. as affected by fertilization, genotype and harvest time, Energy Policy 35 (2007) 6046–6059.

[185] G. Petropoulos, Comparative evaluation of energy crops in Greek space for the production of renewable diesel. M.Sc. Hellenic Open University. Patra. (in Greek), 2019.

[186] J. Mondol and N. Koumpetsos. Overview of challenges, prospects, environmental impacts and policies for renewable energy and sustainable development in Greece, 2013.

[187] G. Arabatzis, Ch. Malesios, An econometric analysis of residential consumption of fuel wood in a mountainous prefecture of Northern Greece, Energy Policy 39 (2011) 8088–8097.

[188] G. Arabatzis, K. Kitikidou, S. Tampakis, K. Soutsas, The fuel wood consumption in a rural area of Greece, Renew. Sustain. Energy Rev. 16 (2012) 6489–6496.

[189] G. Arabatzis, C. Malesios, Pro-environmental attitudes of users and non-users of fuelwood in a rural area of Greece, Renew. Sustain. Energy Rev. 22 (2013) 621–630.

[190] E. Zafeiriou, G. Arabatzis, T. Koutroumanidis, The fuelwood market in Greece: An empirical approach, Renew. Sustain. Energy Rev. 15 (2011) 3008–3018.

[191] K. Moustakas, P. Parmaxidou, S. Vakalis, Anaerobic digestion for energy production from agricultural biomass waste in Greece: Capacity assessment for the region of Thessaly, Energy 191 (2020) 116556.

[192] I. Vakakis, Energy crops and agricultural income, Agric. Farming 8 (2007) 34–42 In Greek.

[193] C. Panoutsou, Bioenergy in Greece: Policies, diffusion framework and stakeholder interactions, Energy Policy 36 (2008) 3674–3685.

[194] E. Zafeiriou, K. Petridis, C. Karelakis, G. Arabatzis, Optimal combination of energy crops under different policy scenarios; The case of Northern Greece, Energy Policy 96 (2016) 607–616.

[195] J.N. Lekakis, M. Kousis, Economic crisis, troika and the environment in Greece, S. Eur. Soc. Polit. 18 (3) (2013) 305–331.

[196] D.M. Knight, S. Bell, Pandora's box: Photovoltaic energy and economic crisis in Greece, J. Renew. Sustain. Energy 5 (3) (2013) 033110.

[197] N.E. Koltsaklis, A.S. Dagoumas, G.M. Kopanos, E.N. Pistikopoulos, M.C. Georgiadis, A spatial multi-period long-term energy planning model: A case study of the Greek power system, Appl. Energy 115 (2014) 456–482.

[198] R.I. Nazli, V. Tansi, H.H. Öztürk, A. Kusvuran, Miscanthus, switchgrass, giant reed, and bulbous canary grass as potential bioenergy crops in a semi-arid Mediterranean environment, Ind. Crops Prod. 125 (2018) 9–23.

[199] I. Pappas, Z. Koukoura, C. Kyparissides, C. Goulas, C. Tananaki, *Phalaris aquatic* L. lignocellulosic biomass as second generation bioethanol feed-

stock. Grassland. A European resource? Proceedings of the 24th General Meeting of the European Grassland Federation, 448–450, 2012.
[200] I.A. Pappas, K.G. Papaspyropoulos, C.N. Karachristos, A.S. Christodoulou, Generating carbon credits from perennial forage species crops in the Mediterranean region: The case of *Phalaris aquatica* L, Fut. Eur.Grasslands (2014) 145–147.
[201] M.B. Villamil, A. Myles, A. Heinz, M.E. Gray, Producer perceptions and information needs regarding their adoption of bioenergy crops, Renew. Sustain. Energy Rev. 16 (2012) 3604–3612.
[202] D.A. Landis, C. Gratton, R.D. Jackson, K.L. Gross, D.S. Duncan, C. Liang, T.D. Meehan, B.A. Robertson, T.M. Schmidt, K.A. Stahlheber, J.M. Tiedje, B.P. Werling, Biomass and biofuel crop effects on biodiversity and ecosystem services in the North Central US, Biomass Bioenergy 114 (2018) 18–29.
[203] United States Energy Information Administration, Renewable energy explained. Available from: http//tonto.eia.doe.gov/energyexplained/index.cfm?page=renewa ble home, 2010.
[204] M.B. Villamil, A.H. Silvis, G.A. Bollero, Potential miscanthus' adoption in Illinois: information needs and preferred information channels, Biomass Bioenergy 32 (2008) 1338–1348.
[205] D.L. Rockwood, D.R. Dippon, Biological and economic potentials of Eucalyptus grandis and Slash Pine as biomass energy crops, Biomass 20 (1989) 155–165.
[206] L.K. Paine, T.L. Peterson, D.J. Undersander, K. Rinner, G.A. Bartelst, S.T.A. Temple, D.W. Sample, R.M. Klemm, Some ecological and socio-economic considerations for biomass energy crop production, Biomass and Bioettergy 10 (4) (1996) 231–242.
[207] M. Downing, R.L. Graham, The potential supply and cost of biomass from energy crops in the Tenessee valey authority region, Biomass Bioenergy 11 (4) (1996) 283–303.
[208] D.S. Zamora, G.J. Wyatt, K.G. Apostol, U. Tschirner, Biomass yield, energy values, and chemical composition of hybrid poplars in short rotation woody crop production and native perennial grasses in Minnesota, USA, Biomass Bioenergy 49 (2013) 222–230.
[209] K.C. Surendra, R. Ogoshi, H.M. Zaleski, A.G. Hashimoto, S.K. Khanal, High yielding tropical energy crops for bioenergy production: Effects of plant components, harvest years and locations on biomass composition, Bioresour. Technol. 215 (2018) 218–229.
[210] G.M. Pedroso, R.B. Hutmacher, D. Putnam, J. Six, C. van Kessel, B.A. Linquist, Biomass yield and nitrogen use of potential C_4 and C_3 dedicated energy crops in a Mediterranean climate, Field Crops Res. 161 (2014) 149–157.
[211] A. Uslu, A.P. Faaij, P.C. Bergman, Pre-treatment technologies, and their effect on international bioenergy supply chain logistics.Techno-economic evaluation of torrefaction, fast pyrolysis and pelletisation, Energy 33 (8) (2008) 1206–1223.
[212] H. Shahrukh, A.O. Oyedun, A. Kumar, B. Ghiasi, L. Kumar, S. Sokhansanj, Techno-economic assessment of pellets produced from steam pretreated biomass feedstock, Biomass Bioenergy 87 (2016) 131–143.
[213] D.F. Lozano-García, J.E. Santibanez-Aguilar, F.J. Lozano, A. Flores-Tlacuahua, GIS-based modeling of residual biomass availability for energy and production in Mexico, Renew. Sustain. Energy Rev. 120 (2020) 109610.
[214] J.A. Honorato-Salazar, J. Sadhukhan, Annual biomass variation of agriculture crops and forestry residues, and seasonality of crop residues for energy production in Mexico, Food Bioprod. Process. 119 (2020) 1–19.

[215] T.F. Cardoso, M.D.B. Watanabe, A. Souza, M.F. Chagas, O. Cavalett, E.R. Morais, L.A.H. Nogueira, M.R.L.V. Leal, O.A. Braunbeck, L.A.B. Cortez, A. Bonomi, A regional approach to determine economic, environmental and social impacts of different sugarcane production systems in Brazil, Biomass Bioenergy 120 (2019) 9–20.

[216] L.C. de Freitas, S. Kaneko, Ethanol demand in Brazil: Regional approach, Energy Policy 39 (2011) 2289–2298.

[217] A.T. Furtado, M.I.G. Scandiffio, L.A.B. Cortez, The Brazilian sugarcane innovation system, Energy Policy 39 (2011) 156–166.

[218] J. Goldemberg, Ethanol for a sustainable energy future, Science 315 (2007) 808–810.

[219] CONAB, Companhia Nacional de Abastecimento, Perfil do Setor do Aη car e do Ethanol no Brasil, Safra 2014/15, Brasvlia, (2017) 1–64.

[220] CONAB, Cana-de-açúcar-Brasil: Série Histórica de Produção de Cana-de-Açúcar: Safras 2005/06 a 2015/16 em kg/ha, (2016).

[221] A.L. de Carvalho, R.S.C. Menezes, R.S. Nobrega, A. Pinto, S. de, J.P.H.B. Ometto, C. von Randow, A. Giarolla, Impact of climate changes on potential sugarcane yield in Pernambuco, northeastern region of Brazil, Renew. Energy 78 (2015) 26–34.

[222] C.P. Barros, L.A. Gil-Alana, P. Wanke, Ethanol consumption in Brazil: Empirical facts based on persistence, seasonality and breaks, Biomass Bioenergy 63 (2014) 313–320.

[223] S.L.M. Salles-Filho, P.F. Drummond de Castro, A. Bin, C. Edquist, A.F. Portilho Ferro, S. Corder, Perspectives for the Brazilian bioethanol sector: The innovation driver, Energy Policy 108 (2017) 70–77.

[224] R.C. Leite, C. de, M.R.L.V. Leal, L.A.B. Cortez, W.M. Griffin, M.I.G. Scandiffio, Can Brazil replace 5% of the 2025 gasoline world demand with ethanol?, Energy 34 (2009) 655–661.

[225] W.A. Araújo, Ethanol industry: surpassing uncertainties and looking forward, in: S.L.M. Salles-Filho, L.A.B. Cortez, J.M.F.J. Silveira, S.C. Trindade (Eds.), Global Bioethanol: Evolution, Risks, and UncertaintiesElsevier, London, 2016, pp. 1–33.

[226] EIA. Independent statistics and analysis, 2014.

[227] X. Li, Y. Huang, J. Gong, X. Zhang, A study of the development of bio-energy resources and the status of eco-society in China, Energy 35 (11) (2010) 4451–4456.

[228] X. Chen, Economic potential of biomass supply from crop residues in China, Applied Energy 166 (2016) 141–149.

[229] H.P. Zhong, Y.Z. Yue, J.W. Fan, Characteristics of crop straw resources in China and its utilization, Res. Sci. 25 (4) (2003) 62–67.

[230] J.J. Huang, W.D. Han, The current research and perspective utilization on the energy tree species in China, Guangdong Forest. Sci. Technol. 4 (2006) 24–28.

[231] J. Lin, X.W. Zhou, K.X. Tang, F. Chen, A survey of the studies on the resources of Jatropha curcas, J. Trop. Subtrop. Bot. 12 (3) (2004) 285–290.

[232] T. Wang, A survey of the woody plant resources for biomass fuel oil in China, Sci. Technol. Rev. 5 (2005) 23–29.

[233] Y.J. Chen, H. Meng, The present situation and countermeasures for bioenergy development in China, Chinese High Technol. Lett. 17 (12) (2007) 1312–1316.

[234] N.Y. Li, L. Zhi, G.S. Wang, H.J. Li, J.Y. Hong, Status and industrialization development of wood biomass energy in three north regions, Sci. Soil Water Conserv. 5 (4) (2007) 70–74.

[235] S.A.U. Rehman, Y. Cai, M. Nafees, G. Das Walasai, N.H. Mirjat, W. Rashid, Optimal biomass power plants locations with assigned biomass supply areas (red areas). Source: Overcoming electricity crisis in Pakistan: An overview of

the renewable energy status and development in Pakistan, J. Environ. Account. Manag. 5 (4) (2017) 357–383.
[236] M. Irfan, Z.-Yu. Zhao, M.K. Panjwani, F.H. Mangi, H. Li, A. Jan, M Ahmad, A. Rehman, Assessing the energy dynamics of Pakistan: Prospects of biomass energy, Energy Reports 6 (2020) 80–93.
[237] Pakistan Economic Survey 2017-18, (2018). Ministry of Finance. Government of Pakistan. Available at: http://www.finance.gov.pk/survey/chapters_19/2-Agriculture.pdf.
[238] N.A. Khan, H. Dessouky, Prospect of biodiesel in Pakistan, Renew. Sustain. Energy Rev. 13 (2009) 1576–1583.
[239] M. Kashif, M.B. Awan, S. Nawaz, M. Amjad, B. Talib, M. Farooq, A.S. Nizami, M. Rehan, Untapped renewable energy potential of crop residues in Pakistan: Challenges and future directions, J. Environ. Manag. 256 (2020) 109924.
[240] K. Ahmadi, H. Gholizadeh, H.R. Ebadzadeh, R. Hosseinpour, H. Abdi Shah, A. Kazimian, and M. Rafiei, Agricultural Ministry of Iran, Agricultural Statistics Volume I: Crop Products, Agriculture Ministry, p. 125. Iran, 2018.
[241] S.H. Samadi, B. Ghobadian, M. Nosrati, Prediction and estimation of biomass energy from agricultural residues using air gasification technology in Iran, Renew. Energy (2020).
[242] M. Hiloidhari, D. Das, D.C. Baruah, Bioenergy potential from crop residue biomass in India, Renew. Sustain. Energy Rev. 32 (2014) 504–512.
[243] P. Dwivedi, J.R.R. Alavalapati, Economic feasibility of electricity production from energy plantations present on community-managed forestlands in Madhya Pradesh, India, Energy Policy 37 (2009) 352–360.
[244] B. Patel, A. Patel, B. Gami, P. Patel, Energy balance, GHG emission and economy for cultivation of high biomass verities of bamboo, sorghum and pearl millet as energy crops at marginal ecologies of Gujarat state in India, Renew. Energy 148 (2020) 816–823.
[245] M. Siriwardhana, G.K.C. Opathella, M.K. Jha, Bio-diesel: Initiatives, potential and prospects in Thailand: A review, Energy Policy 37 (2009) 554–559.
[246] C. Pichalai, Thailand's energy conservation program and the role of new and renewable energy in energy conservation, J. R. Inst. Thailand 32 (1) (2007) 125–133.
[247] J.B. Gonsalves, An assessment of the biofuels industry in Thailand. United Nations Conference on Trade and Development, 2006.
[248] H. Chin, W.W. Choong, S. Rafidah, W. Alwi, A.H. Mohammed, Issues of social acceptance on biofuel development, J. Clean. Prod. 71 (2014) 30–39.
[249] S. Sohni, N.A.N. Norulaini, R. Hashim, S.B. Khan, W. Fadhullah, A.K. Mohd Omar, Physicochemical characterization of Malaysian crop and agro-industrial biomass residues as renewable energy resources, Ind. Crops Prod. 111 (2018) 642–650.
[250] A. Buchwal, S. Weijers, D. Blok, B. Elberling, Temperature sensitivity of willow dwarf shrub growth from two distinct High Arctic sites, Int. J. Biometeorol. 63 (2019) 167–181.
[251] S. Kivinen, E. Koivisto, S. Keski-Saari, L. Poikolainen, T. Tanhuanpaa, A. Kuzmin, A. Viinikka, R.K. Heikkinen, J. Pykala, R. Virkkala, P. Vihervaara, T. Kumpula, A keystone species, European aspen (*Populus tremula L.*), in boreal forests: Ecological role, knowledge needs and mapping using remote sensing, Forest Ecol. Manag. 462 (2020) 118008.
[252] R. van den Driessche, W. Rude, L. Martens, Effect of fertilization and irrigation on growth of aspen (*Populus tremuloides* Michx.) seedlings over three seasons, Forest Ecol. Manag. 186 (2003) 381–389.
[253] A.J. Franzluebbers, N. Nazih, J.A. Stuedemann, J.J. Fuhrmann, H. Schomberg, P.G. Hartel, Soil carbon and nitrogen pools under low- and high-endophyte-infected tall fescue, Soil Sci. Soc. Am. J. 63 (1999) 1687–1694.

[254] FAO, Grassland Index. A searchable catalogue of grass and forage legumes. FAO, Rome, Italy. Available from: https://web.archive.org/web/20170120044942/ http://www.fao.org/ag/AGP/AGPC/doc/GBASE/default.htm, 2015.

[255] B.G. Cook, B.C. Pengelly, S.D. Brown, J.L. Donnelly, D.A. Eagles, M.A. Franco, J. Hanson, B.F. Mullen, I.J. Partridge, M. Peters, and R. Schultze-Kraft, Tropical forages. CSIRO, DPI & F(Qld), CIAT and ILRI, Brisbane, Australia, 2005. Available from: http://www.tropicalforages.info/.

[256] J.A. Duke, Handbook of Energy Crops. New CROPS web site, Purdue University, 1983. Available from: https://hort.purdue.edu/newcrop/duke_energy/dukeindex.htm.

[257] P.H.A.U. de Silva, R.W. Snaydon, Chromosome number in *Cynodon dactylon* in relation to ecological conditions, Ann. Bot. 76 (5) (1995) 535–537.

[258] J.A. Rueda, E. Ortega-Jimenez, A. Hernandez-Garay, J.F. Enríquez-Quiroz, J.D. Guerrero-Rodríguez, A.R. Quero-Carrillo, Growth, yield, fiber content and lodging resistance in eight varieties of Cenchrus purpureus (Schumach.) Morrone intended as energy crop, Biomass Bioenergy 88 (2016) 59–65.

[259] J.E. Knoll, W.F. Anderson, R. Malik, R.K. Hubbard, T.C. Strickland, Production of napiergrass as a bioenergy feedstock under organic versus inorganic fertilization in the southeast USA, Bioenergy Res. 6 (2013) 974–983.

[260] D. Ogle, M. Majerus, and L. St. John, Plants for saline to sodic soil conditions. Plant Materials Tech Note No. 9. USDA NRCS, Boise, Idaho, 2008. Available from: http://www.id.nrcs.usda.gov/programs/tech_ref.html#TechNotes.

[261] J. Retana, D.R. Parker, C. Amrhein, A.L. Page, Growth and trace element concentrations of five plant species grown in a highly saline soil, J. Environ. Qual. 22 (1993) 805–811.

[262] M. Sakellariou-Makrantonaki, D. Kalfountzos, P. Vyrlas, Water saving and yield increase of sugar beet with subsurface drip irrigation, Global Nest Int. J. 4 (2–3) (2002) 85–91.

[263] S. Kiymaz, A. Ertek, Yield and quality of sugar beet (*Beta vulgaris* L.) at different water and nitrogen levels under the climatic conditions of Kırsehir, Turkey, Agric. Water Manage. 158 (2015) 156–165.

[264] M.M.S. Ouda, Effect of nitrogen and sulphur fertilizers levels on sugar beet in newly cultivated sandy soil, Zagazig J. Agric. Res. 29 (2002) 33–50.

[265] M.F. Fathy, A. Motagally, K.K. Attia, Response of sugar beet plants to nitrogen and potassium fertilization in sandy calcareous soil, Int. J. Agric. Biol. 11 (6) (2009) 695–700.

[266] A. Herrera, T.L. Dudley, Invertebrate community reduction in response to *Arundo donax* invasion at Sonoma Creek, Biol. Invasions 5 (2003) 167–177.

[267] CABI, *Arundo donax* (giant reed). In: Invasive Species Compendium. CAB International, Wallingford, UK, 2020. Available from: https://www.cabi.org/isc/datasheet/1940.

[268] Global Invasive Species Database, Species profile: Arundo donax. (2020). Available from: http://www.iucngisd.org/gisd/species.php?sc=112, Accessed 20.05.20.

[269] S.L. Cosentino, D. Scordia, E. Sanzone, G. Testa, V. Copani, Response of giant reed (*Arundo donax* L.) to nitrogen fertilization and soil water availability in semi-arid Mediterranean environment, Eur. J. Agronom. 60 (2014) 22–32.

[270] W.H. Verheye, Soils, plant growth and crop production-Volume II, Eolss Publishers, Co Ltd, U.K, 2010, pp. 216–222.

[271] Klebesadel, L.J., & Dofing, S.M. (1991). Reed canarygrass in Alaska: influence of latitude-of-adaptation on winter survival and forage productivity, and ob-

servations on seed productivity. University of Alaska, School of Agriculture and Land Resources Management, Agricultural and Forestry Experiment Station, Bulletin 84, Fairbanks.

[272] GISD (Global Invasive Species Database), Species profile: *Phalaris arundinacea*. Invasive species specialist group, gland, Switzerland, 2017. Available from: http://www.iucngisd.org/gisd/speciesname/Phalaris+arundinacea.

[273] K.A. Amouzou, J.P.A. Lammers, J.B. Naab, C. Borgemeister, P.L.G. Vlek, M. Becker, Climate change impact on water- and nitrogen-use efficiencies and yields of maize and sorghum in the northern Benin dry savanna, West Africa, Field Crops Res. 235 (2019) 104–117.

[274] D. Xiao, D.L. Liu, B. Wang, P. Feng, H. Bai, J. Tang, Climate change impact on yields and water use of wheat and maize in the North China Plain under future climate change scenarios, Agric. Water Manag. 238 (2020) 106238.

[275] I.M. Azare, I.J. Dantana, M.S. Abdullahi, A.A. Adebayo, M. Aliyu, Effects of climate change on Pearl millet (*Pennisetum glaucum* [L.R. Br.]) production in Nigeria, J. Appl. Sci. Environ. Manag. 24 (1) (2020) 157–162.

[276] N. Bassam, Handbook of Bioenergy Crops. A Complete Reference to Species Development and Applications. Energy Crops Guide, Earthscan publishing, London, Washington D.C, 2010, pp. 319–333.

[277] A. Nishad, A.N. Mishra, R. Chaudhari, R.K. Aryan, P. Katiyar, Effect of temperature on growth and yield of rice (*Oryza sativa* L.) cultivars, Int. J. Chem. Studies 6 (5) (2018) 1381–1383.

[278] R. Kaur, Climate change impacts on rice (Oryza sativa) productivity and strategies for its sustainable management, Indian J. Agric. Sci. 89 (2) (2019) 171–180.

[279] B. Sliz-Szkliniarz, Energy planning in selected European regions Methods for evaluating the potential of renewable energy sources, Scientific Publishing, Karlshruhe, German, 2012, pp. 32.

[280] J.C. Clifton-Brown, I. Lewandowski, Overwintering problems of newly established *Miscanthus* plantations can be overcome by identifying genotypes with improved rhizome cold tolerance, New Phytol. 148 (2000) 287–294.

[281] R. Pyter, E. Heaton, F. Dohleman, T. Voigt, S. Long, Agronomic experiences with *Miscanthus* x *giganteus* in Illinois, USA, in: J.R. Mielenz (Ed.), Biofuels: Methods and Protocols, Human Press, NY, USA, 2009, pp. 41–52.

[282] L. Ercoli, M. Mariotti, A. Masoni, E. Bonari, Effect of irrigation and nitrogen fertilization on biomass yield and efficiency of energy use in crop production of Miscanthus, Field Crops Res. 63 (1999) 3–11.

[283] Sh.F. Huseynli, I.B. Mirjalally, R.R. Efendiyeva, J.G. Gurbanova, S.F. Hajiyeva, *Medicago sativa* L.'s sowing periods and biological characteristics, Global J. Biol. Agric. Health Sci. 5 (3) (2016) 57–59.

[284] P.C. Friesen, M. de Melo Peixoto, D.K. Lee, R.F. Sage, Sub-zero cold tolerance of *Spartina pectinata* (prairie cordgrass) and *Miscanthus* × *giganteus*: candidate bioenergy crops for cool temperate climates, J. Exp. Bot. 66 (14) (2015) 4403–4413.

[285] S.E. Travers, Z. Tang, D. Caragea, K.A. Garrett, S.H. Hulbert, J.E. Leach, J. Bai, A. Saleh, A.K. Knapp, P.A. Fay, J. Nippert, P.S. Schnable, M.D. Smith, Variation in gene expression of *Andropogon gerardii* in response to altered environmental conditions associated with climate change, J. Ecol. 98 (2) (2010) 374–383.

[286] K.J. Mitchell, Growth of pasture species under controlled environment. 1. Growth at various levels of constant temperature, New Zealand J. Sci. Technol. 38 (2) (1956) 203–215.

[287] D. Thorogood, Perennial ryegrass (*Lolium perenne* L.). In: M.D. Casler, R.R. Duncan (Eds.), Turfgrass Biology Genetics and Breeding. John Wiley and Sons, New Jersey, USA, (2003) 75–105.

[288] M. Cool, D.B. Hannaway, C. Larson, D. Myers, Perennial ryegrass (Lolium perenne L.), forage fact sheet, Oregon State University0, Oregon, USA, 2004 http://forages.oregonstate.edu/php/fact_sheet_print_grass.php?SpecID=6&use=.

[289] R.A. Waller, P.W.G. Sale, Persistence and productivity of perennial ryegrass in sheep pastures in south-western Victoria: a review, Austr. J. Exp. Agric. 41 (1) (2001) 117–144.

[290] S.L. Harris, E.R. Thom, D.A. Clark, Effect of high rates of nitrogen fertiliser on perennial ryegrass growth and morphology in grazed dairy pasture in northern New Zealand, New Zealand J. Agric. Res. 39 (1) (1996) 159–169.

[291] L. Gilbert, *Phacelia tanacetifolia*: A brief overview of a potentially useful insectary plant and cover crop. Small Farm Success Project. Fact Sheet Number 2a, 2003.

[292] J. Sáez-Bastante, P. Fernández-García, M. Saavedra, L. López-Bellido, M.P. Dorado, S. Pinz, Evaluation of *Sinapis alba* as feedstock for biodiesel production in Mediterranean climate, Fuel 184 (2016) 656–664.

[293] W.-L. Chen, W.-J. Yang, H.-F. Lo, D.-M. Yeh, Physiology, anatomy, and cell membrane thermostability selection ofleafy radish (*Raphanus sativus* var. *oleiformis* Pers.) with different tolerance under heat stress, Sci. Horticultur. 179 (2014) 367–375.

[294] Z. Li, C. Fontanier, B.L. Dunn, Physiological response of potted sunflower (*Helianthus annuus* L.) to precision irrigation and fertilizer, Sci. Horticultur. 270 (2020) 109417.

[295] V. Tilvikiene, Z. Kadziuliene, A. Raila, E. Zvicevicius, I. Liaudanskiene, Z. Volkaviciute, L. Pociene, *Artemisia dubia* wall-A novel energy crop for temperate climate zone in Europe. 23rd European Biomass Conference and Exhibition, 1–4 June 2015, Vienna, Austria, 2015.

[296] Ž. Kadziuliene, V. Tilvikiene, I. Liaudanskiene, L. Pociene, Ž. Černiauskiene, E. Zvicevicius, A. Raila, *Artemisia dubia* growth, yield and biomass characteristics for combustion, Zemdirbyste Agric. 104 (2) (2017) 99–106.

[297] J. Gominho, H. Pereira, An overview of the research on the pulping aptitude of *Cynara cardunculus* L. First World Conference on Biomass for Energy and Industry, Sevilla, Spain, 5–9 June, pp. 1187–1190, 2000.

[298] J. Gominho, A. Lourenço, M.D. Curt, J. Fernández, H. Pereira, *Cynara cardunculus* in large scale cultivation. A case Study in Portugal, Chem. Eng. Trans. 37 (2014) 529–534.

[299] D. Sekella, Cold hardiness of 5 Eucalypts in Northern California, Pacific Hortic. 64 (1) (2003).

[300] J.K. Hasey, J.M. Connor, Eucalyptus shows unexpected cold tolerance, Calif. Agric. 44 (2) (1990) 25–27.

[301] W.H. Maes, A. Trabucco, W.M.J. Achten, B. Muys, Climatic growing conditions of *Jatropha curcas* L, Biomass Bioenergy 33 (2009) 1481–1485.

[302] T.A. Shestakova, J. Voltas, M. Saurer, R.T.W. Siegwolf, A.V. Kirdyanov, Warming effects on *Pinus sylvestris* in the cold–dry Siberian forest–steppe: Positive or negative balance of trade?, Forests 8 (2017) 490.

[303] S.R. McGreevy, A. Shibata, Mobilizing biochar. A multistackholder scheme for climate-friendly foods and rural sustainable development, in: T.J. Goreau, R.W. Larson, J. Campe (Eds.), Geotherapy. Innovative methods of soil fertility, restoration, carbon sequestration and reversing CO_2 increase, CRS Press, Taylor and Francis group, 2015, pp. 270.

[304] A. Fernández-Cancio, R.M. Navarro Cerrillo, R. Fernández, P. Gil Hernández, E. Manrique Menéndez, C. Calzado Martínez, Climate classification of *Abies pinsapo* Boiss. Forests in Southern Spain, Investigación Agraria: Sistemas y Recursos Forestales 16 (3) (2007) 222–229.

[305] R.M. Navarro, M.J. Retamosa, J. Lopez, A. del Campo, C. Ceaceros, L. Salmoral, Nursery practices and field performance for the endangered Mediterranean species *Abies pinsapo* Boiss, Ecol. Eng. 27 (2006) 93–99.
[306] J.N. Owens, L.M. Chandler, J.S. Bennett, T.J. Crowder, Cone enhancement in abies amabilis using GA4/7. Fertilizer, firding and tenting, Forest Ecol. Manag. 154 (2001) 227–236.
[307] X. Feng, G. Liu, J.M. Chen, M. Chen, J. Liu, W.M. Ju, R. Sun, W. Zhou, Net primary productivity of China's terrestrial ecosystems from a process model driven by remote sensing, J. Environ. Manag. 85 (2007) 563–573.
[308] X. Feng, P. An, X. Li, K. Guo, C. Yang, X. Liu, Spatio temporal heterogeneity of soil water and salinity after establishment of dense-foliage *Tamarix chinensis* on coastal saline land, Ecol. Eng. 121 (2018) 104–113.
[309] X. Feng, X. Liu, X. Zhang, J.S. Li, Growth dynamic of *Tamarix chinensis* plantations in high salinity coastal land and its ecological effect, Sabkha Ecosyst. 6 (2019) 113–124.
[310] L. Xie, B. Wang, M. Xin, Q. Wei, W. Wang, X. He, J. Wang, X. Shi, X. Sun, Impacts of coppicing on *Tamarix chinensis* growth and carbon stocks in coastal wetlands in northern China, Ecol. Eng. 147 (2020) 105760.
[311] J. Xia, X. Zhao, J. Ren, Y. Lang, F. Qu, H. Xu, Photosynthetic and water physiological characteristics of *Tamarix chinensis* under different groundwater salinity conditions, Environ. Exp. Bot. 138 (2017) 173–183.
[312] Q. Meng, Y. Niu, X. Niu, R.H. Roubin, J.R. Hanrahan, Ethnobotany, phytochemistry and pharmacology of the genus *Caragana* used in traditional Chinese medicine, J. Ethnopharmacol. 124 (2009) 350–368.
[313] W.G. Taylor, D.H. Sutherland, K.W. Richards, H. Zhang, Oleanane triterpenoid saponins of *Caragana arborescens* and their quantitative determination, Ind. Crops Prod. 77 (2015) 74–80.
[314] K.-H. Chung, Transesterification of Camellia japonica and Vernicia fordii seed oils on alkali catalysts for biodiesel production, J. Ind. Eng. Chem. 16 (2010) 506–509.
[315] W. Chen, F. Wu, J. Zhang, Potential production of non-food biofuels in China, Renew. Energy 85 (2016) 939–944.
[316] K.A. Langeland, H.M. Cherry, C.M. McCormick, K.A. Craddock Burks, Identification and biology of non-native plants in Florida's natural areas, 2nd edition., University of Florida, Gainesville, Florida, USA:, 2008, pp. 210.
[317] PROTA, PROTA4U web database. In: G.J.H. Grubben, O.A. Denton (Eds.), Plant Resources of Tropical Africa, Wageningen, Netherlands, 2014. Available from: http://www.prota4u.org/search.asp.
[318] D.-G Guo, X.-Y Zhang, H.-B Shao, Z.-K Bai, L.-Y Chu, T. -L Shangguan, K Yan, L.-H Zhang, G. Xu, J.-N. Sun, Energy plants in the coastal zone of China: Category, distribution and development, Renew. Sustain. Energy Rev. 15 (2011) 2014–2020.
[319] Bostid, Firewood Crops: Shrub and Tree Species for Energy Production, 1983.
[320] R. Wang, M.A. Hanna, W.-W. Zhou, P.S. Bhadury, Q.-A. Chen, B. Song, S. Yang, Production and selected fuel properties of biodiesel from promising non-edible oils: *Euphorbia lathyris* L., *Sapium sebiferum* L. and *Jatropha curcas* L, Bioresour. Technol. 102 (2011) 1194–1199.
[321] S. Nijjer, W.E. Rogers, C.-T.A. Lee, E. Siemann, The effects of soil biota and fertilization on the success of *Sapium sebiferum*, Appl. Soil Ecol. 38 (2008) 1–11.
[322] M. Tang, P. Zhang, L. Zhang, M. Li, L. Wu, A potential bioenergy tree: *Pistacia chinensis* bunge. 2012 International Conference on Future Energy, Environment, and Materials, Energy Proc. 16 (2012) 737–746.

[323] L. Lu, D. Jiang, J. Fu, D. Zhuang, Y. Huang, M. Hao, Evaluating energy benefit of *Pistacia chinensis* based biodiesel in China, Renew. Sustain. Energy Rev. 35 (2014) 258–264.
[324] T.J. Mwine, P. van Damme, *Euphorbia tirucalli* L. (Euphorbiaceae): The miracle tree: Current status of available knowledge, Sci. Res. Essays 6 (23) (2011) 4905–4914.
[325] C. Orwa, A. Mutua, R. Kindt, R. Jamnadass, A. Simons, Agroforestree Database: a tree reference and selection guide version 4.0, 2009. Available from: http://www.worldagroforestry.org/af/treedb/.
[326] R.S. Troup, H.B. Joshi, The silviculture of Indian trees. Vol IV., Leguminosae, Controller of Publications, Delhi, India, (1983).
[327] R.K. Luna, Plantation trees, International Book Distributors, Delhi, India:, (1996).
[328] A.J. Nath, G.W. Sileshi, A.K. Das, Bamboo based family forests offer opportunities for biomass production and carbon farming in North East India, Land Use Policy 75 (2018) 191–200.
[329] S. Gantait, B.R. Pramanik, M. Banerjee, Optimization of planting materials for large scale plantation of *Bambusa balcooa* Roxb.: Influence of propagation methods, J. Saudi Soc. Agric. Sci. 17 (2018) 79–87.
[330] C.I. Ogbonnaya, M.C. Nwalozie, H. Roy-Macauley, D.J.M. Annerose, Growth and water relations of Kenaf (*Hibiscus cannabinus* L.) under water deficit on a sandy soil, Ind. Crops Prod. 8 (1998) 65–76.
[331] S.L. Falasca, A.C. Ulberich, S. Pitta-Alvarez, Possibilities for growing kenaf (*Hibiscus cannabinus* L.) in Argentina as biomass feedstock under dry-subhumid and semiarid climate conditions, Biomass Bioenergy 64 (2014) 70–80.
[332] C. Petrini, R. Bazzocchi, P. Montalti, Yield potential and adaptation of kenaf *(Hibiscus cannabinus)* in north-central Italy, Ind. Crops Prod. 3 (1994) 11–15.
[333] World Energy Council, World energy resources bioenergy 2016. Technical Report. World Energy Council. Available from: https://doi.org/10.1016/0165-232X(80)90063-4, 2016.
[334] IEA, Renewables information 2017 final edition. Technical Report, Paris: International Energy Agency, 2017.
[335] REN21, Renewables 2017. Global Status Report 2017. Technical Report, Renewable Energy Policy Network for the 21st Century, Paris, 2017. Available from: https://doi.org/10.1016/j.rser.2016.09.082.
[336] A. Al Nouss, G. McKay, T. Al-Ansari, Enhancing waste to hydrogen production through biomass feedstock blending: A techno-economic-environmental evaluation, Appl. Energy 266 (2020) 114885.
[337] IEA, World energy outlook 2016—Part B special focus on renewable energy. Technical Report. Paris: International Energy Agency, 2016. Available from: https://doi.org/10.1787/20725302.
[338] J. Goffe, J.-H. Ferrasse, Stoichiometry impact on the optimum efficiency of biomass conversion to biofuels, Energy 170 (2019) 438–458.
[339] G. Butera, S.H. Jensen, R.O. Gadsboll, J. Ahrenfeldt, L.R. Clausen, Flexible biomass conversion to methanol integrating solid oxide cells and Two Stage gasifier, Fuel 271 (2020) 117654.
[340] U. Henriksen, J. Ahrenfeldt, T.K. Jensen, B. Gobel, J.D. Bentzen, C. Hindsgaul, L.H. Sorensen, The design construction and operation of a 75 kW two-stage gasifier, Energy 31 (2006) 1542–1553.
[341] J.D. Bentzen, U.B. Henriksen, C. Hindsgaul, P. Brandt, Optimized two-stage gasifier. 1st World Conference on Biomass for Energy and Industry. Proceedings of the Conference held in Sevilla, Spain, 5–9 June 2000, London, James & James (Science Publishers) Ltd, 2000.

[342] S. Li, H. Zheng, Y. Zheng, J. Tian, T. Jing, J.–S. Chang, S.–H. Ho, Recent advances in hydrogen production by thermo-catalytic conversion of biomass, Int. J. Hydrogen Energy 44 (2019) 14266–14278.
[343] M. Shahbaz, T. Al-Ansari, M. Aslam, Z. Khan, A. Inayat, M. Athar, S.R. Naqvi, M.A. Ahmed, G. McKay, A state of the art review on biomass processing and conversion technologies to produce hydrogen and its recovery via membrane separation, Int. J. Hydrogen Energy 45 (2020) 15166–15195.
[344] J. Ren, Y.–L. Liu, X.–Y. Zhao, J.–P. Cao, Biomass thermochemical conversion: A review on tar elimination from biomass catalytic gasification, J. Energy Inst. 93 (2020) 1083–1098.
[345] H.C. Ong, W.-H. Chen, Y. Singh, Y.Y. Gan, C. Chen, -Y. Showg, A state-of-the-art review on thermochemical conversion of biomass for biofuel production: A TG-FTIR approach, Energy Conver. Manag. (2020) 112634.
[346] A. Hirano, K. Hon-Nami, S. Kunito, M. Hada, Y. Ogushi, Temperature effect n continuous gasification of microalgal biomass: theoretical yield of methanol production and its energy balance, Catal. Today 45 (1998) 399–404.
[347] M.J.A. Tijmensen, A.P.C. Faaij, C.N. Hamelinck, M.R.M. Van Hardeveld, Exploration of the possibilities for production of Fischer Tropsch liquids and power via biomass gasification, Biomass Bioenergy 23 (2002) 129–152.
[348] K.J. Ptasinski, Thermodynamic efficiency of biomass gasification and biofuels conversion, Biofuels Bioprod. Biorefin. 2 (2008) 239–253.
[349] K. Kumabe, S. Fujimoto, T. Yanagida, M. Ogata, T. Fukuda, A. Yabe, T. Minowa, Environmental and economic analysis of methanol production process via biomass gasification, Fuel 87 (2008) 1422–1427.
[350] M. Higo, K. Dowaki, A Life Cycle Analysis on a Bio-DME production system considering the species of biomass feedstock in Japan and Papua New Guinea, Appl. Energy 87 (2010) 58–67.
[351] J. Sheehan, A. Aden, K. Paustian, K. Killian, J. Brenner, M. Walsh, R. Nelson, Energy and environmental aspects of using corn stover for fuel ethanol, J. Ind. Ecol. 7 (2004) 117–146.
[352] W.D. Huang, Y.H.P. Zhang, Energy efficiency analysis: biomass-to-wheel efficiency related with biofuels production, fuel distribution, and powertrain systems, PLoS One 6 (2011) 1–10.
[353] X. Li, J. Xia, X. Zhao, Y. Chen, Effects of planting Tamarix chinensis on shallow soil water and salt content under different groundwater depths in the Yellow River Delta, Geoderma 335 (2019) 104–111.
[354] T. Kirch, P.R. Medwell, C.H. Birzer, P.J. van Eyk, Small-scale autothermal thermochemical conversion of multiple solid biomass feedstock, Renew. Energy 149 (2020) 1261–1270.

2

Examining low carbon energy technologies and their contribution as sustainable energy systems

4

Public attitudes toward the major renewable energy types in the last 5 years: A scoping review of the literature

Evangelia Karasmanaki, Georgios Tsantopoulos
Department of Forestry and Management of the Environment and Natural Resources, Democritus University of Thrace, Orestiada, Greece

Chapter outline

1 Introduction

During the first years of implementation, renewable energy seemed to be universally approved as it lacked the severe and, in some cases, irreversible impacts of the traditional fossil fuel-based energy system. Environmentalists and scientists supported fervently the adoption of renewables using the argument that they can prevent resource depletion and environmental degradation. Yet, the increasing number of renewable energy projects that have failed [1] shows that the advantages of renewables do not suffice to prevent public opposition.

The public concerns, which induce opposition to renewables typically, revolve around four areas. The first area of concern focuses on the visual impact of the facility on the natural landscape, which is disrupted by the addition of the facility [2,3]. Second, the

Low Carbon Energy Technologies in Sustainable Energy Systems. http://dx.doi.org/10.1016/B978-0-12-822897-5.00004-3

local public often complains about the noise pollution caused during the construction of the project as well as during the operation of the facility. Noise disturbance is pronounced especially in the case of wind energy projects with affected residents expressing concerns about both the actual noise and the infrasound of wind turbines [4,5]. Another frequent concern that sparks conflict is the impact of these projects on the sustainability of the local natural environment and animal species [6]. Beside these concerns, which focus mainly on the technical and physical nature of the projects and could perhaps be overcome with technological progress, place attachment, the strong bond between a place and a person, can induce residents to react against new energy applications [7,8]. Moreover, the public opposes projects, if the planning and decision-making procedures are perceived as unfair and if local actors are not invited to participate in these processes [9,10]. As it can be seen, the concerns of the public are multi-faceted, but, most importantly, reveal that the public plays a critical role in the development of renewable energy projects. This may also suggest that public attitudes to RES consist a determinant of the widespread implementation of renewable energy and the achievement of energy policy goals [11,12].

With a multitude of policies and directives pursuing the transition from fossil fuels to renewable energies, the European Union is often described as a leader in energy policy [13]. In particular, the Energy Strategy and Energy Union require safe, competitive, and sustainable energy within the EU and define challenging objectives for the deployment of renewable energy. For instance, renewable energy should account for as high as 32% in the EU energy mix by 2030 [14]. The achievement of this objective requires all Member States to deploy renewable energy and decrease their reliance on fossil fuels (both indigenous and imported). In retrospect, the foundations of the European Union renewable energy policy were laid in 1997 when the European Council and the European Parliament established the "White Paper for a Community Strategy and Action Plan" and when the percentage of renewable energy accounted for 6% of gross internal energy consumption [15]. About a decade later, the European Commission (EC) established "the Europe 2020 Strategy," a 10-year strategy that set out the vision of a social market economy and sought to transform the Union into a smart, sustainable and inclusive economy. In the context of this strategy and in order to achieve the ambitious energy and climate change goals, the EC established the "2020 climate and energy package" which involved a set of binding regulations to make sure that the following targets would be met: (1) 20% reduction in greenhouse gas emissions, (2) 20% of gross final energy consumption from renewables, and (3) 20% enhancement in energy efficiency [16]. Over the longer term, the EU has ratified the Paris Agreement, which aims at keeping the global temperature rise below 2°C and keeping it to 1.5°C. In this context, the European Union developed its 2050 long-term strategy referred to as "A clean Planet for all" which visualizes a prosperous, modern, competitive, and climate-neutral European economy by 2050. To manifest this vision, the 2050 long-term strategy demonstrates the ways in which Europe can become the leader of climate neutrality by making investments in realistic technologies, empowering its citizens, and aligning action in important areas, such as finance, research, industrial policy while providing social fairness for a just transition [17]. Simultaneously, the strategy assesses the portfolio of available options for member states, businesses, and citizens to determine how they can contribute to the modernization of the economy and improve the life quality of Europeans. Furthermore, the strategy tries to ensure that this transition is socially just and that the competitiveness of the European economy and industry in global markets is improved [17].

Given that public acceptability can curtail the exploitation of renewables, it is necessary to evaluate current knowledge of public attitudes in order to provide the basis from which further research work in this field can be undertaken. In addition, understanding public attitudes to RES and their influence on the development of renewable energies will potentially guide future implementation strategies. Therefore, the aim of this chapter is to provide a review of research into public attitudes toward renewable energy in general as well as toward specific renewable types. The focus is placed on recent empirical research, which was carried out in countries belonging to the European Union during the period 2015–20; that is, the period that followed the Paris Agreement and the ambitious energy targets that were set in response to the agreement. Moreover, it was deemed appropriate to use a 5-year time frame in order to capture the current trend because attitudes toward renewables tend to shift due to the public's exposure to varying information and political debates on the energy industry. It should also be noted that this work faced the problem of defining who "the public" is. Although there is a tendency to regard "the public" as an amorphous mass, the "public opinion" often takes two forms in the literature on public attitudes to RES. That is, there are the "representative" national views, which often perceive matters in an abstract and distanced manner and there are also the "local" views, which are founded on the possible or actual experience with specific renewable energy projects [18]. In this work, we include both forms of "the public" in order to form a complete picture on public attitudes. Finally, the format of the scoping review was chosen as it enables the researchers to map a wide range of evidence and to sum research findings produced by studies of potentially widely varying designs.

2 Methodology

To perform this review, the framework proposed by Arksey and O'Malley [19] was followed. Arksey and O'Malley [19] suggest that the aim of a scoping review is to put together the materials, which map an area of research. In contrast to other types of reviews, the scoping study method is driven by the requirement to identify all relevant literature regardless of study design. It is likely that as familiarity with the literature is increased, researchers will wish to redefine search terms and carry out more sensitive searches of the literature. To that end, the researchers might not want to set strict limitations on search terms, identification of relevant studies, or study selection at the beginning. Therefore, the procedure is not linear but iterative, while it requires researchers to deal with each stage of the process in a reflexive way and, if needed, repeat steps to ensure that the literature is covered in a comprehensive way. Hence, such a review attempts to collect evidence from the literature in order to answer a specific research question with the aim of generating a synthesis based on existing knowledge. However, this type of review does not to intend to critique the methodology of the reviewed literature works. Moreover, researchers can carry out a scoping study to investigate the extent, range, and nature of research activity, detect the value of performing a complete systematic review, summarize, and disseminate existing research findings, or recognize gaps in the extant literature [19]. In an effort to provide guidance to authors undertaking scoping studies, Arksey and O'Malley [19] proposed a methodological framework but also prompted the authors, who would follow their methodology, to adjust their framework in order to improve the methodology. In this chapter, we utilize the Arksey and O'Malley framework but build on the existing methodological framework.

With this information in mind, we now describe the stages of the framework we adopted for performing a scoping study (Fig. 4.1).

Following the above framework, in the first stage, we stated our research question, which was: "What are the public attitudes toward the major renewable energy types in the last 5 years?" Employing this question as a guide of our search of the relevant literature, the database *Scopus* was searched using the search terms: public attitudes; social acceptance of renewable energy; public opinion on

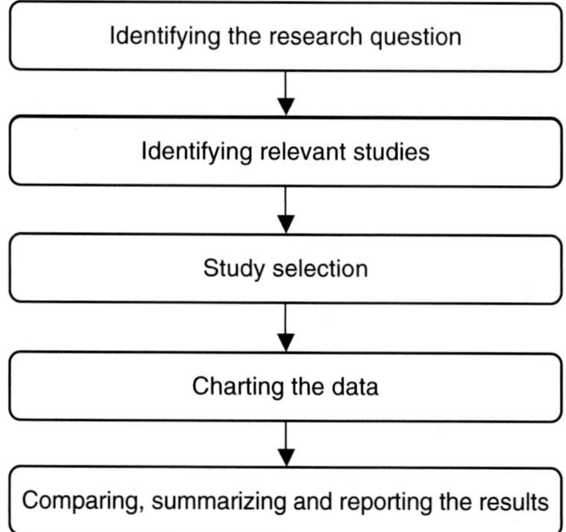

Figure 4.1. Proposed research framework.

renewable energy; public perceptions of renewable energy; public attitudes to wind energy; public attitudes to solar energy; public attitudes to hydropower; public attitudes to geothermal energy. Articles were limited to studies that focused on the public as their sample while studies with groups like tourists, policymakers, local authorities or students as sample were excluded. In addition, only studies carried out in EU countries and written in the English language from 2015 to 2020 were included. Then, in stage 3 a repetitive process of study selection was conducted. After the potential abstracts were identified from the searches, each researcher reviewed the abstract and reported whether it should be included in the review.

3 Results

The search, which was conducted in February 2020, identified 217 relevant papers in the database of which 29 articles were chosen to be fully reviewed. Based on this full-text review, 21 articles finally met all the criteria for inclusion in the scoping review. The selected articles were studies, which investigated the attitudes of the public toward specific renewable types or renewable energy in general in EU countries during the period 2015–20. Of these 21 articles, many studies examined attitudes and acceptance toward renewable energy in general and a considerable number of studies explored attitudes to wind and solar energy. It is interesting to observe that there was a scarcity of articles examining public attitudes to biomass, hydropower while there were no studies on geothermal or other less commonly known renewable types (such as wave and tidal energy) (Table 4.1).

Table 4.1 Summary of the articles selected for inclusion in the scoping review.

References	Study aims	Study area	Sample characteristics and study design	Main findings
Ólafsdóttir and Sæþórsdóttir (2019) [31]	Evaluation of attitudes of local residents and tourism service providers towards the first wind farm	Southern Iceland	$n = 178$ questionnaire respondents Interviewees: 23	The majority was familiar with the two existing experimental wind turbines (WTs), and nearly 1 in 10 residents stated they could see them from their residence. Most residents were positive to WTs, while less than half were positive to WTs in Iceland. Male respondents were significantly more positive. Most perceived WTs as a positive addition to Iceland's energy production. The respondents were less enthusiastic about the proposed "Búrfell Wind Farm." Male respondents are less negative than female ones. The visual impact of WTs caused uneasiness and the most negative aspect of WTs would be the obstruction of the mountain view of the surrounding mountains.
Roddis et al. (2019) [20]	Analysis of 25 "waves" of the UK PAT.	UK	$n = 52{,}525$ Face-to-face in-home interviews, analysis with Mann-Kendall tests and ordinal regression	Solar was the most favored energy source in Great Britain between July 2012 and April 2018. Solar was followed by RE (in general), wave and tidal, offshore wind, onshore wind, and biomass. Support for all RES is increasing over time. Rural dwellers were more supportive to RES compared to urban residents. Younger people, women, and those with higher climate concern were more likely to support RES.
Seidl et al. (2019) [28]	Investigation of the perceived responsibilities and the intended technology uptake of DESs (Distributed energy systems)	Switzerland, Germany, and Austria	$n = 2104$ Web-based survey and mixed-method approach with ANOVA, Bonferroni post-hoc test) and linear regression analysis	An openness to engage with DESs was recorded. In all countries, the participants perceived that the national government and large energy supply utilities were responsible for energy transitions, whereas SMEs, households, municipalities, landowners and property owners were the least responsible. The support for DESs was irrespective of different scales of rationale. The acceptance of DES was driven by the perceived opportunities, whereas the variables *general attitude to innovations*, *knowledge about energy themes*, and *potential behavior toward DES participation*, account only for minor percentages of variance.

Liebe and Dobers (2019) [26]	Examination of factors affecting acceptance of hypothetical new power plants in respondents' proximity and investigation of protest intentions.	Germany	$n = 3400$ Online survey, regression models	The respondents preferred solar over wind power, followed by biogas and natural gas-fired plants. While the protest potential was low, it was greatest for natural gas, followed by biomass, wind, and solar. Wind and solar energy were viewed as costlier than biomass and natural gas. Solar and wind energy were seen as less harmful than biomass and natural gas. Climate change concern has a positive effect on the acceptance of wind power plants and solar fields but not of biogas plants. Respondents who perceived that new plants should be built in someone else's vicinity were also more likely to disapprove new power plants in their vicinity. The higher the place attachment, the stronger the intention to oppose new plants. Protest intentions were stronger for older respondents and weaker for the higher educated and urban dwellers.
Zaunbrecher et al. (2018) [30]	Assessment of attitudes and preferences for electricity mix scenarios.	Germany	$n = 176$ Online survey	The electricity mix with the highest share of PV and the lowest share of biogas ("PV-Mix Pasture") was the most favored, whereas the most negatively rated scenario was "Wind-Mix Forest." The scenario "PV-Mix Pasture" was also perceived as most likely to be supported. Its impact on health was ranked lowest and it seen as the quietest and safest. It was perceived as having the most positive effect on the image of a region and for providing the most added value for the local community.
Sposato and Hampl, (2018) [24]	Investigation of respondents' perceptions of various renewable energy-related issues	Austria	$n = 1000$ Quantitative data via questionnaire and multiple regression analysis	The more positive respondents perceive RE, the more likely they are to accept RES in their community. Respondents who rated positively arguments for switching to RES expressed greater acceptance of RES in their vicinity whereas individuals who were skeptical to the future viability of RE were less likely to approve it. Age, communitarianism-egalitarianism and gender were associated with RES acceptance, indicating that older and female participants were less supportive to WTs and photovoltaics in their vicinity. Wind power was less accepted than photovoltaics.
Ntanos et al. (2018) [22]	Analysis and evaluation of the use of renewable energy sources and their contribution to citizens' life quality.	Greece	$n = 400$ Principal Components Analysis and Logit Regression	Respondents knew little about hydrodynamic, geothermal, and biomass-based sources of energy but were informed about wind and solar power. Most respondents used at least one type of RES (mostly solar water heaters) and 11% had installed solar PVs. These results resonated with their knowledge about RES types since solar power was the most familiar and commonly used renewable type. Subsidies were the most favored incentive whereas credit provision was the least favored. Most perceived that RES improve life quality and environmental protection was seen as the most important parameter followed by reduced oil dependence.

(*Continued*)

Table 4.1 Summary of the articles selected for inclusion in the scoping review. (*Cont.*)

References	Study aims	Study area	Sample characteristics and study design	Main findings
Sánchez-Pantoja et al. (2018) [40]	Examination of citizens' feelings when they see Building-integrated photovoltaics (BIP) and BAPV (Building-applied photovoltaics)	France and Spain	Sample 1: $n = 87$ Sample 2: $n = 253$ sample 3: $n = 165$ Self-assessment Manikin	BIPV technology, which requires greater attention in the design and a higher initial economic investment, is more accepted by respondents with higher values for both hedonic valence and level of emotional intensity. The high cost or intense technological difficulties could be significant factors that result in a BAPV system being chosen rather than a BIPV installation. Age influences preferences with younger respondents accepting more BAPV technology and older people being skeptical about the lack of a subtle integration of the system within the building envelope.
Koirala et al. (2018) [29]	Investigation of willingness to participate in community energy systems (CESs) can be predicted using demographic, socio-economic, socio-institutional, and environmental factors.	Netherlands	$n = 599$ Methods: web-based digital survey and factor analysis and multi-variate regression analysis	The respondents find solar panels less disturbing than the WTs while the noise of WTs is perceived as the most disturbing. Eighty per cent were aware of local energy projects such as CESs, 53% were willing to participate and 8% were willing to steer different activities of CES. The perceived barriers to participation are lack of time, financial resources, technical expertise and ownership of a PV installation. The willingness to participate in CESs is driven by concern about the environment and climate change, community trust, renewables acceptance, energy independence, and community resistance. Community trust factor is the most important and statistically significant predictor of willingness to participate in CESs followed by community resistance, energy independence, and environmental concern factor as well as education, energy-related education, and awareness about local energy initiatives.
Ntanos et al. (2018) [21]	Examination of public perceptions about RES and willingness to pay for a greater expansion of RES into the Greek energy mix.	Greece	$n = 400$ Face-to-face questionnaire completion.	60% of respondents used RE technologies with most users having installed solar water heaters, while another 13% were using photovoltaic systems. Only four RES users have invested in wind energy systems and two in geothermal energy. The Internet was the most favored information source followed by television. Also, 54% and 42% were adequately informed about solar power and wind power, respectively. On the contrary, respondents were inadequately informed about hydropower, geothermal, and biomass applications.

Hai et al. (2017) [38]	Understanding the patterns of intention-behavior gap in solar energy adoption.	Finland	*n* = 25 Qualitative semi-structured interviews	Solar energy was appreciated and fully supported by all respondents. All respondents expressed pro-environmental interest in terms of their positive thinking about the environment. Despite the stated positive attitude, what the respondents thought and what they actually did were contradicting. The availability and variability of sunlight created confusion and skepticism about the practicality of solar power in Finland. Moreover, having two or more sources of energy connection would not be affordable. Regarding investment in solar energy, financial consideration was important in decision-making. The absence of incentives for individual house owners and a proper system of selling produced solar electricity as well as not owning the house affected non-adopters.
Moula et al. (2017) [41]	Assessment of potential customers' views on biofuels.	Finland	*n* = 90 Survey questionnaire Content analysis	Half participants perceived that biofuel production affects food prices and would not purchase biofuels derived from food crops. Only 60% would switch to biofuels, while limited information about biofuels prevented them from choosing biofuels. Moreover, car owners preferred hydrogen, followed by electricity, and hybrid fuels.
Sonnberger and Ruddat (2017) [32]	Investigation of public acceptance of wind farms. Examination of the role of trust in key actors, the general attitude toward the energy transition in Germany.	Germany	*n* = 2,009 The survey was carried out as a CATI survey (CATI = Computer Assisted Telephone Interview).	The respondents perceived offshore wind farms as the most acceptable siting, suggesting that they are more likely to accept wind farms when they are far away from their residence and, perhaps, not visible. Positive attitude to the energy transition affects positively the socio-political acceptance of onshore and offshore wind farms and the local acceptance of wind farms. Trust in large energy companies had a significant influence on the acceptance of wind farms. Perceived risks associated with wind turbines will have a negative effect on the acceptance of wind farms. However, the influence is stronger for onshore wind farms, compared to offshore wind farms. Perceived benefits and from wind energy and fairness have a positive effect on the acceptance of wind farms.
Flacke and De Boer (2017) [25]	Analysis of whether the COLLAGE tool helps to increase community engagement in renewable-energy projects and planning.	Germany	Workshops using the interactive planning support tool "COLLAGE"	Most participants had a generally positive attitude to RE and almost half of them had some kind of RE installed already - mostly solar panels. Most participants considered that RE could decrease GHG emissions and ensure future energy supply. Areas suitable for solar farms were the green areas along the highway, the abandoned airport, and a contested area. Most participants preferred solar options over wind because of the lower visual impact on the landscape, while some other participants perceived that a combination of both is the most logical option. Finally, 75% stated that they had learned something about RES during the workshop.

(*Continued*)

Table 4.1 Summary of the articles selected for inclusion in the scoping review. (*Cont.*)

References	Study aims	Study area	Sample characteristics and study design	Main findings
Petrakopoulou (2017) [23]	Investigation of key factors to energy projects. Evaluation of social awareness and opinions.	Greece	$n = 183$ Descriptive statistics and one-to-one discussions were planned and realized	Most were positive to RES and thought their broader use can reduce the environmental impact of traditional fuels. High awareness levels were reported for RE technologies (such as solar heaters), but lower for relatively unknown energy sources (wave energy, fuel cells). Respondents were more open to new installations of solar, wind and geothermal energy, but rejected nuclear power and coal plants. The major barriers to RE implementation were aesthetically problematic plant design, the lack of political will, the lack of public awareness and of experts and the absence of a legal framework. Policy driven by an environmentally friendly framework was acknowledged as the most important followed by policies ensuring cheaper electricity generation and business motivation toward a green economy.
Frantál et al. (2017) [42]	Examination of the factors shaping landscape perceptions and attitudes toward renewable energy developments.	Iceland	Principal Component Analysis (PCA).	One half of the respondents perceived WTs as incompatible with the landscape, while about one-third perceived them as compatible. Also, almost half of the respondents would rather reject the project, while only about one third would approve it. People who perceive the landscape as more open, homogenous, industrial, alien and resilient, viewed the wind farm as compatible with the landscape. Conversely, those who perceive the landscape as less open, more diverse, pastoral, familiar and vulnerable considered the project as incompatible with the landscape. Male participants were more likely to perceive the proposed wind farm as compatible with the landscape than females. Older people tolerated more the project than younger persons. The aspects of potential visual impact of the wind farm on the local landscape and the need for the production of more energy emerged as the main points of contention.
Kastner and Matthies (2016) [39]	Measurement of internal factors in the form of value orientations.	Germany	$n = 232$ Online survey and discrete choice experiment, data analyzed with mixed logit models	Eco-social value orientations affected positively solar thermal energy investments as opposed to conservative value orientations and hedonistic value orientations which were both found to have a negative influence. Moreover, energy cost saving potentials, carbon dioxide saving potentials, guarantee extent, and trustworthiness of recommendations were found to have a positive effect on investments in solar thermal energy systems. The influence of investment costs was negative.

Wilson and Dyke (2016) [33]	Investigating reasons to support or oppose wind power proposals pre- and post-installation.	England	$n = 52$ interviewees A multi-method approach was adopted that used both quantitative and qualitative data	Perceptions of the community about WTs changed and became more positive. Out of 52 respondents, about a third still have negative perceptions while most respondents now have "neutral" attitudes, while the remainder either "like" them or are "very positive" toward them. This change in opinion was expressed by many living near the wind farm, suggesting that the sense of place attachment has now been transferred to WTs. Only 18% would support other wind power proposals and very few said it would depend on proximity.
Lindén et al. (2015) [34]	Measurement of NIMBYism and using that variable as an informative measure of public opinion.	Finland	n=3459 Mixed effects ordinal regression model (with logit link and multinomial error structure).	Gender and age were strong determinants of NIMBYism with men and older people being less likely to express a higher level of NIMBYism. Individuals with no previous experience about wind tended to express NIMBYism. Population density reduced the level of NIMBYism, supporting the community attachment hypothesis. For the general attitude towards wind power, gender and age again were strong determinants with men being less supportive to wind power compared to females, while older individuals expressed more positive attitudes. Population density was not statistically significant, suggesting that community attachment may account for NIMBYism only, but not for the general attitude toward wind power.
Michel et al. (2015) [37]	Examination of the opinions of residents and tourists on photovoltaic installations	Switzerland	n=352 Questionnaires Factor analysis and regression	The new installations were not perceived as a drastic intrusion into the landscape, because the view was already affected by the avalanche barriers, which are accepted because of their vital protective function. No significant difference was found between residents' and tourists' evaluation of the new photovoltaic installations. However, conceptions related to place played an important role in the evaluation of possible photovoltaic sites.
de Sousa and Kastenholz (2015) [35]	Investigation of the potential for integrating wind energy production in the rural tourism experience in Portugal	Portugal	On-site semi-structured interviews with open questions	Residents expressed favorable opinions on wind energy mainly due to possible economic benefits, such as income from the lease of land and overall local development. They referred more to community's benefits and less to societal benefits. Visual/esthetic impacts of wind farms on the landscape were the most negative consequences. Noise disturbance was another negative aspect of wind farms commonly mentioned. Positive impacts involved improved air quality, saving the depletable fossil fuel resources. Negative impacts that the respondents mentioned were to the disturbance of birds. As for effects on tourism, many participants perceived no (or neglectable) impacts.

3.1 Attitudes to renewable energy sources in general

A significant number of studies included in this review examined public attitudes to renewable energy sources in general and covered topics relating to their use. These studies have provided a picture of how the public responds to renewable energy and, in specific, they gave information on the public's most preferred renewable types, factors affecting the public support for renewables, preferences for policy schemes, barriers to the deployment of renewables, willingness to participate in distributed and community energy systems as well as attitudes to different renewable energy scenarios.

The large-scale study of Roddis et al. [20] has revealed how British people respond to renewables and the factors affecting their support for them. In particular, the place of residence was found to be a very influential factor with rural dwellers being significantly more supportive to renewables compared to their urban counterparts. This higher support was ascribed to the former group's greater familiarity with RES applications, which are more common in rural areas. However, the support levels decreased in rural areas where RES installations were intense, indicating that the perceived distributional injustice can reduce support levels. Beside the factor of residence, age, gender and environmental concern also affected attitudes with younger individuals, women and people with concerns about the climate being significantly more likely to support RES as opposed to older individuals, men and people with fewer climate concerns who expressed a higher likelihood to support conventional energy sources.

Similar positive attitudes to renewables were also recorded among the Greek public. Ntanos et al. [21] examined attitudes towards RES in Attica, the country's biggest conurbation. The respondents in their majority perceived that RES can improve life quality due to their ability to minimize environmental degradation, which is caused by the consumption of conventional fuels. It was, moreover, indicated that they had a low level of knowledge about geothermal, hydropower and biomass whereas they were considerably more knowledgeable about wind and solar energy. In addition, the majority was using solar water heaters. Interestingly, the non-users thought that high cost, lack of information and lack of confidence were major barriers to RES adoption but generally tended to overlook the benefits of RES. As for the policy schemes, which are introduced to promote RES, subsidies provided for the installation of the systems were mostly favored while credit provision was the least favored measure. Another study [22] indicated that in Nikaia, another large urban area in Greece, the residents were adequately informed about solar and wind energy but were insufficiently informed about hydropower, geothermal and biomass. Furthermore, they preferred the Internet as a source to learn about RES. Meanwhile, in the study of Petrakopoulou [23], the majority of the inhabitants in the island of Skyros in Greece expressed a positive attitude to renewable energy and acknowledged that environmental protection ought to be the most significant criterion when assessing an energy technology. The islanders favored solar, wind and geothermal energy and rejected coal-fired electricity and nuclear power. In addition, they perceived that RES can bring notable benefits and, most importantly, recognized that renewables can create new jobs and exploit domestic resources for energy production. As for the barriers to RE implementation, the islanders were quite skeptical to the unaesthetic design of many renewable plants and also recognized that the lack of political will, lack of public and expert awareness and the absence of a legal framework obstruct the deployment of renewable energy.

Given that solar and wind energy are often the renewable types with which the public is mostly familiar, Sposato and Hampl [24] compared the acceptance of the two renewable types among Austrians and indicated that wind energy was less accepted than solar energy. In addition, individuals who expressed a general positive attitude to renewable energy were more likely to accept this type of plants in their vicinity whereas people who were skeptical to the future viability of renewables were less likely to accept them. The acceptance in both cases was affected by respondents' age, gender, and communitarianism-egalitarianism beliefs.

In their study Flacke and De Boer [25] performed a workshop using the interactive planning support tool "COLLAGE." According to their findings, most respondents exhibited a positive attitude to RE and perceived that RE could lower GHG emissions and secure future energy supply. In their opinion, the most appropriate areas for the installation of solar farms would be the green spaces along the highways, the abandoned airports, and any type of contested area.

The intention to protest against various hypothetical energy installations was examined by Liebe and Dobers [26]. Although the citizens expressed an overall low intention to protest, the greatest protest intention was reported for natural gas followed by biomass, wind energy and solar. Moreover, they considered that solar and wind energy consist less detrimental energy types than biomass and natural gas. Simultaneously, participants' concerns about climate change affected positively the acceptance of solar and wind applications, however, not of biogas plants. Further analysis revealed that NIMBY (Not In My Back Yard) beliefs, place attachment and old age are statistically important and can affect positively protest intention.

A distributed energy system (DES) may be described as a multi-input and multi-output energy system which is operated at a local level and involves a broad range of technologies, like energy storage units, combined cooling systems, power and heating subsystems as well as renewable energy subsystems [27]. As this system involves a varying degree of renewable energy generation, it was considered appropriate to include the study of Seidl et al. [28] in which the intention to engage in DESs was examined among Swiss, German, and Austrian citizens. Citizens in these three countries were found to be quite open to these systems but considered, at the same time, that national governments in cooperation with energy utilities should bear full responsibility for the transition to such energy systems. Conversely, municipalities, households, media, and landowners were viewed as the least responsible for undertaking this task.

Another study included in this review examined attitudes to community energy systems. These systems are broadly acknowledged as a remarkable approach to promote renewable energies while meeting sustainably the energy needs of a community. Koirala et al. [29] found that 80% of citizens in the Netherlands were aware of and interested in community energy systems but recognized that participation in these systems was hindered by lack of time, financial constraints, existing PV installation, limited trust in the neighborhood, inability to develop such systems and lack of technical expertise. Despite these barriers, over half participants (by 53%) were willing to participate and this willingness was mostly influenced by community trust followed by overall acceptance of RES, concerns about the environment and climate change as well as factors, such as community resistance and energy independence. However, willingness to participate could not be predicted by age, gender, house ownership, income, and community type.

A diversified energy system is often regarded as a safe path to achieve energy security while deploying renewables. The research work of Zaunbrecher et al. [30] investigated how the German public perceives different energy scenarios in which a certain renewable type has a higher share than the other energy types in the mix. It was shown that the most preferred scenario was the "PV-Mix Pasture" in which a high amount of the calculated renewable energy would be produced on both public and private roofs (with solar energy accounting for 60% of energy supply) and therefore the effect on the landscape would be negligible. In the same scenario, wind energy (which accounted only for 10% of energy supply as only one wind turbine would be installed) and biogas (which accounted for 30%) would be installed only on meadows and pastures, whereas forests would not be affected at all. In contrast to the "PV-Mix Pasture," the "Wind-Mix Forest" was the scenario that received the lowest acceptance. In this scenario, large numbers of wind applications would be installed in the forest resulting in high land-use and permanent disruption of ecological processes.

3.2 Attitudes to wind energy

The results of the studies included in our scoping study confirm that wind energy remains a controversial renewable type mainly due to its visual and audio impact on the local area. In specific, six studies focused explicitly on wind energy applications with two of them having been conducted in Iceland while the remaining studies were conducted in Germany, England, Finland, and Portugal.

In Iceland, the study of Ólafsdóttir and Sæþórsdóttir [31] revealed that most residents living in municipalities, which were adjacent to a wind farm, were positive to wind turbines in general with male respondents exhibiting a more positive attitude compared to female participants. Moreover, the majority (by 81.6%) thought that wind energy can benefit the country's energy production. However, the respondents expressed a lower level of enthusiasm when they were asked specifically about the proposed wind farm. This reduced support was associated with their fear that the wind turbines would obstruct the view of the surrounding mountains and generate noise pollution. Interestingly, the respondents were somewhat indifferent to the height of the turbines.

A year prior to the above study, Frantàl et al. [42] explored how perceptions of the characteristic landscapes in Iceland affect residents' views on wind farms. Residents' opinions were divided with one-half stating that wind turbines would be utterly incompatible with the Icelandic landscape and only one-third of participants perceiving them as compatible. As with previous studies, gender was again an influential variable since more male respondents perceived wind turbines as a compatible addition to the landscape than female participants. Age was another significant variable with older participants being slightly more favorable to the proposed project than younger individuals. Moreover, the visual impact and the need for energy production influenced participants' attitudes. Conversely, profession and personal background did not have a notable effect on respondents' attitudes.

Sonnberger and Ruddat [32] examined the local acceptance of a wind farm in Germany and it was found that only 35% of the respondents would accept a wind farm at a distance of 500 meters from their residence. The acceptance rate, however, increased to 66% for offshore and to 52% for onshore wind farms indicating that offshore wind farms were the most acceptable form of wind energy. Moreover, the same study examined the factors affecting the acceptance of wind farms and it was shown that trust in key actors and perceived fairness had a positive effect on local acceptance of wind farms while perceived risks outplayed the perceived benefits.

Another research work that is worthwhile to discuss would be the study of Wilson and Dyke [33] who focused on the perceptions of a wind farm in Cornwall. This study is particularly interesting because the research team made a meaningful comparison of Cornish residents' perceptions before and after the installation of wind farms in an area, which due to its scenic landscapes consists one of the leading holiday destinations in the United Kingdom. During the planning stages, the residents reacted negatively to the proposal for a wind project in their vicinity. These negative responses were driven by concerns about the environment and wildlife, the visual and audio impact of the wind turbines on the local area, as well as concerns about possible effects on traffic. However, residents' attitudes changed appreciably after the wind farm had been installed with many of the previous objectors now "favoring," "tolerating" or even being "strongly for" the wind farm. What is more, the acute concerns about the visual intrusion of the wind turbines into the Cornish landscape which had, inter alia, triggered the opposition were now milder, and many "objectors" stated that they were now "used to seeing them" in their vicinity. Only 6 out of 52 interviewees were still skeptical to the effect of the wind farm on the landscape, however, these were respondents who had unfavorable attitudes to wind energy in general or resided in locations that wind turbines were acutely visible. In addition, the notable change in perceptions after installation suggests that the wind turbines have may become a local landmark and may have been integrated into residents' sense of place attachment. That being said, the Cornish people would not support a new wind energy project while almost all of them thought that the wind farm had not brought any benefits to the local area.

As it is often indicated in the literature on attitudes to wind energy, the NIMBY[a] theory has consistently been used to explain residents' opposition to wind projects in their vicinity. In an effort to scrutinize possible NIMBYism, Lindén et al. [34] found that in Finland gender and age functioned as determinants of NIMBYism with men and older people being less likely to express NIMBYism.

[a] NIMBY: acronym for "Not In My Back Yard" to describe the opposition of residents to a proposed facility or project in their local area.

Moreover, people who had no prior experience with wind energy tended to express NIMBYism, whereas those residing in areas with high population density displayed a lower level of NIMBYism. Regarding the general attitude toward wind energy, gender and age were significant variables and, in specific, male individuals were less supportive while older people expressed more positive attitudes to this type of renewable energy.

Finally, de Sousa and Kastenholz [35] examined the attitudes of the residents in the historic village of Linhares da Beira in Portugal that is located close to a wind farm, which is visible from the village. The residents expressed an overall favorable attitude to wind energy, despite their stated limited knowledge about it. Moreover, the respondents recognized that wind energy projects can bring notable economic benefits to the local area while only a minority thought that wind farms could have a negative effect on tourism.

3.3 Attitudes to solar energy

In this scoping study, solar energy emerged as the public's most preferred renewable type [20,22,24,25,26,29]. As solar energy applications require investments and involve some visual impact [36], there has been substantial research effort to explore public attitudes towards solar energy in the European Union.

Indicatively, Sanchez-Pantoja et al. [40] captured how French and Spanish citizens perceived two different solar applications: building-integrated photovoltaics (BIP), photovoltaic materials which are installed in the place of conventional building materials in parts of the building envelope, and building-applied photovoltaics (BAP) which are the common applications of panels on rooftops and plots of land. The former type of photovoltaics, which consists the most expensive option, was the most preferred type despite the higher cost indicating a clear preference for the less noticeable photovoltaic systems. Preferences were affected by respondents' age with older participants being more negative to the lack of a seamless integration of the system into the building envelope which is stronger in BAPs.

At the same time, in the study of Michel et al. [37] respondents were required to evaluate potential sites for photovoltaic installation in the Swiss Alps. Industrial sites and modern buildings followed by agricultural buildings and avalanche barriers were considered as the most suitable installation sites. This suggests that solar panels on buildings or locations, which bear no symbolic meaning, were preferred over solar applications on buildings with a symbolic meaning such as historic buildings.

In the study of Hai et al. [38], all Finnish interviewees motivated by their strong environmental values exhibited a positive attitude to solar energy and all of them expressed their full support for it. The stated high level of support, however, did not correspond to their behavior since they did not purchase or invest in solar energy. The reason for this intention-behavior gap was traced in respondents' concerns about various issues related to solar energy applications. Such issues included the possible deficiencies due to the limited availability of sunlight in Finland, investment costs, costs of connection to the grid, the lack of any proven cost-effective mechanism for strong solar system for the sale of produced solar electric power. In addition, the Finnish respondents perceived that the absence of a rigid political commitment and initiative as well as the lack of provided incentives would lead to the inability to raise awareness about solar power or motivate people. Moreover, not owning a house was perceived as a major obstacle to adopt solar power.

Acknowledging that the intention to invest is a determinant of the wider development of solar energy, Kastner and Matthies [39] explored the effect of value orientations on investment intention among German houseowners. Interestingly, respondents with strong conservative value orientations were more skeptical to investment costs but less sensitive to the reliability of recommendations and the potential of cost saving. Conversely, individuals with acute eco-social value orientations were more mindful of carbon dioxide saving potentials and recommendation reliability while investment costs played a relatively insignificant role in their investment decision. As for the factors affecting the intention to invest, the potential of saving energy costs together with guarantee period and the reliability of recommendations had a positive impact whereas investment costs exerted a negative influence on investment decisions.

4 Discussion and conclusions

The range of research works on public attitudes toward renewable energy sources included in this review allows us to infer that overall the public in EU countries exhibits a positive attitude to renewable energy and is mostly familiar with wind and solar energy but has a clear preference for solar energy. In addition, the visual effect of wind farms on the landscape is still a point that attracts controversy.

Despite the wide range of studies on attitudes, a profound understanding of the dynamics of public acceptance has not been achieved. This can perhaps be ascribed to the methodologies the researchers used to examine public attitudes in the period 2015–20. In terms of data collection, most of the reviewed studies employed face-to-face interviews and on-site or telephone assisted questionnaire surveys, whereas only four studies employed web-based surveys [26,29,30,39]. Regarding data analysis, most studies conducted statistical analyses which are frequently used in social psychology such as principal component analysis [21,42] and various types of regression analysis [20,24,26,28,29,34,39]. However, only one study included in this scoping review moved away from these somewhat "conventional" methodologies and employed a different technique to evaluate respondents' attitudes. In specific, Flacke and De Boer [25] used the interactive planning support tool "COLLAGE" and managed to gain a deeper understanding of attitudes to renewables while the respondents became more knowledgeable about the complexities surrounding the installation of renewables through their participation in the study. It could perhaps be stated that although "conventional" methodologies have served as valuable tools to shed light on public attitudes during the early stages of renewable energy deployment, more innovative methodologies, like workshops using the COLLAGE tool, should be used in order to capture ways of thinking about energy technologies. In other words, the focus should shift toward the identification of thinking patterns, which shape how individuals perceive energy technologies. It is also necessary to probe the symbolic meaning that the siting of energy projects holds for individuals. For instance, Michel et al. [37] have concluded that the construction of solar facilities at sites, which bear a symbolic meaning, can decrease public acceptance.

As discussed at the outset of this chapter, the European Union is often characterized as a global leader in energy policy and sets the most ambitious energy goals including 32% of energy from renewables by 2030 [14]. Wind energy has contributed and can further contribute to fulfilling such targets, but the public acceptance of wind energy projects can inhibit this progress. The public opposition stems often from the aesthetic intrusion of wind turbines into the natural landscape [31,33,42]. Yet, turbine sizes have been dramatically increasing over the last years and given that the public has already been reacting to the visual impact of the previous structures, the increasing size of turbines could be translated into further conflicts in the years to come. To overcome these challenges, efforts should be made to increase citizens' trust in key actors and enhance the perceived fairness of procedures as perceived fairness has been found to be a determinant of public acceptance [9,10,32]. This could be achieved by including public representatives and local groups in every stage of the planning and decision processes while ensuring that their voices are heard and valued. Another recommendation to overcome conflicts would be the provision of notable financial benefits (such as lower electricity costs) to the affected residents who "bear the cost" of the project that is implemented in their "backyard."

Another point that is worthwhile to discuss would be the clear preference for solar energy which was recorded in a significant number of studies [20,22,24,25,26,29]. This preference can be ascribed not only to the public's higher familiarity with solar installations which have become very common throughout Europe, but also to the EU policies which have consistently been applied to increase private small-scale investments and public investments in solar photovoltaics. Hence, the applied policies and the provided incentives appear to have contributed both to the development and the popularity of solar energy.

The review has also shown that not only wind applications but also installations of other renewable types can receive negative responses [37]. For this reason, it is critically important to avoid focusing only on wind energy but to examine public attitudes to all renewable types so that our knowledge about public attitudes to renewable energy becomes comprehensive. Interestingly, this review revealed that the public exhibited limited preference and knowledge about renewable types other than wind and solar energy [22] while there was a scarcity of studies examining public attitudes explicitly toward bioenergy, hydropower and geothermal energy in the period 2015–20. Hence, a significant literature gap, which ought to be filled by future research, was highlighted in this scoping study.

Regarding the overall support for renewable energy, there is evidence that environmental awareness plays a critical role in the acceptance of renewables. In specific, individuals with environmental values and concerns about environmental problems are more likely to take pro-environmental decisions and support renewable energy [20,38]. Hence, efforts aiming at raising awareness about the environment and renewable energy are worthwhile and could yield benefits to the renewable sector. The first step in this direction would be to continue examining the public attitudes and based on these findings to design strategies which would enhance public awareness about renewables. An effective strategy would be the conduct of environmental education programs and information campaigns in a systematic and frequent manner as well as the provision of information on renewable energy through the Internet as this is a favored information source [22].

To conclude, instead of perceiving public attitudes as a barrier to the development of renewable energy, it is preferable to seek to understand the dynamics of public involvement in renewable energy. To that end, innovative research methodologies aiming at capturing the ways of thinking can shed light on the shaping of attitudes and expand our knowledge.

References

[1] M. Wolsink, Near-shore wind power—Protected seascapes, environmentalists' attitudes, and the technocratic planning perspective, Land Use Policy 27 (2) (2010) 195–203.

[2] M.J. Pasqualetti, Opposing wind energy landscapes: a search for common cause, Ann. Assoc. Am. Geogr. 101 (4) (2011) 907–917.

[3] M. Wolsink, Wind power and the NIMBY-myth: institutional capacity and the limited significance of public support, Renew. Energy 21 (1) (2000) 49–64.

[4] F. Crichton, G. Dodd, G. Schmid, G. Gamble, K.J. Petrie, Can expectations produce symptoms from infrasound associated with wind turbines?, Health Psychol. 33 (4) (2014) 360.

[5] E. Pedersen, K. Persson Waye, Perception and annoyance due to wind turbine noise—a dose–response relationship, J. Acoust. Soc. Am. 116 (6) (2004) 3460–3470.

[6] W. Krewitt, J. Nitsch, The potential for electricity generation from on-shore wind energy under the constraints of nature conservation: a case study for two regions in Germany, Renew. Energy 28 (10) (2003) 1645–1655.

[7] P. Devine-Wright, Rethinking NIMBYism: The role of place attachment and place identity in explaining place-protective action, J. Commun. Appl. Soc. Psychol. 19 (6) (2009) 426–441.

[8] M. Lewicka, Place attachment: How far have we come in the last 40 years?, J. Environ. Psychol. 31 (3) (2011) 207–230.

[9] C. Gross, Community perspectives of wind energy in Australia: The application of a justice and community fairness framework to increase social acceptance, Energy Policy 35 (5) (2007) 2727–2736.

[10] J. Firestone, W. Kempton, M.B. Lilley, K. Samoteskul, Public acceptance of offshore wind power: does perceived fairness of process matter?, J. Environ. Plan. Manage. 55 (10) (2012) 1387–1402.

[11] E. Karasmanaki, G. Tsantopoulos, Exploring future scientists' awareness about and attitudes towards renewable energy sources, Energy Policy 131 (2019) 111–119.

[12] H. Karlstrøm, M. Ryghaug, Public attitudes towards renewable energy technologies in Norway. The role of party preferences, Energy Policy 67 (2014) 656–663.

[13] J. Li, J. Shi, Z. Wang, New progress of renewable energy policy in Europe and its reference for China, Kezaisheng Nengyuan/Renew. Energy Resour. 25 (3) (2007) 1–3.

[14] Council of the European Union. Renewable Energy: Council Confirms Deal Reached with the European Parliament, 2018. Available from: https://www.consilium.europa.eu/en/press/press-releases/2018/06/27/renewable-energy-council-confirms-deal-reached-with-the-european-parliament/.

[15] European Commission. Energy for the Future: Renewable Sources of Energy: White Paper for a Community Strategy and Action Plan 1997; COM (97) 599: Communication from the Commission of the European Communities, 1997.

[16] European Commission. 2020 by 2020: Europe's climate change opportunity. Communication from the Commission to the European Parliament, the Council, The European Economic and Social Committee, and the Committee of the Regions, 2008. Available from: https://eur-lex.europa.eu/legal-content/EN/TXT/?uri=CELEX:52008DC0030.

[17] European Commission. Communication from the Commission. A clean planet for all. A European strategic long-term vision for a prosperous, modern, competitive and climate neutral economy. COM (2018) 773 final, 2018. Available from: https://eur-lex.europa.eu/legal-content/EN/TXT/?uri=CELEX:52018DC0773.

[18] G. Walker, Renewable energy and the public, Land Use Policy 12 (1) (1995) 49–59.
[19] H. Arksey, L. O'Malley, Scoping studies: towards a methodological framework, Int. J. Soc. Res. Methodol. 8 (1) (2005) 19–32.
[20] P. Roddis, S. Carver, M. Dallimer, G. Ziv, Accounting for taste? Analysing diverging public support for energy sources in Great Britain, Energy Res. Soc. Sci. 56 (2019) 101226.
[21] S. Ntanos, G. Kyriakopoulos, M. Chalikias, G. Arabatzis, M. Skordoulis, S. Galatsidas, et al. A social assessment of the usage of renewable energy sources and its contribution to life quality: The case of an Attica urban area in Greece, Sustainability 10 (5) (2018) 1414.
[22] S. Ntanos, G. Kyriakopoulos, M. Chalikias, G. Arabatzis, M. Skordoulis, Public perceptions and willingness to pay for renewable energy: A case study from Greece, Sustainability 10 (3) (2018) 687.
[23] F. Petrakopoulou, The social perspective on the renewable energy autonomy of geographically isolated communities: evidence from a Mediterranean island, Sustainability 9 (3) (2017) 327.
[24] R.G. Sposato, N. Hampl, Worldviews as predictors of wind and solar energy support in Austria: Bridging social acceptance and risk perception research, Energy Res. Soc. Sci. 42 (2018) 237–246.
[25] J. Flacke, C. De Boer, An interactive planning support tool for addressing social acceptance of renewable energy projects in the Netherlands, ISPRS Int. J. Geo-inf. 6 (10) (2017) 313.
[26] U. Liebe, G.M. Dobers, Decomposing public support for energy policy: What drives acceptance of and intentions to protest against renewable energy expansion in Germany?, Energy Res. Soc. Sci. 47 (2019) 247–260.
[27] M. Di Somma, B. Yan, N. Bianco, G. Graditi, P.B. Luh, L. Mongibello, et al. Multi-objective design optimization of distributed energy systems through cost and exergy assessments, Appl. Energy 204 (2017) 1299–1316.
[28] R. Seidl, T. Von Wirth, P. Krütli, Social acceptance of distributed energy systems in Swiss, German, and Austrian energy transitions, Energy Res. Soc. Sci. 54 (2019) 117–128.
[29] B.P. Koirala, Y. Araghi, M. Kroesen, A. Ghorbani, R.A. Hakvoort, P.M. Herder, Trust, awareness, and independence: Insights from a socio-psychological factor analysis of citizen knowledge and participation in community energy systems, Energy Res. Soc. Sci. 38 (2018) 33–40.
[30] B.S. Zaunbrecher, B. Daniels, M. Roß-Nickoll, M. Ziefle, The social and ecological footprint of renewable power generation plants. Balancing social requirements and ecological impacts in an integrated approach, Energy Res. Soc. Sci. 45 (2018) 91–106.
[31] R. Ólafsdóttir, A.D. Sæþórsdóttir, Wind farms in the Icelandic highlands: Attitudes of local residents and tourism service providers, Land Use Policy 88 (2019) 104173.
[32] M. Sonnberger, M. Ruddat, Local and socio-political acceptance of wind farms in Germany, Technol. Soc. 51 (2017) 56–65.
[33] G.A. Wilson, S.L. Dyke, Pre-and post-installation community perceptions of wind farm projects: the case of Roskrow Barton (Cornwall, UK), Land Use Policy 52 (2016) 287–296.
[34] A. Lindén, L. Rapeli, A. Brutemark, Community attachment and municipal economy: Public attitudes towards wind power in a local context, Environ. Sci. Policy 54 (2015) 10–14.
[35] A.J.G. de Sousa, E. Kastenholz, Wind farms and the rural tourism experience–problem or possible productive integration? The views of visitors and residents of a Portuguese village, J. Sustain. Tourism 23 (8–9) (2015) 1236–1256.

[36] E. Karasmanaki, S. Galatsidas, G. Tsantopoulos, An investigation of factors affecting the willingness to invest in renewables among environmental students: A logistic regression approach, Sustainability 11 (18) (2019) 5012.
[37] A.H. Michel, M. Buchecker, N. Backhaus, Renewable energy, authenticity, and tourism: social acceptance of photovoltaic installations in a Swiss alpine region, Mountain Res. Dev. 35 (2) (2015) 161–170.
[38] M.A. Hai, M.M.E. Moula, U. Seppälä, Results of intention-behaviour gap for solar energy in regular residential buildings in Finland, Int. J. Sustain. Built Environ. 6 (2) (2017) 317–329.
[39] I. Kastner, E. Matthies, Investments in renewable energies by German households: A matter of economics, social influences and ecological concern?, Energy Res. Soc. Sci. 17 (2016) 1–9.
[40] N. Sánchez-Pantoja, R. Vidal, M.C. Pastor, Aesthetic perception of photovoltaic integration within new proposals for ecological architecture, Sustain. Cities Soc. 39 (2018) 203–214.
[41] M.M.E. Moula, J. Nyári, A. Bartel, Public acceptance of biofuels in the transport sector in Finland, Int. J. Sustain. Built Environ. 6 (2) (2017) 434–441.
[42] B. Frantál, T. Bevk, B. Van Veelen, M. Hărmănescu, K. Benediktsson, The importance of on-site evaluation for placing renewable energy in the landscape: A case study of the Búrfell wind farm (Iceland), Moravian Geogr. Rep. 25 (4) (2017) 234–247.

5

Understanding willingness to pay for renewable energy among citizens of the European Union during the period 2010–20

Evangelia Karasmanaki

Department of Forestry and Management of the Environment and Natural Resources, Democritus University of Thrace, Orestiada, Greece

Chapter outline

1 Introduction

While the concerns over the environmental impact of conventional fuel usage are growing, the idea of generating clean, inexhaustible and sustainable power from renewable energy sources sparks interest around the world. That is because renewable energy sources produce less harmful emissions and even when calculating the "life cycle" emission rates (the emissions produced at every stage of a facility's life: manufacturing, installation, operation, decommissioning), the emissions of renewable energy compared to fossil fuels are still significantly lower [1]. In addition, the deployment of renewables can bring considerable advantages such as diversification of energy supply, promotion of locally

Low Carbon Energy Technologies in Sustainable Energy Systems. http://dx.doi.org/10.1016/B978-0-12-822897-5.00005-5

produced power, reduction in the reliance on imported fuels, and creation of new job positions [2,3].

Having recognized the catalytic role of the renewable sector in the effort to tackle climate change, the European Union was one of the first to establish binding policies seeking to minimize climate-harming emissions through the development of renewable energy. For this reason, the EU is considered one of the leaders in renewable energy while energy policy lies at the heart of the political and economic integration process [4]. A landmark policy on renewable energy was the Renewable Energy Directive 2009/28/EC, which mandated particular levels of renewable energy use within the European Union by the year 2020 [5]. In addition, through the implementation of the "Europe 2020" strategy, the European Union sought to achieve sustainable growth and development, enhance its competitiveness in the global market and become one of the strongest as well as most knowledge-based economies in the world [6]. The "Europe 2020" strategy defined the widely known "20-20-20" targets which were interrelated and mutually supporting each other. In specific, they involved a 20% reduction in GHG emissions compared to the levels recorded in 1990, a 20% share of renewable energy in gross final energy consumption as well as a 20% reduction in energy consumption compared to a 2020 business-as-usual projection [7]. As early as in 2017, there were indications that the EU would successfully meet most of these ambitious targets as the emissions produced from international aviation and indirect carbon dioxide emissions, had decreased by 21.7% compared to the 1990 levels. What is more, the EU was already on track to achieve the renewable energy target it had set for the year 2020 [8]. In view of this progress, it is not surprising that the EU has set even more challenging targets for the period from 2021 to 2030. The key targets involve a 40% reduction in greenhouse gas emissions compared to the 1990 levels, a 32% share for renewables as well as a minimum of 32% enhancement in energy efficiency [9].

Meanwhile, the workings of the European Parliament and Council Directives had previously resulted in the liberalization of EU energy markets (Directives 96/92/EC, 2003/54/EC, 96/92/EC) which means that the European energy market is, at least theoretically, free and subject to competition and consumers can benefit from the most affordable energy options while companies' costs are decreased [10]. The market liberalization suggests also that the diffusion of renewable energy lies, to some extent, in the hands of consumers. From this perspective, the investigation of consumers' willingness-to-pay for renewable energy assumes a great importance and becomes a significant tool in order to promote renewable energy [11,12]. That is, if suppliers and policymakers are informed about the consumer surplus (the difference between the actual price they pay and the price which they would be willing to pay), it is feasible to design marketing strategies that will promote the purchase of renewable energy or to calculate with greater accuracy the size of subsidies.

Realizing how important it is to learn more about willingness-to-pay for renewable energy, the number of studies on this topic has been following a rising trend over the past years leading to an abundance of data and making it difficult to form a precise picture of estimations of willingness-to-pay and the key factors affecting it. Hence, this chapter seeks to review the existing research works examining willingness-to-pay among EU citizens and identify the common variables across a varied set of WTP studies to create a basis for comparison. It also highlights areas for future studies and makes specific policy recommendations. The observations made in this review study may be particularly useful to those involved in the marketing of green electricity offers and policymakers who bear the responsibility for the policy development of renewables.

2 Methodology

The present review was carried out in March of 2020 and the scoping review method was chosen. The objective of a scoping review is to collect the materials, which map a specific area of research. Here, willingness-to-pay for renewable energy was the area of research. To carry out the scoping review, we followed the six-stage framework proposed by Arksey and O'Malley [13]. According to the framework, the researchers first identify a research question, which serves as a guide for the subsequent stages. Hence, the research question was stated as "How willing is the European public to pay for renewable energy and what affects this willingness or unwillingness"? Next, the database *Scopus* was searched using the terms: willingness to pay; willingness to pay for renewable energy; factors affecting willingness to pay for renewable energy; green consumerism; willingness to pay for green energy; willingness to pay for green products; willingness to pay for clean energy. Articles were limited to research works that included social groups (citizens, customers, university students, tourists and so forth) as sample in European countries and were published in English from 2010 to 2020.

In the first search, 118 documents were identified as meeting all the above search criteria. Out of the 118 documents, 6 were excluded since they were not written in English. Next, the titles and abstracts of the remaining documents were carefully read (n = 112) and 75 were excluded because they were not related to the field of willingness-to-pay or because they examined willingness-to-pay for other services and goods (such as electric vehicles) or because they were not research articles (such as reviews and perspective papers).

3 Results

After excluding 81 studies, a total of 37 publications finally met all criteria and were included in this review. Fig. 5.1 displays the number of studies on willingness-to-pay conducted in countries of the EU for the period 2010–20. It can be seen that the number of publications increased over this 10-year period and the highest number of WTP studies can be seen in 2015 and 2019. From 2012 to 2013, the number of publications stood still at four and then decreased abruptly to reach two in 2014 and increased sharply the following year. From 2016 to 2017, the number reached again two and in 2018 it reached four.

Regarding the countries where the studies were conducted, Fig. 5.2 shows that the majority of studies was conducted in Germany followed by Poland and Great Britain, whereas the lowest number of studies was recorded for Switzerland, Cyprus, and Ireland.

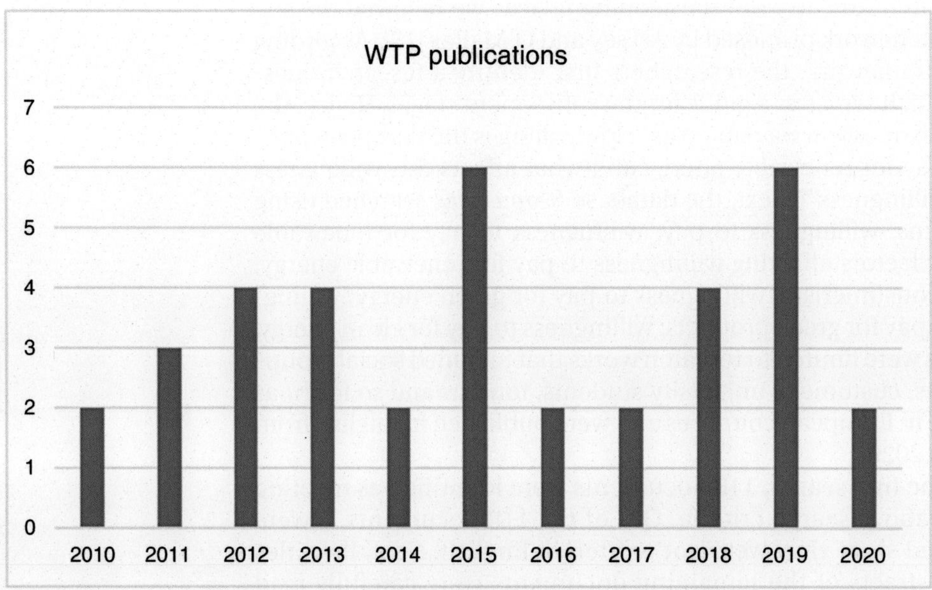

Figure 5.1. WTP publications in EU countries from 2010 to 2020.

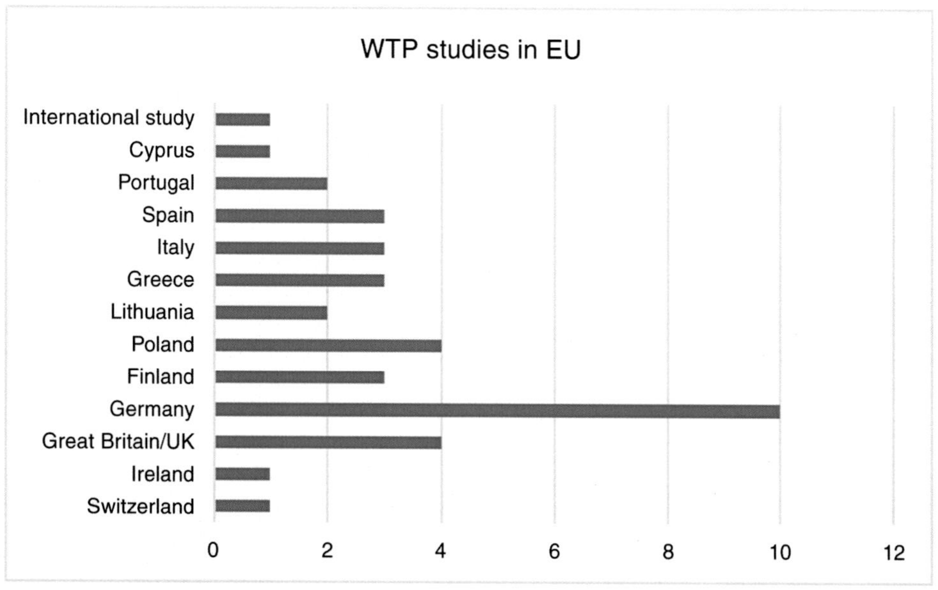

Figure 5.2. Number of WTP studies in EU countries.

In our data chart (Table 5.1), we included information on authorship and date of publication, survey design, sample size, methodology, and major findings.

3.1 Estimations of willingness to pay for renewable energy in European Union

The research works that fell within the scope of this review indicated that citizen majorities in many EU countries expressed a willingness to pay an amount of money in order to facilitate the development of renewable energy sources.

As Fig. 5.2 shows, most WTP studies in the European Union were conducted in Germany during the period 2010–20. It is important to note that Germany has been pursuing the so-called "Energiewende", the planned massive transition to a low-carbon and nuclear-free energy system, which has set the ambitious aim to reduce the share of fossil fuels from 80% to 20% in the energy supply by the year 2050 [14,15]. With this aim in mind, it is quite relevant to observe how the German consumers respond to proposed renewable energy payment schemes. In 2019, the average WTP was estimated at 23.8 eurocents per kWh for a tariff that would involve 20% renewable energy while this rate would rise to 28.3 eurocents per kWh for a tariff with 100% renewable energy [16]. In 2018, a large-scale study had indicated that Germans would pay more

Table 5.1 Summary of included studies.

References	Location	Sample; Methodology	Main findings
Ghesla et al. (2020) [37]	Switzerland	$n = 1362$ Field study and survey-based approach	Being informed before the decision, perceiving the choice as complex, and delaying the decision predict an opt-out of the default contract. Households with a low socio-demographic status are more likely to remain with the default while those who know the share of RE in the electricity market are more likely to opt out. The more respondents are informed, the more they opt out of the default.
Merk et al. (2019) [16]	Germany and Great Britain	$n = 1300$ respondents in each country vignette study	Respondents were willing to pay a price premium for an increasing share of renewables in the electricity mix. In Germany, the average WTP for a tariff with 20% RES was 23.8 eurocents per kWh and increased to 28.3 eurocents per kWh for a tariff with 100% RES. In Great Britain, the respective percentages were 16.1 eurocents per kWh and 18.4 eurocents per kWh.
Paatero et al. (2019) [22]	Finland	$n = 1352$ web-based questionnaire survey	445 of 1352 interviewees would pay extra for a RES-based energy system, but 340 of 575 respondents (59%) would pay only 5%–10% more, 167 interviewees (29%) 10%–20% more and 52 interviewees (9%) 20%–30% more. The WTP extra for renewable energy is pronounced for 25–34-year-olds, academically educated, leaders and experts and freelancers. High annual income is positively related to WTP.
Meloni et al. (2019) [61]	Italy	$n = 242$ Correlational field study, Principal Component Analysis, and Scale Reliability Analysis	A relationship between stress and identity for both WTP for green energy and non-littering behavior occurred which could be explained by the argument that these place-related behavioral responses to the place-related environmental stress are intertwined with the identification with the place itself
Kowalska-Pyzalska (2018) [20]	Poland	$n = 502$ Contingent valuation, data collection through standardized telephone survey	WTP for green electricity increases with consumers' material well-being, their pro-environmental attitudes and peers' support, with education and knowledge of green electricity tariffs and, to some extent, with the type of accommodation. WTP for green electricity decreases with age (older consumers) and with the satisfaction with the current energy supplier.
Ortega-Izquierdo et al. (2019) [31]	Spain	$n = 1250$ Telephone interviews	57% of respondents with RES knowledge would pay more, 35% wouldn't pay more and 8% of respondents did not answer. Young people, those with university education, rural dwellers and people with income higher than the average are in general, more willing to pay.

Gargallo et al. (2020) [27]	Spain	$n = 231$	Most of the respondents with WTP would pay between €100 and €500. More than 17% are willing to pay more than €2000. Those who have a higher level of education are willing to pay more.
Zawojska et al. (2019) [62]	Poland	$n = 801$ Interview-based survey	Increasing policy-consequentiality is associated with increasing WTP for renewable energy
Botelho et al. (2018) [28]	Portugal	Discrete-choice experiment	Respondents were willing to pay to reduce the environmental impacts. Respondents were, on average, willing to pay 1.6 Euros for electricity based on hydropower relative to photovoltaic, and 0.6 to have wind generated electricity relative to photovoltaic.
Knoefel et al. (2018) [17]	Germany	$n = 2000$ Discrete choice experiment and kernel density estimates	High additional WTP for all attributes, which, in total, amount to more than a quarter of the overall electricity price per kilowatt hour. Male respondents were willing to pay less than female respondents: female respondents were willing to pay 1.2 Eurocents and 0.8 Eurocents more than male respondents, respectively.
Kowalska-Pyzalska (2018) [21]	Poland	$n = 502$ Contingent valuation method and standardized telephone survey, t-test	42% would not pay more for electricity than they pay now. 13% of the respondents would pay between 1 and 5 PLN, 18% would pay 6–10 PLN, 13% between 11 and 20 PLN, and 8% between 21 and 30 PLN. Only 2% of respondents would pay more than 30 PLN. Age, income, and education were relevant demographics. Age was negatively linked to WTP whereas income and education were positively associated with WTP. Also, peers' support and approval, pro-environmental attitudes and beliefs affect positively WTP.
Ntanos et al. (2018) [29]	Greece	$n = 400$ Questionnaire-based survey and principal component analysis and binary logit model	One-third of respondents favored a further expansion of 10% of RES share in the electricity generation mix, Of these, 28.9% would pay more 6–10 euro in their electricity bill, while 52.4% would pay more than 10 euros per quarter. The mean WTP for a 10% increase of RES penetration in electricity mix, is 26.5 euros per household, quarterly. WTP was positively associated with education, energy subsidies, and state support.
Bartczak et al. (2017) [55]	Poland	800 face-to-face interviews using computer-assisted personal interviewing system	Risk seeking people are more willing to pay for proposed changes in renewable energy development while individuals who are more risk averse require lower compensation before they accept externalities from renewable electricity production.
Karanikola et al. (2017) [63]	Cyprus	$n = 385$ people. Face-to-face interviews	Only 12.2% would not pay for a 20% increase, whereas the other respondents agreed to gradual increase in the price of electricity ranging from 5% to 20%
Rommel et al. (2016) [39]	Germany	$n = 2,000$ Discrete choice experiment	WTP for renewable energy is higher when offered by cooperatives or utilities owned by municipalities. Consumers with experience in switching suppliers have an additional WTP of one eurocent per kWh for cooperatives and two Eurocents for public enterprises.

(*Continued*)

Table 5.1 Summary of included studies. (*Cont.*)

References	Location	Sample; Methodology	Main findings
Uehleke (2016) [36]	Germany	Contingent valuation survey and median regression	Determinants of WTP are personal norm, climate change scepticism, agreement with the German Renewable Energy Law and individualistic cognition. Income and gender also predict WTP.
Caporale and De Lucia (2015) [26]	Italy	$n = 375$ Choice experiment market	Most respondents (60%) were willing to pay to subsidize electricity production from RES. 45% of respondents would pay 60 Euros/year. Only 28% would pay 90 Euros/year.
Akcura (2015) [32]	United Kingdom	Web-based surveys with contingent valuation method. Data analysis with fractional logit regression	While UK households were more inclined to pay under a mandatory scheme, the amount of payment was higher under a voluntary payment option. Women were willing to pay less than men. Age is a significant factor in how much to contribute and older respondents have a lower WTP. Lower income groups were less willing to pay compared to the highest income earners.
Vecchiato and Tempesta (2015) [25]	Italy	$n = 509$ Choice experiment	86% of the respondents were willing to pay more for RES. WTP for photovoltaics is 3.4 times than for forest biomass and 5.4 times than for agricultural biomass. 86% would pay 2 €/month for a quota of electricity produced from RE.
Štreimikienė and Baležentis (2015) [33]	Lithuania	n = 99 Questionnaire surveys with focus group approach and regression analysis	Information and environmental awareness are crucial in WTP for renewables while those having higher income and higher education are willing to pay more for renewable electricity. Gender and age did not have significant impact on WTP. The percentage of respondents who would pay an additional sum of 3.5 EUR per month for a 5% RES increase, rose from 50% before the campaign to 67% after the campaign.
Štreimikienė (2015) [24]	Lithuania	$n = 1002$ The survey was conducted at home of respondents.	Most of Lithuanian consumers (81%) would not pay more for electricity produced from renewables. Only 18% of respondents would pay more for green electricity.
Mahieu et al. (2015) [38]	International survey	Large web-based survey	As nuclear plants get older, respondents living near them are more willing to support RE. Participation in an environmental organization has a positive effect on WTP for RE, while females report lower WTP than males. The mean WTP of the sample for RE was 3.36 %.
Ribeiro et al. (2014) [34]	Portugal	$n = 3646$ Survey, t-test	WTP was high among those considering RES as increasing the costs of the electricity, but few respondents believed that RES increase the electricity bill. Females were more willing to pay. Education is significant in two cases, and respondents with a lower educational degree are more willing to pay.

Sagebiel et al. (2014) [15]	Germany	$n = 287$ Questionnaire and choice experiment. Conditional logit and latent class logit	Consumers expressed high WTP for RE. There is also a considerable WTP for transparent pricing, participation in decision-making, and local suppliers, democratic decision-making and electricity produced by cooperatives. For an increase from zero to a 33% share of RE, respondents would pay 9.21 Eurocent. For an increase from zero to 66%, the WTP is 14.4 Eurocent.
Kosenius and Ollikainen (2013) [23]	Finland	$n = 947$	The residents in Eastern Finland were inclined to accept a 261**€** increase in their annual electricity bill to replace the current energy mix with the wood-based bioenergy.
Kontogianni et al. (2013) [30]	Greece	$n = 312$ Open-ended contingent valuation survey and data analysis with non-parametric statistical analysis	The majority of respondents (by 76.9%) would pay to promote RES, while 23.1% were unwilling to pay. Five out of the 240 respondents were willing to pay only if they agreed with the location of RES investments. On average they would pay in one-shot payment per household **€**180 (for onshore wind) and **€**137 (for photovoltaic) for the realization of a 10% substitution in the primary energy mix of the island.
Kaenzig et al. (2013) [18]	Germany	$n = 414$ Choice experiments and conjoint analysis	The German electricity customers would pay about 16% for electricity from RES. Electricity mixes with renewable energy are preferred over those with high shares of non-renewable energy sources.
Moula et al. (2013) [60]	Finland	Multiple choice questionnaire	38% of respondents would pay up to 5% more for using green energy, 33% prefer the cheapest possible choice and about 29% would pay up to 10% of extra costs.
Gracia et al. (2012) [56]	Spain Spain in 2010	$n = 400$ Face-to-face questionnaire, discrete choice experiment	Most consumers were not willing to for increases in the shares of renewable in their electricity mix. For two of the three renewable sources considered (wind and biomass) an increase of the renewable mix would require a discount. Instead, they were willing to pay for increases in the share of both solar power and locally generated power. feed-in tariffs.
Kraeusel and Möst (2012) [11]	Germany	$n = 130$ Choice experiment multiple regression analysis	WTP for carbon capture storage technology is lower than for RE. The WTP for CCS power is thus 7 times lower than for green electricity or in other words, it is nearly nothing.
Soliño et al. (2012) [53]	Spain	Choice experiment and open-ended contingent valuation survey	Galician inhabitants would enhance their well-being when a program of replacement of fossil fuels with biomass in electricity generation processes should be applied. WTP is between 0.94 and 1.13 times up annual version for the non-random parameters and between 2.59 and 2.59 for random parameters.
Kostakis and Sardianou (2012) [35]	Greece	$n = 400$ Binary logistic regression models	Marital status and educational level are not statistically significant factors in WTP. Information dissemination and environmental awareness affect the willingness to pay for RES.

(*Continued*)

Table 5.1 Summary of included studies. (*Cont.*)

References	Location	Sample; Methodology	Main findings
Diaz-Rainey and Ashton (2011) [12]	United Kingdom	$n = 1800$ Telephone interview	42% of respondents would pay a premium of between 5% and 10% for green electricity. WTP was positively associated with higher income and younger age. Respondents who perceive that their actions do not affect the environment were less willing to pay. The desire for information on environmental matters and whether customers consider the environment in their personal decision-making, membership of an environmental group were all positively associated with WTP.
Grösche and Schröder (2011) [59]	Germany	$n = 2948$ Hedonic regression framework	WTP increases for the share of renewable fuels in the electricity mix while WTP tends to fall for the nuclear share.
Claudy et al. (2011) [54]	Ireland	Double-bounded contingent valuation method computer-assisted questionnaire telephone interviews (CATI)	WTP varies significantly among the four technologies. WTP for solar water heater is the lowest, at about €2380. Those who perceive that investments in microgeneration technologies will make them independent from conventional fuels and energy suppliers have a greater WTP. Homeowners with high to medium levels of education favor solar panels. Urban dwellers have a higher WTP for solar water heaters; rural dwellers have a higher WTP for solar panels and lower WTP for micro wind turbines.
Gerpott and Mahmudova (2010) [19]	Germany	$n = 238$ Telephone interviews based on a standardized questionnaire Binary logistic (ordinal) regression analysis	53.4% would pay a mark-up for green electricity. 26.1% report a price tolerance equal to a 5%–10% increase in their current electricity bill. Binary logistic and ordinal regression analyses indicate that price tolerance for green electricity is particularly influenced by attitudes towards environmental issues and toward one's current power supplier, perceptions of the evaluation of green energy by an individual's social reference groups, household size and current electricity bill level.
Scarpa and Willis (2010) [57]	United Kingdom	$n = 1279$ questionnaire with assisted personal interview Choice experiment, conditional and mixed logit model	Consumers were WTP GBP2.91 ± GBP0.30 in capital costs to reduce annual fuel bills by GBP1. This value also applied to solar electricity, solar thermal, and wind power.

than a quarter of the overall electricity price per kilowatt hour [17]. A former study had indicated that the German respondents would pay an additional amount of 3.72 eurocents per kWh on the condition that the renewable energy would be offered by regional suppliers [15]. Moreover, Kaenzig et al. [18] had found that the German consumers were more willing to pay for energy mixes in which renewables would account for high shares but were unwilling to pay for energy mixes that would be based on conventional energy sources. A year before, Kraeusel and Möst [11] reported that German respondents would pay a premium of 10.85% every month if this amount would be used to achieve a 10% increase of the RES share in the overall energy mix. Remarkably, a survey conducted in 2010 had indicated that over half respondents (by 53.4%) were already willing to pay for green electricity [19]. Therefore, it can be observed that throughout the decade 2010–20 the German consumers have been steadily expressing their willingness to pay for renewable energy but at varying WTP levels.

In contrast to the high WTP of their German counterparts, consumers in Poland, the eighth largest economy in the European Union, expressed quite low levels of WTP. In specific, a recent survey indicated that the Poles would pay only an additional amount of 3.5 USD for green electricity [20]. In addition, a year prior to this study, the same researcher had indicated that 42% of Polish respondents were unwilling to pay any amount of money, while only 8% would pay an additional 21–30 PLN (about 5.29 to 7.56 USD) [21]. Likewise, consumers in Finland would pay merely 10%–20% more for renewable energy [22]. However, a former study had revealed that if renewable energy was produced from wood-based bioenergy, WTP in Finland was estimated as high as a 261 euros increase in the annual electricity bill [23]. Quite low levels of WTP were also recorded among Lithuanians, as the strong majority of respondents were unwilling to pay any amount of money for electricity from renewable energies [24].

Southern European Union countries reported varying WTP levels. In Italy, the overwhelming majority of Italian consumers were found to be willing to pay "more" for green electricity, however, this WTP was translated into low sums of money as Italians would pay merely two euros more for an energy contract of renewable electricity [25]. Nevertheless, higher WTP levels were reported in another study, which indicated that 45% of Italian consumers would be willing to pay 60 euros per year (6 euros per month) and 19% up to 120 euros per year (10 euros per month) in order to subsidize renewable electricity [26]. In addition, most Spanish respondents were willing to pay between one and 500 euros and, remarkably, 17% would pay more than 2000 euros, which is higher than the minimum salary in Spain [27]. In Portugal, on average the consumers would pay 1.6 euros for hydropower-produced electricity and 0.6 eurocents for wind-based electricity [28]. Likewise, a significant share of Greek respondents (by 28.9%) would pay an additional amount of 6–10 euros in their electricity bill, while the average WTP was estimated at 26.5 euros per household for a 10% growth in the RES share in the Greek electricity mix [29]. Moreover, a previous study had indicated that the majority of Greek households were willing to pay some amount of money for renewables [30].

In view of the above WTP estimations, it may be inferred that there is a wide market for expanding "green" electricity contracts as there is empirical evidence that in many EU countries consumers are willing to pay some amount to facilitate the adoption of RE. However, in order to understand differences in willingness-to-pay, it is important to examine the factors, which determine individuals' willingness to pay.

3.2 Factors affecting the willingness or unwillingness to pay for renewable energy

As willingness-to-pay for renewable energy is multifaceted, many research works have sought to detect those variables, which affect individuals' willingness or unwillingness to pay for renewable energies.

According to the reviewed research works, individuals' demographic characteristics and, most importantly, age, gender, education level and income status exert a significant influence on WTP. In terms of income, high-income earners were found to be more willing to pay for RES compared to those with lower income who were less willing to pay [20–22,31–33]. Moreover, education level is an influential factor with a positive effect on WTP as university graduates expressed greater WTP levels than those with a lower education level [21,22,27,31,33]. In contrast to income and education level which have a clear positive effect on WTP, the research findings concerning age and gender were conflicting. Regarding age, on the one hand, there are studies indicating that younger individuals were more willing to pay for RES than their older counterparts [12,21,22,31]. On the other hand, however, Ribeiro et al. [34] and Kostakis and Sardianou [35] found that older people expressed higher WTP. Conflicting findings were also recorded for gender with the studies of Akcura [32] and Kostakis and Sardianou [35] indicating that male respondents were more willing to pay and the study of Knoefel and Sagebiel [17] showing a lower WTP for male respondents.

The attitudes and the awareness of individuals about the environment consist another important factor,with a clear positive impact on willingness-to-pay. That is because concerns about the environment can extend to pro-environmental behaviors such as paying for renewables. The influence of environmental attitudes was confirmed as a factor affecting positively WTP in a considerable number of the studies reviewed in this chapter [12,19–21,36].

Information about renewable energy was another factor exerting influence on willingness-to-pay with those being sufficiently informed about renewable energy sources or the available options to pay for RES (such as "green" tariffs and contracts) being more inclined to pay for renewable energy sources [20,37]. What is more, Štreimikienė and Baležentis [33] compared WTP levels for a focus group before and after an information campaign on renewable energy and found that the share of participants who would pay an additional amount of money for RES rose from 50% to 67%.

Moreover, the preference for specific renewable types can affect WTP. For instance, Italian respondents expressed a significantly higher WTP for photovoltaics than other renewable types [25] while Finnish respondents would pay considerable amounts of money for wood-based bioenergy [23].

Apart from the above, willingness-to-pay has been found to be associated with some diverse factors. In specific, individuals who participated in environmental organizations [38], discussion groups and decision-making processes [30] reported higher willingness-to-pay. Additionally, German respondents expressed greater WTP levels for renewable energy, which is provided by cooperatives or utilities owned by municipalities rather than companies owned by investors [39] while they preferred renewable energy which is locally produced [15]. In addition, the process of switching energy suppliers affects the decision of consumers and makes them hesitant whereas those having gone through this experience had a higher WTP for renewables [39].

The above factors affecting willingness-to-pay can help marketers and policy designers to build the profile of consumers and correspond to their expectations and needs while the ways in which "green" products and services could be promoted are highlighted.

3.3 Methodologies employed in WTP studies

At this point it is relevant to pay attention to the research methodology used in WTP studies. As observed, stated preference methods are the most widely used methodologies to calculate values like WTP. In most cases, a stated preference method assumes the simulation of a market in which a good, a service or an amount of goods, is provided at a specific price. This market is described in the questionnaire, which is administered to a sample of the population under study, and then the participants complete the questionnaire by "stating" their preferences. Once the necessary data are collected, researchers conduct statistical analyses to measure the representative maximum WTP of the participants [40].

Stated preference methods may be categorized into two broad groups: the contingent valuation method (CVM), which has many variations [41], and choice modeling (CM), which encompasses choice experiments, as well as contingent rating and ranking [42–44]. Contingent valuation is a survey-based technique to calculate stated preferences and, in specific, it can assess non-market goods and services [45,46]. First, the study participants are introduced to a hypothetical scenario of a certain environmental situation and then are required to express their willingness to pay in order to help avoid a welfare loss or to ensure a welfare gain. The questions that respondents answer in contingent valuation surveys can be either open-ended or closed-ended. The open-ended form of question asks the participant to state the maximum amount of money which he or she is willing to pay for the service in question whereas in the closed-ended form, the participants are provided with specified amounts of money to choose from [20,41,45–47]. Regardless of the question format, a drawback of this method would be the substantial "hypothetical bias" which occurs when participants responding to hypothetical scenarios overstate their preferences. Moreover, hypothetical bias may occur in the case that the respondents are not certain about the amount of money they would pay since they might not have considered the presented hypothetical environmental situation before [48,49].

Beside the contingent valuation method, the method of choice experiment (CE) has been widely used in research on environmental economics to calculate consumers' willingness to pay for renewable energy sources. The CE draws on the consumer theory of Lancaster [50] as well as the random utility model (RUM), which was introduced by McFadden [51]. In essence, choice experiments are a combination of these two theories and expand the dichotomous contingent valuation method. Choice experiments consist a valuable tool to measure the economic value of non-market goods, their shadow-prices (monetary values of difficult-to-estimate costs in the absence of valid market prices) as well as their attributes [52]. Regarding the practical application of choice experiments, the data are gathered by means of a survey. A classic choice experiment imitates real-life market scenarios and presents to participants a hypothetical market whose distinctive characteristics are described in full detail. Once respondents are familiar with the hypothetical market, they are required to choose among products or services, which are differentiated by a specified set of characteristics, such as the price of an electricity contract, the type of energy and so forth. Often, the respondent is able not to select any of the proposed products/services [25]. Moreover, during the application of the choice experiment respondents are required to act as in a real market and as if they were actually purchasing the proposed product or service. For this reason, respondents are encouraged to include any budget constraints during the experiment [25].

We can now turn our attention to the present review on WTP for renewables. As shown in Table 5.1, a significant number of WTP studies used either the contingent valuation method [20,21,30,32,36,53,54] or the choice experiment method [11,15, 17,18,25,26,28,39,55,56,57] to determine the effect of different factors on WTP. It could be said that the contingent valuation method and the choice experiment are the most prominent techniques used in current willingness-to-pay studies and their application has managed to translate willingness-to-pay into monetary values.

While WTP calculations can inform marketers and policymakers about the size of the market for renewable energy sources, the factors affecting this willingness-to-pay can guide policymakers in designing strategies, which will ensure the success of renewables in the market. In terms of methodology, regression models consist the most widely used tools to analyze social data, and detect factors affecting willingness-to-pay. In specific, the most typical form is linear regression, which takes on a continuous dependent variable. At present, limited forms of linear regression are used in WTP studies including the regression types of binary, multinomial, censored, and so forth [58]. In the studies included in this review, regression models were used in an appreciable of studies [11,19,32,33,35,36,59].

Based on the literature works presented in this review and with a focus on the techniques used to estimate and analyze willingness-to-pay and the factors affecting it, it may be inferred that contingent valuation, choice experiments and regression models have offered valuable insights into WTP and paved the way for future research.

4 Discussion

Overall, this review study has confirmed that EU citizens have expressed some degree of willingness-to-pay for renewable energy in the decade 2010–20. This willingness could perhaps be attributed to the wider EU energy policy framework, which has been consistent in promoting the diffusion of RES through the provision of subsidies, incentives and the funding of large-scale investments [3].

As stated previously, in liberalized markets consumers can choose renewable energy providers, contracts and tariffs suggesting that it is totally up to their will whether they will support renewables. Hence, the main question is what induces individuals to actually pay for renewables. A response has been provided by a substantial number of studies concluding that willingness-to-pay has a clear positive relationship with pro-environmental attitudes [12,19–21,36]. From this perspective, it may be argued that a key strategy to ensuring payments for renewables is to raise public environmental awareness. To that end, policymakers in cooperation with educators and other bodies involved in the implementation of policies and programs, should organize environmental education programs especially in countries in which citizens have expressed their unwillingness to pay or exhibited quite low WTP levels [2,20,33].

Willingness-to-pay was also affected by households' ability to afford payments for renewable energy. Although low-income earners' unwillingness to pay is understandable and expected, there is evidence that sometimes low-income households wish to pay in order to contribute to the development of renewables but are prevented from the unaffordable available options [60]. To achieve a wider diffusion of renewables across EU, however, it is of great significance that citizen majorities will pay some premium regardless of their income status. For this reason, there should be some affordable "green" tariffs or contracts, which will enable also low-income households to contribute to the diffusion of renewable energy. In this way, not only larger shares of citizens would make contributions, but renewable electricity payments would be viewed as "socially just."

Unwillingness-to-pay was negatively related to low levels of information on green electricity contracts and renewable energy in general whereas citizens with higher information levels were more willing to pay [20,33,37]. An effective way to enhance citizens' WTP would be, therefore, to provide consumers with adequate information both on the benefits of renewable energies and the available renewable electricity contracts and tariffs. To that end, information campaigns and discussion groups could be held in citizens' proximity frequently and systematically. In addition, since individuals hesitate to opt for renewable electricity contracts due to the uncertainty they feel about the process and the cost of switching energy suppliers [39], policymakers should ensure that all relevant processes are simple, fast, unbureaucratic, and cost-effective. Moreover, it is critical to ensure that customers can switch energy providers without having to pay penalties or exit fees and without any disruption to their energy supply.

Another point requiring discussion would be that citizens tend to be more willing to pay for renewable energy, which is produced by local energy cooperatives or offered by utilities owned by municipalities rather than large investor-owned companies [39]. Therefore, the establishment of energy cooperatives in the EU countries could increase the share of individuals who pay for renewable electricity.

Finally, a significant number of WTP studies was conducted in Germany and southern EU countries, such as Italy, Greece, and Spain. This may imply that the interest in investigating WTP is greater in countries where massive energy transitions requiring citizens' acceptance and support are being implemented [14,15].

5 Conclusions

Payments for renewables made by citizens are of pivotal importance to achieve further increase in the share of renewable energy in EU countries' energy mix and, to that end, willingness-to-pay studies can guide the efforts of policymakers to achieve greater payment rates. As indicated by the studies included in this review, considerable levels of willingness-to-pay were expressed by the citizens in many European Union countries while there is a clear preference for lower premiums and tariffs showing that more affordable renewable electricity contracts would encourage more citizens to pay for renewables.

Willingness-to-pay in EU countries seems to be affected by individuals' demographics and, in specific, by their age, gender, education level, and income status. Moreover, individuals with pro-environmental attitudes were found to be more inclined to pay for renewable energy suggesting that it would be worthwhile to take the necessary steps in order to raise public awareness about the environment and environmentally friendly behaviors such as choosing renewable electricity. In addition, providing citizens with information on the benefits of renewable energy and the available "green" electricity contracts together with more practical measures, such as simplifying the process of supplier switching, could yield greater willingness to pay for renewables.

Finally, research works estimating and analyzing willingness-to-pay should be conducted regularly so that policymakers and "green" energy suppliers can evaluate and adjust the provided contracts and services according to the preferences and expectations of the public.

References

[1] IPCC Summary for policymakers, in: O. Edenhofer, R. Pichs–Madruga, Y. Sokona, K. Seyboth, P. Matschoss, S. Kadner, et al. (Eds.), IPCC Special Report on Renewable Energy Sources and Climate Change Mitigation, Cambridge University Press, Cambridge, United Kingdom and New York, NY, USA, 2011.

[2] E. Karasmanaki, G. Tsantopoulos, Exploring future scientists' awareness about and attitudes towards renewable energy sources, Energy Policy 131 (2019) 111–119.

[3] G. Tsantopoulos, G. Arabatzis, S. Tampakis, Public attitudes towards photovoltaic developments: Case study from Greece, Energy Policy 71 (2014) 94–106.

[4] J.H. Matlary, Energy Policy in the European Union, Macmillan, Basingstoke, (1997).

[5] European Union Directive 2009/28/EC of the European Parliament and of the Council of 23 April 2009 on the promotion of the use of energy from renewable sources and amending and subsequently repealing Directives 2001/77/EC and 2003/30/EC, Off. J. Eur. Union 5 (2009) 2009.

[6] A. Fedajev, D. Stanujkic, D. Karabašević, W.K. Brauers, E.K. Zavadskas, Assessment of progress towards "Europe 2020" strategy targets by using the MULTIMOORA method and the Shannon Entropy Index, J. Clean. Product. 244 (2020) 118895.

[7] European Commission, Taking stock of the Europe 2020 strategy for smart, sustainable and inclusive growth, COM(2014) 130 final, Brussels, 2014.

[8] Eurostat. Europe 2020 indicators - climate change and energy, 2019. Available from: https://ec.europa.eu/eurostat/statisticsexplained/index.php/Europe_2020_indicators__climate_change_and_energy#The_EU_is_on_track_to_achieving_its_GHG_emission_reduction_target_for_2020.

[9] European Commission. 2030 climate & energy framework, 2020. Available from: https://ec.europa.eu/clima/policies/strategies/2030_en.

[10] L. Ilie, A. Horobet, C. Popescu, Liberalization and regulation in the EU energy market, 2007. Available from: https://mpra.ub.uni-muenchen.de/6419/1/MPRA_paper_6419.pdf.

[11] J. Kraeusel, D. Möst, Carbon capture and storage on its way to large-scale deployment: Social acceptance and willingness to pay in Germany, Energy Policy 49 (2012) 642–651.

[12] I. Diaz–Rainey, J.K. Ashton, Profiling potential green electricity tariff adopters: green consumerism as an environmental policy tool?, Bus. Strategy Environ. 20 (7) (2011) 456–470.

[13] H. Arksey, L. O'Malley, Scoping studies: towards a methodological framework, Int. J. Soc. Res. Methodol. 8 (1) (2005) 19–32.

[14] O. Renn, J.P. Marshall, Coal, nuclear and renewable energy policies in Germany: From the 1950s to the "Energiewende", Energy Policy 99 (2016) 224–232.

[15] J. Sagebiel, J.R. Müller, J. Rommel, Are consumers willing to pay more for electricity from cooperatives? Results from an online Choice Experiment in Germany, Energy Res. Soc. Sci. 2 (2014) 90–101.

[16] C. Merk, K. Rehdanz, C. Schröder, How consumers trade off supply security and green electricity: Evidence from Germany and Great Britain, Energy Econ. (2019) 104528.

[17] J. Knoefel, J. Sagebiel, Ö. Yildiz, J.R. Müller, J. Rommel, A consumer perspective on corporate governance in the energy transition: Evidence from a discrete choice experiment in Germany, Energy Econ. 75 (2018) 440–448.

[18] J. Kaenzig, S.L. Heinzle, R. Wüstenhagen, Whatever the customer wants, the customer gets? Exploring the gap between consumer preferences and default electricity products in Germany, Energy Policy 53 (2013) 311–322.

[19] T.J. Gerpott, I. Mahmudova, Determinants of price mark up tolerance for green electricity–lessons for environmental marketing strategies from a study of residential electricity customers in Germany, Bus. Strategy Environ. 19 (5) (2010) 304–318.
[20] A. Kowalska-Pyzalska, Do consumers want to pay for green electricity? a case study from Poland, Sustainability 11 (5) (2019) 1310.
[21] A. Kowalska-Pyzalska, An empirical analysis of green electricity adoption among residential consumers in Poland, Sustainability 10 (7) (2018) 2281.
[22] J.V. Paatero, M.E. Moula, K. Alanne, Occupants' acceptability of zero energy housing in Finland, Int. J. Sustain. Energy 38 (6) (2019) 542–560.
[23] A.K. Kosenius, M. Ollikainen, Valuation of environmental and societal trade-offs of renewable energy sources, Energy Policy 62 (2013) 1148–1156.
[24] D. Štreimikienė, The main drivers of environmentally responsible behaviour in Lithuanian households, Amfiteatru Econ. 17 (2015) 1023–1035.
[25] D. Vecchiato, T. Tempesta, Public preferences for electricity contracts including renewable energy: A marketing analysis with choice experiments, Energy 88 (2015) 168–179.
[26] D. Caporale, C. De Lucia, Social acceptance of on-shore wind energy in Apulia Region (Southern Italy), Renew. Sustain. Energy Rev. 52 (2015) 1378–1390.
[27] P. Gargallo, N. García-Casarejos, M. Salvador, Perceptions of local population on the impacts of substitution of fossil energies by renewables: A case study applied to a Spanish rural area, Energy Reports 6 (2020) 436–441.
[28] A. Botelho, L. Lourenço-Gomes, L.M.C. Pinto, S. Sousa, M. Valente, Discrete-choice experiments valuing local environmental impacts of renewables: Two approaches to a case study in Portugal, Environ. Dev. Sustain. 20 (1) (2018) 145–162.
[29] S. Ntanos, G. Kyriakopoulos, M. Chalikias, G. Arabatzis, M. Skordoulis, Public perceptions and willingness to pay for renewable energy: A case study from Greece, Sustainability 10 (3) (2018) 687.
[30] A. Kontogianni, C. Tourkolias, M. Skourtos, Renewables portfolio, individual preferences and social values towards RES technologies, Energy Policy 55 (2013) 467–476.
[31] M. Ortega-Izquierdo, A. Paredes-Salvador, C. Montoya-Rasero, Analysis of the decision making factors for heating and cooling systems in Spanish households, Renew. Sustain. Energy Rev. 100 (2019) 175–185.
[32] E. Akcura, Mandatory versus voluntary payment for green electricity, Ecol. Econ. 116 (2015) 84–94.
[33] D. Štreimikienė, A. Baležentis, Assessment of willingness to pay for renewables in Lithuanian households, Clean Technol. Environ. Policy 17 (2) (2015) 515–531.
[34] F. Ribeiro, P. Ferreira, M. Araújo, A.C. Braga, Public opinion on renewable energy technologies in Portugal, Energy 69 (2014) 39–50.
[35] I. Kostakis, E. Sardianou, Which factors affect the willingness of tourists to pay for renewable energy?, Renew. Energy 38 (1) (2012) 169–172.
[36] R. Uehleke, The role of question format for the support for national climate change mitigation policies in Germany and the determinants of WTP, Energy Econ. 55 (2016) 148–156.
[37] C. Ghesla, M. Grieder, R. Schubert, Nudging the poor and the rich–A field study on the distributional effects of green electricity defaults, Energy Econ. 86 (2020) 104616.
[38] P.A. Mahieu, H.P.P. Donfouet, B. Kriström, Determinants of willingness-to-pay for renewable energy: does the age of nuclear power plant reactors matter? Revue d'économie politique 125 (2) (2015) 299–315.
[39] J. Rommel, J. Sagebiel, J.R. Müller, Quality uncertainty and the market for renewable energy: Evidence from German consumers, Renew. Energy 94 (2016) 106–113.

[40] J. Mogas, P. Riera, R. Brey, Combining contingent valuation and choice experiments. A forestry application in Spain, Environ. Resour. Econ. 43 (4) (2009) 535–551.

[41] R.C. Mitchell, R.T. Carson, Using surveys to value public goods: the contingent valuation method, Resourc. Future (1989).

[42] J. Bennett, R. Blamey (Eds.), The Choice Modelling Approach to Environmental ValuationEdward Elgar Publishing, 2001.

[43] J. Bennett, V. Adamowicz, Some fundamentals of environmental choice modelling, Choice Model. Appr. Environ. Valuat. (2001) 37–69.

[44] N. Hanley, S. Mourato, R.E. Wright, Choice modelling approaches: a superior alternative for environmental valuatioin?, J. Econ. Surveys 15 (3) (2001) 435–462.

[45] I.J. Bateman, R.T. Carson, B. Day, M. Hanemann, N. Hanley, T. Hett, ..., R. Sugden, Economic valuation with stated preference techniques: a manual, Edward Elgar, (2002).

[46] K. Arrow, R. Solow, P.R. Portney, E.E. Leamer, R. Radner, H. Schuman, Report of the NOAA panel on contingent valuation, Feder. Regist. 58 (10) (1993) 4601–4614.

[47] R.T. Carson, Contingent valuation: A practical alternative when prices aren't available, J. Econ. Perspect. 26 (4) (2012) 27–42.

[48] J.A. List, J.F. Shogren, Calibration of the difference between actual and hypothetical valuations in a field experiment, J. Econ. Behav. Org. 37 (2) (1998) 193–205.

[49] H.R. Neill, R.G. Cummings, P.T. Ganderton, G.W. Harrison, T. McGuckin, Hypothetical surveys and real economic commitments, Land Econ. 7 (1994) 145–154.

[50] K.J. Lancaster, A new approach to consumer theory, J. Polit. Econ. 74 (2) (1966) 132–157.

[51] D. McFadden, Conditional logit analysis of qualitative choice behavior, in: P. Zarembka (Ed.), Frontiers in Econometrics, Academic Press, New York, 1973 1974.

[52] R. Scarpa, A. Alberini (Eds.), Applications of Simulation Methods in Environmental and Resource Economics, Vol. 6, Springer Science & Business Media, 2005.

[53] M. Soliño, B.A. Farizo, M.X. Vázquez, A. Prada, Generating electricity with forest biomass: Consistency and payment timeframe effects in choice experiments, Energy Policy 41 (2012) 798–806.

[54] M.C. Claudy, C. Michelsen, A. O'Driscoll, The diffusion of microgeneration technologies–assessing the influence of perceived product characteristics on home owners' willingness to pay, Energy Policy 39 (3) (2011) 1459–1469.

[55] A. Bartczak, S. Chilton, M. Czajkowski, J. Meyerhoff, Gain and loss of money in a choice experiment. The impact of financial loss aversion and risk preferences on willingness to pay to avoid renewable energy externalities, Energy Econ. 65 (2017) 326–334.

[56] A. Gracia, J. Barreiro-Hurlé, L.P. Pérez, Can renewable energy be financed with higher electricity prices? Evidence from a Spanish region, Energy Policy 50 (2012) 784–794.

[57] R. Scarpa, K. Willis, Willingness-to-pay for renewable energy: Primary and discretionary choice of British households' for micro-generation technologies, Energy Econ. 32 (1) (2010) 129–136.

[58] A. DeMaris, Regression With Social Data: Modeling Continuous and Limited Response Variables, Vol. 417, John Wiley & Sons, 2004.

[59] P. Grösche, C. Schröder, Eliciting public support for greening the electricity mix using random parameter techniques, Energy Econ. 33 (2) (2011) 363–370.

[60] M.M.E. Moula, J. Maula, M. Hamdy, T. Fang, N. Jung, R. Lahdelma, Researching social acceptability of renewable energy technologies in Finland, Int. J. Sustain. Built Environ. 2 (1) (2013) 89–98.
[61] A. Meloni, F. Fornara, G. Carrus, Predicting pro-environmental behaviors in the urban context: the direct or moderated effect of urban stress, city identity, and worldviews, Cities 88 (2019) 83–90.
[62] E. Zawojska, A. Bartczak, M. Czajkowski, Disentangling the effects of policy and payment consequentiality and risk attitudes on stated preferences, J. Environ. Econ. Manage. 93 (2019) 63–84.
[63] P. Karanikola, S. Tampakis, F. Florou, Z. Tampakis, Residents' Information About Renewable Forms of Energy in the Island of Cyprus. In: HAICTA, pp. 462–468, 2017.

6

Linking energy homeostasis, exergy management, and resiliency to develop sustainable grid-connected distributed generation systems for their integration into the distribution grid by electric utilities

Fernando Yanine[a], Antonio Sanchez-Squella[b], Aldo Barrueto[b], Sarat Kumar Sahoo[c]

[a]*Faculty of Engineering, Universidad Finis Terrae, Santiago, Chile;* [b]*Department of Electrical Engineering, Universidad Tecnica Federico Santa María (UTFSM), Santiago, Chile;* [c]*A Constituent College of Biju Patnaik Technological University, Parala Maharaja Engineering College, Department of Electrical Engineering, Govt. of Odisha, Berhampur, Odisha, India*

Chapter outline

Low Carbon Energy Technologies in Sustainable Energy Systems. http://dx.doi.org/10.1016/B978-0-12-822897-5.00006-7

1 Introduction

The general concepts of energy homeostasis and homeostaticity in the control and energy management of electric power systems in general and in distributed energy systems in particular: How to engineer energy homeostasis and homeostaticity in distributed generation systems

Ever since Cannon [1,2] first introduced the concept, attention on homeostasis has been focused chiefly on its role in medicine and biology to find cures for diseases like diabetes and obesity for example. However, homeostasis applications in designing and engineering sustainable energy systems (SES) are also part of its focus and scientific work scope, realizing that exergy and sustainability are both directly linked to homeostasis mechanisms present in all living systems. This is so because exergy is part of such mechanisms and expresses the capacity of the system to do useful work at any point in time, but it also expresses the potential of the system to cause change, to react, to do work, and thus, it implies a measure of resourcefulness of the system itself. Exergy is also a measure of the available energy in the system, capacity of an energy system to to bring about change. In thermodynamics, the exergy of a system is the maximum useful work possible that the system can do during a given process that brings the system into equilibrium with a heat (energy) reservoir or surrounding. Hence, exergy is the capacity of a system to be able to bring itself into efficient equilibrium between the system's supply and the demand for energy. This entails performing energy balance which is closely linked to the system's energy efficiency. Moreover, the systemic capacity of the energy system is enhanced, thus increasing the exergy level of the system, whenever energy efficiency (EE) and thriftiness are combined with reactive and predictive homeostasis, as it is proposed here, in the form of a generic homeostatic control (HC) system. The HC system was designed for the energy system being modeled and simulated in this study, along with presenting the theoretical and empirical foundations that sustain it [2a, 2b]. The HC system proposed here is the result of an ongoing joint research and development effort between ENEL and the corresponding author's industry and research team. The HC system is to be built in 2017 to operate ENEL's utility-run microgrid project to be implemented in a number of buildings, whereby customers and the utility are expected to contribute to potentiating each other's role in this new energy scenario.

In today's electric utilities market two main concerns arise. First there is the need to advance toward further electric power systems (EPS) decentralization, incorporating more renewables and furthering the opportunity to incorporate distributed generation (DG) solutions to the energy grid and to offer more flexible, personalized and cost effective services. This in tandem with the industry transformation that is taking place worldwide toward the adoption of Smart Grid technologies, as is the case in Chile where new legislation is emerging to force the industry to open up and become much more competitive than in the past.

The second big concern in Chile is the increasing threat from climate change and the ever more frequent problem of quakes and fires, which greatly affect the traditional EPS infrastructure [3–6]. Environmental challenges like natural disasters and hazardous climatic events are becoming more severe and recurrent in many parts of the world including Chile, and they are here to stay, affecting millions.

Nowhere is the matter being taken more seriously than in the United States, where the US Senate was expected to take up microgrid policy in early 2016 to tackle, among others, the issue of electric power grid readiness and resilience [7–9]. In North America, for example, large-scale power outages spanning several urban and semi-rural areas are not new. Still fresh in people's memory is Hurricane Sandy, known as "Superstorm Sandy" [10–13]. This natural event was the deadliest and most destructive hurricane of the 2012 Atlantic hurricane season, and the second-costliest hurricane in United States history [10–13] with damages estimated as of 2015 to have been about $75 billion (2012 USD), a total surpassed only by Hurricane Katrina [14,15]. This monstrous calamity caused unprecedented infrastructure damages including major power outages. Yet there were other power disruptions as predecessors of Sandy, among them the Northeast Blackout of 2003, the Hurricanes Katrina and Rita in 2005 and Hurricane Irene and the Northeast's freak Halloween snowstorm in 2011[14]. After each of these events, more consensuses were built among public opinion, local authorities and power industry experts that something new had to be done fast to strengthen the power distribution grid against such recurrent catastrophes.

Even in the best-case scenarios where everything seems to go smooth, the electricity grid may experience short-term, temporary changes in overall capacity that may adversely impact power supply and cause severe problems. Thus electric utilities must be prepared to face such disruptions and to account for power plant malfunctions or a transmission lines that suddenly go out of service, resulting in a power outage [16,17]. Also rapid and unexpected increases or decreases in electricity demand can cause abrupt and unforeseen changes in frequency and voltage, affecting hundreds of thousands of customers. Particularly electric utilities in Chile are becoming increasingly concerned about such threats and the power disruptions the poise. Even small voltage variations that were common in the past are now unacceptable, since changes in the law now obligate electric power suppliers to indemnify their customers for any damage to their customers' goods. Appliances, PCs, and a variety of infrastructure damage, can be at risk. They are also accountable for damages and can be forced to respond for financial losses to any residential or commercial customer that may be affected by a power malfunction or unplanned power outage. Therefore, the need to become much more resilient, robust and at the same time flexible, has made them refocus their priorities, and embrace more localized, smaller EPS employing renewables.

Electric utilities are also seeking to increase their ability to attenuate or ameliorate peak demand hurdles, smoothing out peaks and troughs in their energy supply in order to respond more effectively and proactively to changing energy needs and to environmental disruptions. In the case of Chile, a developing country with abundant wind in its northern territory and having one of the best solar irradiations in the world, has incorporated better legislation in recent years to make the energy grid more competitive. This has brought important changes to the country's EPS industry scenario, with a lot more renewables coming into the market, such as eolic energy parks, solar power plants and other renewables such as biomass and others which are expected to come into the generation market in the coming years. No doubt their contribution will further reinforce power supply and also make a better case for building energy hubs, thus providing more alternatives to local communities. This will also give more power to the regions in Chile where these resources are being exploited, in opposition to the past when the central government monopolized all the decisions and decided purely on a centralized and highly hierarchized political criterion with little regard for regional interests.

With storage technologies maturing, such as lithium batteries (lithium is quite abundant in Chile and there are plans to exploit this technology in the coming years) or hydrogen production, not only for electricity but also for all electric public transportation; it is evident that their integration with the current EPS infrastructure is a matter of time. Yet this would only be possible with a smart grid transformation. ENEL Chile knows this and is keenly aware of the need for such industry changes in a country like Chile, with several conditions that favor such changes. Thus ENEL Chile is aggressively moving in that direction. ENEL and other utilities are also considering high power DC grids to face mounting costs of building new AC transmission lines, which generate tremendous costs and chaos to local communities.

In Chile in particular, being such a long country, rich in natural and renewable energy resources, and with huge hurdles to advance transmission lines, there is the need to explore more flexible, small and medium-range scale DC solutions, as well as other medium range power supply alternatives that can operate as energy hubs. Particularly in the far south and north of Chile, where there are growing plans to integrate electric power services with neighboring countries and to provide electricity to vast zones in those countries, such as Peru and Argentina, it is all too clear for utilities like ENEL that the future of electric power generation, distribution, and consumption will rely chiefly on regional integration, a decisive use of large-scale renewables like solar, small hydro and wind and also large size energy storage units that can further their ability to deliver energy efficiently, safely, and sustainably.

The work is divided into five sections. Section 1 serves as the introduction, emphasizing the need for SES in light of ongoing changes in the electric power systems (EPS) industry in Chile and the threats that climate change and the energy crisis poise. The section also underlines the hurdles of the EPS decentralization and the roadblocks for adopting and engineering SES and particularly sustainable hybrid energy systems (SHES) such as the microgrid concept. Section 2 presents a brief review of the HC systems literature and then points to the need to incorporate homeostasis-based control systems in the design of SES, particularly in line with the opportunity that electric power industry icons like ENEL are soon to present to the market for the incorporation of such systems. Section 3 presents a homeostasis-based power and energy management system for a sustainable hybrid energy system (SHES) like the microgrid. The new homeostasis-based power and energy management and control system being proposed here compares more traditional control methodology with the new theoretical approach being presented. Section 4 deals with SHES and the need to view and manage sustainable hybrid energy systems (SHES) as living open systems. This section offers a discussion section as well to enhance our insight and to enlighten the subject's industry spectrum in light of its modern day transformation, and where Chile is no stranger. Section 5 offers conclusions.

1.1 Climate change and the energy crisis

Building the case for sustainable energy systems (SES) in the electric utilities' landscape

Like the US, Chile is no stranger to these harsh scenarios either, and has had its share of disasters too. The country is "sitting on a hot seat" so to speak, with earthquakes, volcano eruptions, and rain floods becoming increasingly present in the collective consciousness of its people. Such events are simply not uncommon but are becoming prevalent not only in Chile but in many parts of the world, with climate change and harsher weather on the rise. The difference is that in today's 21st century world, much of the fragile living systems and economic sustainability depend on modern utilities' infrastructure of which roads, electric power transmission and distribution networks, and telecommunications are a vital part, yet increasingly vulnerable when faced with these such phenomena [3,16,17].

On September 17, 2015 a powerful 8.3-magnitude earthquake struck off Chile's coast causing havoc and chaos in an otherwise tranquil Wednesday afternoon [4,5]. Unlike its predecessor of 2010, the natural disaster triggered an immediate tsunami alert and coastal evacuations were readily executed yet utility infrastructure was compromised, particularly electricity. The tremendous earthquake that struck Chile in 2010 [6] was much worse and found the country largely unprepared. It occurred on February 27, 2010 at 3:34 a.m., off the coast of south-central Chile, taking everyone by surprise. The 8.8 magnitude earthquake had its epicenter some 200 miles (325 km) southwest of the country's capital, Santiago, causing widespread damage on land and initiating a tsunami that devastated some coastal areas of the country. Together, the earthquake and tsunami were responsible for more than 500 deaths and caused major damage to infrastructure [6]. Yet, despite these and other natural disasters, the country remains largely unprepared against massive telecomm and electric power systems brake-down [3].

The problem lies in the high concentration of centralized electric power and communication systems—a model than once proved efficient and secure but that it is no longer. To further compound the risks posed by environmental threats, there is also the lack of adequate technologies and back-up/emergency power systems for disaster recovery, something that even extends to the armed forces of the country today. There is really no energy sustainability roadmap for the country whatsoever, aside from a timid effort to advance in the incorporation of non-conventional renewable energies. Environmental policy is also weak and short sighted owing to strong corporate interests and lobby to keep the status quo in order to continue superseding environmental management to corporate economic interests.

The flaws that are built into the very fabric of our presently centralized power systems were on full display in the aftermath of the February 27, 2010 earthquake in Chile [6]. Nowhere it becomes more evident that hugely centralized power generation and distribution systems are extremely vulnerable and ineffective to disruptions from natural disasters, human error or other calamities than in a situation like this. Thus, the large power and telecommunication networks that once proved very efficient and secure are now at the center of discussion fueling the need for decentralization, further upgrades in technology for the national energy grid matrix and the rapid growth of distributed generation (DG). Hence it makes sense to follow other nations example seeking more decentralized, diversified and DG-oriented energy matrix, a solution that is notoriously much better suited to withstand these disasters [7–28].

1.2 Electric power systems (EPS) decentralization for growing environmental threats

The hurdles of electric power systems (EPS) decentralization and the roadblocks for adopting and engineering SES

The saying is clear: a chain is only as strong as its weakest link, where the weakest link, figuratively speaking, applies to a system's characteristic or technical feature that makes it quite vulnerable in terms of its design, rather than the link of an actual chain. Due to its geography and utility infrastructure design and operational conditions, Chile is a country that is quite susceptible to be struck by natural disasters including landslides, floods and earthquakes, which can seriously impair its utility infrastructure let along roads and transportation. Such events can cause major damage, producing havoc and mayhem all around, compromising the operation of key infrastructure like the power grid. Therefore, there is a clear need—as it has been already understood and acted upon in North America—to develop better, more resilient and robust approaches to enable today's EPS infrastructure to successfully withstand and overcome such adverse conditions [7–10,17,18].

The weakest link in the case of Chile's electric power distribution system is its inability to adequately sort out these events, as it was designed for normal conditions, without the level of stress and severity being imposed on the system by such scenarios. For this very reason, DG solutions ought to be designed around the idea of flexibility, resourcefulness, and independence, all common features of distributed control systems. These solutions may take several forms, sometimes with autonomous control coexisting with other forms of control like the traditional centralized control, but they all point to the same goal. This way, if a sudden power failure were to occur, like a distribution line being brought down or a power transformer being lost as a result of large violent lightning storm or wind gusts, the result would be widespread shutdown. A utility service supply disruption would impact an entire region, with long periods of limited or no electricity or water for the population until the damaged is repaired and service is brought back up again.

Although of great concern for millions of people, particularly for those countries where the technology is currently being used, the nuclear energy issue is still a double edged knife, with disastrous implications to humanity should an accident or negligent act were to occur again (like the disasters as a result of the Fukushima nuclear power plant accident in Japan or the Chernobyl nuclear power plant meltdown in the old Soviet Union). Although still quite relevant to energy sustainability and security, the nuclear power issue and its future standing in today's world energy matrix is a case of profound implications on its own right, and would therefore require an entire chapter to discuss it so we are leaving it out. Yet if we are to focus too much on power generation technologies like nuclear, fossil fuels, or hydroelectricity generation, we may be missing the larger picture or at least not giving it its proper place in the scale of concern it deserves. Although economically efficient, traditional centralized EPS (including nuclear energy) are not only vulnerable in regards to natural disasters and other environmental challenges that may threaten our energy supply, presenting very little flexibility and no diversification of energy sources. There is also the fact that EPS' infrastructure vulnerability and its collateral damage manifest themselves in various forms, like the still huge concentration on fossil, non-renewable fuels, the need for safe and steady fuel provisioning, and large hydroelectric projects which require building large dams, inundating vast extensions of fertile land.

Centralized electric power generation and distribution systems as well as large telecommunication networks have, on the one hand, large economies of scale and are very efficient, especially when it comes to serving large interconnected metropolitan areas, as in North America, for example, but that comes at a cost. Their major drawback and weakness are never more evident and alarming than when large power black-outs occur (Chile has had several in the last few years) which leave large populated areas in complete and utter darkness, sometimes for several hours, causing widespread chaos, mayhem and rampant looting all around [4–6]. Their sheer size and highly centralized architecture makes them extremely vulnerable to natural disasters and major accidents due in large part to human error. In this way all the economic gains as a result of high efficiencies, power quality and stability achieved by creating these huge electric power grids are all of a sudden lost when a disaster like Hurricane Sandy or a major earthquake strikes. Centralized EPS are concentrated usually on a few, very large power plants, operating on thermal and hydroelectricity generation for the most part, and distributing power in a radial-type distribution scheme, with each substation supplying electric power to radially connected nearby communities [27,28]. They provide service across a wide range of consumers over vast distances that span hundreds and even thousands of miles, all of which increases the risk of disruption dramatically (generation and distribution power topologies) [27,28].

Hence, the sheer forces involved in just about any natural disaster (whether it is a storm bringing strong winds and snow, flood waters, violent quakes or volcanic eruptions) are no match when it comes to our presently centralized power systems, especially in the case of the two most vulnerable parts of any power system: transmission and distribution. As an example, just one afternoon of strong winds, although rare in the Santiago metropolitan area, can knock dozens of trees and blow away roof tops, disrupting electricity distribution to several areas at once, with fallen trees over power lines, damaged transformers, and other similar havoc that can deprive whole metropolitan areas of power for several hours. All of these are strong arguments in favor of decentralization of power systems and the need for more rapid advancements in DG penetration in the form of SES [3,29–35]. Hopefully, adequate legislation initiatives will be more forthcoming in the years to come, bringing changes that can make the transition to a more secure, robust, resilient, and better prepared EPS come to fruition.

1.3 Role of the microgrid

The shift in microgrid trends, from a more passive role, originally thought to contribute marginally to electrical energy production - in case of an emergency - to a more active industry player

Originally, DG solutions like the microgrid or diesel generator sets were first thought for remote and isolated areas only. They were designed and developed basically as back-up eletric power systems in case of an emergency, like a power outage, or because the grid's power supply was unreliable or simply non-existant, like in many parts of Canada for example, where no transmission lines exist. Indeed, one finds in many places in Latin America and elsewhere in the world that there are still rural and semi-rural areas with weak or no connection to the main grid. However, nowdays electric utilities like ENEL Distribucion in Chile are using microgrids of various sizes and capabilities (https://www.enel.cl/content/dam/enel-cl/sostenibilidad/informes-de-sostenibilidad/enel-distribuci%C3%B3n/2019/IS-EnelDx-2019.pdf) as a means to complement the grid's supply at a larger scale going forward, while advancing renewables as part of their strategic green energy agenda. Thus, microgrids are becoming increasingly more relevant and viable as a solution to both urban and rural areas of all sizes and configurations, just like the North American trend shows [18–28,36–42].

Microgrids are first and foremost local DG solutions that comprise a number of feeders (one or more), servicing a clusters of loads not necessarily grouped together. Some of these loads are more sensitive while others are considered less or non-sensitive, therefore the microgrid system's design and configuration, including the choice of control system, is in part dependent on the role of the microgrid itself, the energy sources employed, and the nature and size of the loads to be serviced. In order to operate autonomously or semi-autonomously so as to reliably supply electricity and heat to a local community whether urban or rural for example, the microgrid may or may not have energy storage systems, depending on budget and system's operation necessities, and may operate connected to the grid if need be and conditions allow so or operate as a standalone system [29–35,43–46]. The degree of resiliency and robustness of the microgrid itself is given by the choice of engineering design and configuration characteristics, both of which depend on technical and budget constraints. Such aspects will influence how much the microgrid, and any other form of DG for that matter, will be prepared to act as a potential solution to endure adverse environmental conditions, while continuing to provide service following a natural disaster or harsh climatic event. Also the question of how microgrids can be designed in a way that can be incrementally integrated to the current EPS will depend on adequate legislation like the net billing law recently passed in Chile [47] as well as finding the right business and operational architectures that can accommodate these new market solutions. Thus in light of the current necessities and changing industry trends, alliances involving communities, power utilities and independent local operators are likely to emerge [23–35,43–46], just like it has occurred in North America, with notable examples in the US and Canada as well as in the UK and other parts of Europe [48–52].

Fig. 6.1 illustrates how HC mechanisms, based on reactive and predictive homeostasis, trigger a system's sustainability stress response (SSSR) that imposes a restraint over the energy

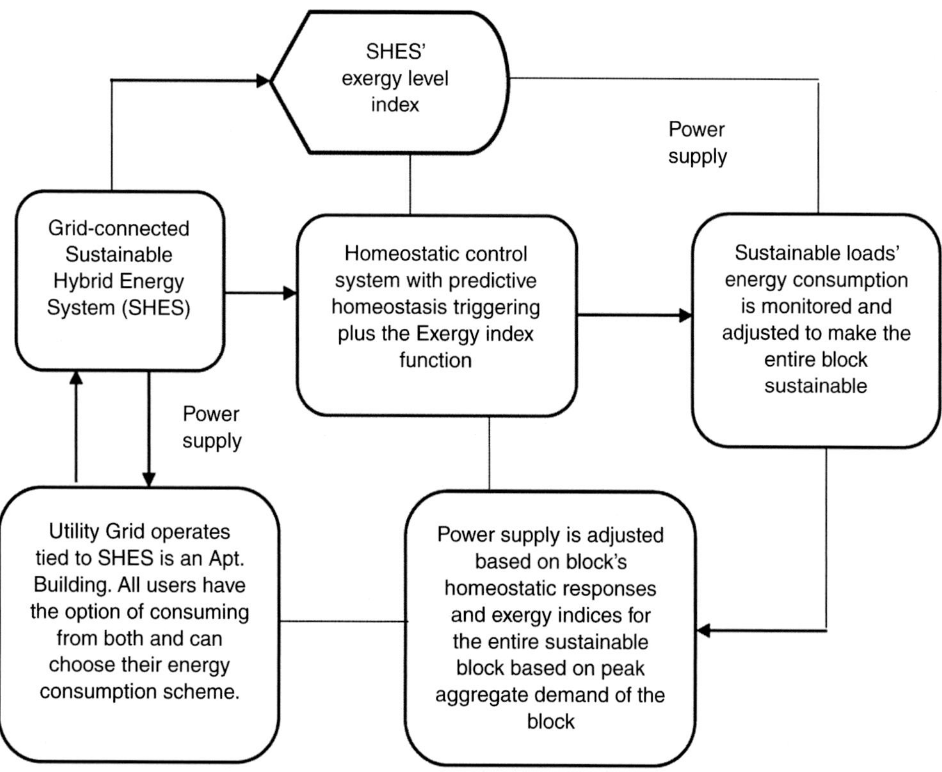

Figure 6.1. The diagram depicts reactive and predictive homeostasis mechanisms engineered in the HC of the SHES (microgrid) to be installed in an apartment building in Santiago, Chile. Source: Own elaboration.

consumption of loads based on the energy supply being available in the green energy system.

In the diagram, there is no energy storage and the Grid operates as back-up. Both HC and artificial intelligence (AI) can play a substantive role in developing SES' capabilities, understanding that homeostasis encompasses both reactive and predictive responses of such systems to changes in internal system's variables as well as changes in the environment (Fig. 6.2).

The interesting Smart Grid concept of energy hubs [36,37] built around the current EPS infrastructure is an idea worth exploring for Chile, where such localized energy hubs could be built incrementally provided adequate legislation and industry incentives in order to transition to a new EPS reality braking away with the old unfit paradigm of today's power long distance transmission and distribution network. SES not only will keep the lights on and basic services running for the residential and critical facilities' loads they serve, but can also act as a power source that can aid the grid in times of trouble. Microgrids constitute a fundamental resource as well as an electric power industry's paradigm change,

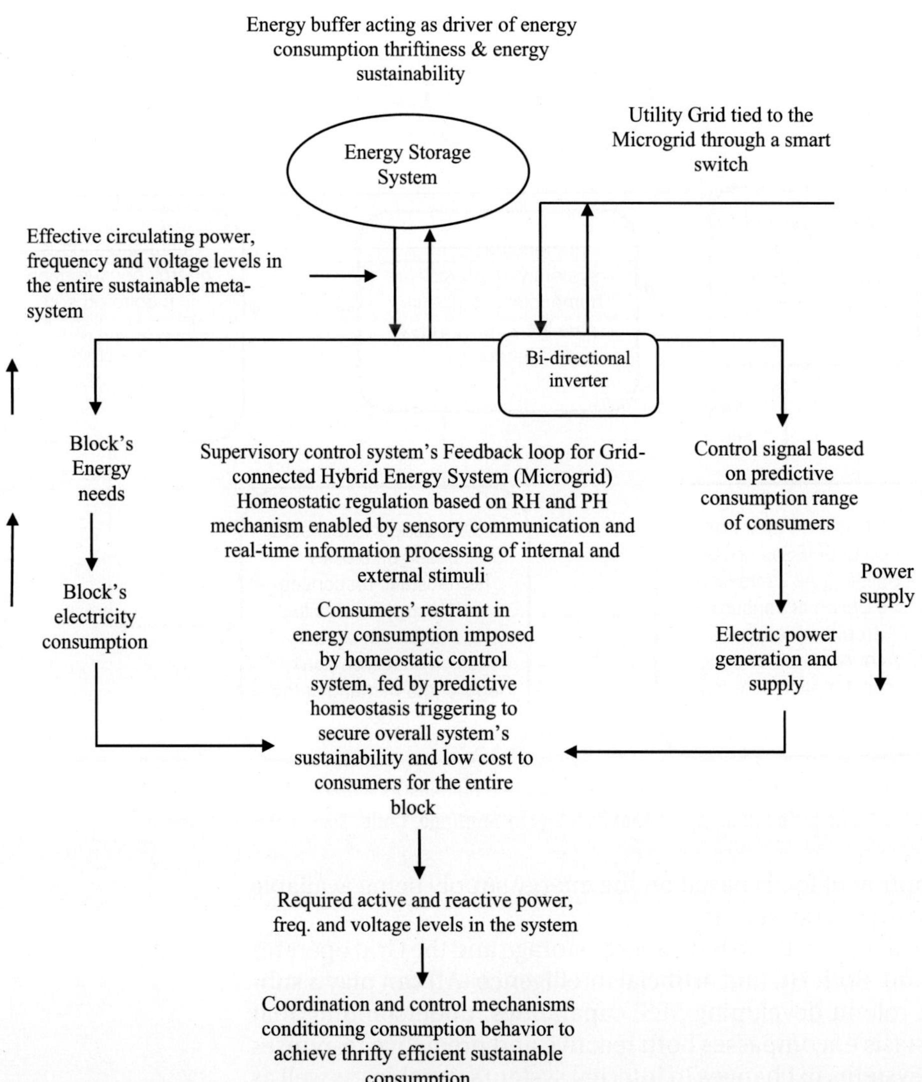

Figure 6.2. The diagram depicts energy homeostasis regulation (HR) mechanisms in operation for a grid-connected microgrid with an energy buffer. The figure is self-explanatory and shows the logic behind the strategies poised by HR mechanisms. Source: Own elaboration.

as demonstrated in New York city for example during Hurricane Sandy, and it is at the forefront of the new energy independence (from the grid-only scenario) trend towards localized energy production and provisioning in the United States, Canada and several places in Europe as well. In fact, North America, along with several European countries and Australia now are seeing microgrids not just as a means towards more energy independence and more resilient and robust EPS, but also as an industry game changer in the face of very pressing energy needs like we see in Chile for the years ahead. Important technological advances in power electronics and renewable energies, the sharp fall in certain energy generation technologies like solar and some wind turbines, along with a plague of very bad weather and natural disasters like torrential rains, earthquakes and landslides have accelerated this change from a more traditional mind frame to a modern, more proactive and customer-conscious and responsive approach on the part of industry agents, local authorities and legislators.

2 Resiliency and energy homeostasis

How to engineer resiliency in sustainable energy systems. Smart energy systems: the need to incorporate homeostasis-based control systems in the design of sustainable energy systems (SES)

Ever since Cannon first formulated the concept of homeostasis over 80 years ago [1,2], attention has largely been focused on the corrective responses initiated after the steady state of the organism is perturbed. However, the concept of homeostasis should be extended not only to include reactive homeostasis but also the precise HC mechanisms that can be designed to enable a sustainable energy system to predict when environmental challenges are approaching or are most likely to occur [29–38,45,46]. Sustainable energy systems encompass both reactive and predictive homeostasis operating recursively and in coordination with one another in the face of an environmental challenge.

On the other hand, HC is a term first introduced by F.C. Schweppe and his group of collaborators at MIT back in 1979 and in early 1980s [53–60] which stems from their highly visionary work and insight regarding both, flexibility and stability in electric power systems (EPS) linked to homeostasis, understood as reaching and maintaining an efficient equilibrium state between energy supply and demand, considering the diverse nature and operation dynamics of the wide variety of industrial and commercial loads [53–60]. Their approach advocated HC of energy supply and demand in an effort to make utilities power supply more efficient, particularly when supplying large industrial and commercial customers. Indeed, Schweppe and his group were much ahead of their time, having had a true insight for what was to come in the years ahead, anticipating also the need for new control technologies and energy management systems to adequately manage the intermittent supply of RETs (mostly wind back then). Homeostasis mechanisms applied to EPS allow utilities to finding ways to tap into a number of opportunities that the great variety of energy consumption patterns of their customers, therefore enabling them to better manage their supply capacity, and at the same time, finding new ways to best fit renewable energy sources and technologies into the current electric power infrastructure [29–35,61–63]. Something in which there are still much to be done, especially in a country like Chile.

2.1 Homeostasis-based control systems in the design of SES

The need to incorporate homeostasis-based control systems in the design of sustainable energy systems: employing reactive and predictive homeostasis to control sustainable energy systems

Reactive homeostasis (RH) in SES, as the name suggests, is a feedback-enabled control mechanism driven by the system's energy generation and supply versus the consumption or expenditure of energy by the loads. This can be engineered in SES by employing sensors, control limit actuators (e.g., set-point fired responses) and AI algorithms that allow the system to make decisions to respond to changes in a predetermined array of systems control variables. Thus, SES take actions to counteract or fend off adverse conditions and noise that may affect the system's normal operation.

On the other hand, predictive homeostasis (PH) mechanisms generate responses well in advance of potential or possible challenges, once the system has reached a threshold signaling a predetermined degree of likelihood that an event will occur. Hence there are a set of precise SES responses that come about in anticipation of predictable environmental challenges. Such PH responses enable the energy system to immediately prepare itself, taking the necessary precautions and actions to adapt and even reconfigure itself if necessary, in order to respond to the challenge ahead of time. Such actions may come in several forms and will depend on the resources and intelligence built into the system, but they are all geared towards making the SES more secured and able to withstand the upcoming challenge by activating its readiness control mechanisms. Actions may come differently in magnitude and timeliness; some may be big and come immediately to adjust parts of the SES operation while others may come in the form of smaller changes in the system, largely as a result of stage-by-stage preparedness protocol building over time. The decision of which changes will occur first, where and how big they will be will be determined by both RH and PH control mechanisms engineered in the SES. Some may come very soon while others may come a longer time in advance of a probable environmental challenge. However, as we all know, systems are prone to internal conflicts when control criteria superimpose on one another generating the wrong response.

Reality is not always adequately interpreted by control and monitoring systems, thus it is possible that, sometimes, there may be signals and perturbations which, in spite of their potential to represent a real challenge or contingency to the energy system, they may not always be correctly anticipated or foreseen in their full magnitude and scope. Also misread signals and misfires may risk a wrong or inadequate response as systems sometimes experience false alarms and misfires. These may occur in part as a result of possible conflicts over certain HC variables that may share common goals and values but different scope of action and logic sequence, depending on the scenario being faced. Such HC variables may involve PH and RH control logic sequence which, if inadequately engineered in the SES may result in inadvertent antagonism that can hurt system's performance [29–32,34,35]. Therefore, careful engineering of such capabilities in SES must account for such conflict of interest and changing scenarios must also be accounted for when establishing set-points [29–35,61–63].

Thus, adequate measures must be engineered in the system's design to prevent PH responses from interfering with RH mechanisms. If these potential conflicts were not accounted for and swiftly overcome, should they arise in the course of events, undesirable conditions may emerge which can compromise the effectiveness of SES readiness mechanisms, risking the very sustainability and efficiency of the system itself.

3 Grid-tied microgrids with and without energy storage

When, where and how to apply each case: The homeostasis-based power and energy management system for SES like the microgrid

In the case of PH the system responses will come as a result of information being processed by the system as the stimulus approaches and is detected by the sensing devices. Here there are both RH and PH sensors and an ample array of control mechanisms ready to act whenever conditions arise. Therefore, energy homeostasis in SES requires a careful equilibrium of such control mechanisms and the coordination of internal and external decision variables—all part of the particular HC strategy designed in the SES—which will stand guard against a variety of adverse conditions and possible challenges. Thus the SES will control the use of its energy resources including the grid and the use of alternative energy sources like energy storage if the grid is off. It will do so recursively and permanently in order to generate and supply enough energy to meet the loads demand, while at the same time signaling to consumers how much energy is the SES capable of supplying. The question of if and how much energy will go into the energy storage system will be determined by the HC system based on the situational awareness and degree of criticality being experienced by the system itself. The HC system will therefore decide when and how much energy to store based on supply surplus and the energy demanded by the loads. Some loads will be more sensitive than others and therefore will occupy a higher hierarchy while others may be spared or serviced partially, as conditions change. Such HC mechanisms will involve both PH and RH operating in unison, determining a generalized state of energy equilibrium between supply and demand, as dynamic scenarios unfold [29–35,44,61–63].

3.1 Building sustainability in energy systems

The role of exergy, exergy management, and how to apply it in on-grid microgrids for buildings and condos, in the quest for higher systems efficiency and exergy levels

Proposition 3.1.1: Energy homeostasis is present in all living organisms. As such it is also present in living systems [1,2,29,31,33,53–60].

Corollary: Energy homeostasis is also present in human living environments and its principles and postulates also apply for electric power supply and consumption in such living environments [29,31,34,35,53–60].

Proposition 3.1.2: Both misuse and excessive use of energy results mainly from inadequate electric power system regulations and the lack of choices and incentives to energy users to behave otherwise. Although environmental and lifestyle factors contribute to excessive use and misuse of energy, homeostatic adaptations to a thriftier and efficient usage of electrical energy is possible and can be induced by voluntary energy consumption restriction while allowing others with greater needs to use such energy with economic benefits for all.

Proposition 3.1.3: Thrifty and efficient use of energy and energy conscious-promoting environmental and systemic factors, regulations and services, along with HC systems employing reactive and predictive homeostasis can be employed to regulate power supply and consumption more efficiently, thus avoiding energy overconsumption and waste. This can contribute to create the conditions for adequate choices and incentives to energy users to behave more thriftily and efficiently in their use of electrical energy, perceiving benefits and incentives for their actions, while allowing electric utilities to profit from such conditions [29–35,53–60].

Corollary: The above propositions are supported by the following mathematical equations which represent the HC model being proposed here, which incorporates both predictive homeostasis (PH) and reactive homeostasis (RH). The expression for the energy equilibrium of the SES is given in terms of the total power supply and the homeostasis regulation mechanisms discussed previously:

$$E_{equilb} = P_{supply}(x)PH(u)RH(v)S(\alpha) = E_{consump}(u,v,\alpha) + \frac{d}{dt}E_{consump}(u,v,\alpha)$$

$$\text{and} \quad P_{supply} = Real\,Power + Reactive\;Power = (P+Q) - Losses \qquad (6.1)$$

Where x represents the internal state of the energy systems at time t_0 and Energy equilibrium E_{equilb} is dependent upon several factors operating adequately in the SES. Both u and v represent the specific predictive and reactive homeostasis variables respectively, which are designed in the HC model. These homeostasis variables are influenced by local factors such as weather, geography and energy use of the particular community or region. They are also dependent on technical factors such as hourly tariffs applied for electricity use during the day with seasonal variations and they can also take into account special tariffs applied to communities with a significant part of their electric power supply coming from renewables, for example in buildings and condos that have this feature in place. These homeostasis variables are designed based on extensive data modeling to incorporate as much accuracy in the system's response as possible.

$S(\alpha)$ is the conditioning function of the SES and operates to alert and condition the HC system's adaptive mechanisms recursively in order to respond to a wide range of stimuli. Its actions are based on the situational awareness and degree of criticality being experienced by the system and the solution incorporates artificial intelligence and intelligent control. Thus $S(\alpha)$ is a function of adverse conditions and environmental challenges being sensed by the energy system and represented by the awareness and criticality variable α. All these three variables: u, v, α are incorporated as key constitutive elements of the intelligent algorithms built in the HC system and as such, are the equivalent of metabolic variables in living organisms like the physiological and endocrine systems' variables that affect the energy expenditure and storage of such systems.

3.2 Sustainability performance indicators

The homeostatic Index and the Grid_frac functions in SHES

The expression $\frac{d}{dt}E_{consump}(u,v,\alpha)$ stands for the rate of change of energy consumption of users, let's say a sustainable block [29–32] somewhere, and is a direct indicator of thriftiness and energy efficiency [32–37,45,46] being built in the SES. It is linked to powerful sustainability performance indicators of SES introduced previously in the literature, such as the homeostatic index H_i and the Grid_frac [31]. The homeostatic index is a powerful new concept previously introduced [31] which measures how much electricity is being drawn by each home from the mains as a percentage of the total electricity (renewable plus non-renewable) being consumed by the home [31]. This is being monitored and recorded in real time and shown to the consumer as a monthly reading or on a daily and hourly basis as preferred [30,31]. Essentially the homeostatic index is a measure of sustainability of energy systems. It shows how thrifty and energy efficient each home is with respect to the power supplied by the SHES (microgrid) and how much power is being drawn from the grid. Both the homeostatic index H_i and the Grid_frac functions are the workhorses that drive the energy system to higher levels of exergy in terms of the amount of energy that the energy source—the sustainable hybrid energy system (SHES)—can indeed deliver at any point in time, as well as the quality of said energy. This is important, not only because of what these two aspects of energy consumption management represent but also because they drive the exergy level being sought in the SES. We must not forget that exergy relates to the amount of energy that is available to be used in the energy system at any time. After the SES and the loads that are being supplied by the SES reach equilibrium, the exergy is zero [31,34,35]. What is different here from the traditional discussion on exergy available in the literature is the fact that exergy is not only an intrinsic quality of the energy source but it is also a variable which depends on the meta-system which includes the energy users in the sustainable block: A value below 1 is considered acceptable yet ideally values closer to 0.50 or below are a truer indicator of a high degree of thriftiness and EE for the home [31]. H_i represents a measure of the energy efficiency and thriftiness of energy consumers.

On the other hand, Grid_frac is an indicator of the fraction of total electricity drawn from the grid per each home. It is also a measure of EE and thriftiness just like the homeostatic index *Hi*. Grid_Frac shows the fraction (in percentage) of the total electricity consumption drawn from the grid by each home in the sustainable block [30,31].

Below is a diagram which illustrates the concepts presented here and which incorporates an Exergy index function which is related to the quality of the energy being produced by the SES and with the amount of thriftiness and EE being exercised by the users which directly impacts on

$E_{consump}(u,v,\alpha)+\frac{d}{dt}E_{consump}(u,v,\alpha)$ determining how much energy is being made available in the SES by the energy users of the sustainable block.

In the model depicted earlier in Fig. 6.3, there is a particular HC and energy management architecture involving

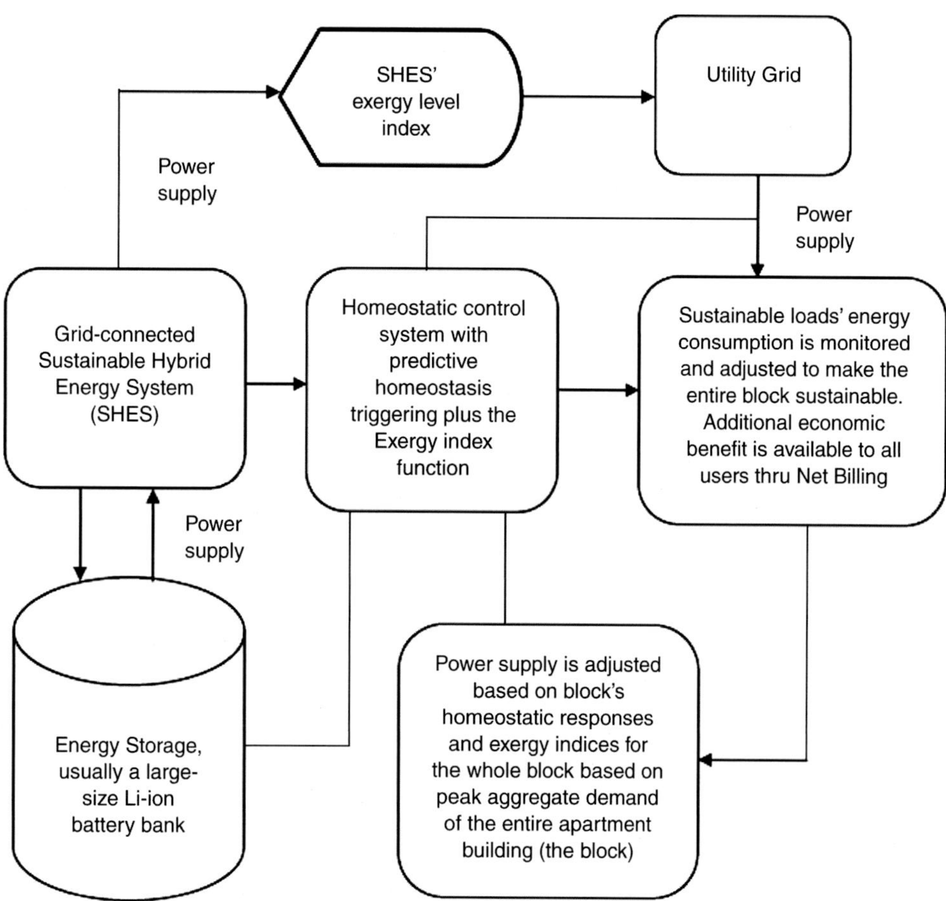

Figure 6.3. The diagram depicts how predictive and reactive homeostasis mechanisms operate in a SHES, as part of the HC of the system, this time with energy storage. Source: Own elaboration.

independent energy sources including an energy storage device and the electric grid. Here HC mechanisms trigger a system's sustainability stress response that imposes a restraint over the energy consumption of the loads in the apartment building based on energy supply being available within the SHES. Both energy and exergy management are built in the SHES to enable resilience and sustainability [32]. The sustainability of the system is in part safeguarded by AI algorithms that make up the autonomous mission control of the SHES (microgrid). There is also an Exergy index function [32] that, like the homeostatic index, is a measure of the quality and efficiency of the energy being generated and utilized by the microgrid, which includes the energy consumers in the sustainable block (the loads). The higher the exergy index of the SHES, the higher its degree of sustainability and energy efficiency and by corollary, the lesser the dependence on the power grid [32]. The power supply by the SHES is adjusted based on the block's homeostasis regulation (HR) and exergy indices for the entire sustainable block based on peak aggregate demand of energy in the system throughout the day.

3.3 Control methods of EPS

More traditional control methods of EPS and how the two may combine

The above control mechanisms come to complement more traditional methods of control for EPS like the well-known droop control method. Traditional electric power generators are engines that drive a generator, whatever the type, providing a constant power output to meet the loads, with a power grid frequency and voltage level very stable for the most part. On the other hand, renewable energy sources like wind and solar systems, which are part of SES solutions like the microgrid, have a variable output, depending on the wind speed or the local solar irradiation of the site. They require electronic inverters to interface with the system before supplying power to the loads. SHES like the microgrid require both, renewables and traditional sources of power generation to keep the system stable (unless connected to the grid) and capable of meeting loads' demand at all times. Depending on the size of these loads and the power consumption profile, there will be a need for more or less engine-type generation combined with renewables and also with the grid when and where it is available.

When it comes to industrial power plants and large microgrids operating grid-tie, the power is mostly supplied by large rotating AC generators turning in synchrony with the frequency of the grid. The electric power grid system operates on a single frequency for all these generators and they must be synchronized in order to keep the system stable. If there is sufficient installed capacity in the system, that is, if there are enough generators to meet the loads at any given point, then the frequency can be maintained at the desired rate (i.e., 50 Hz or 60 Hz depending on country), otherwise the frequency will drop. When operating grid-tie, the phase angle of the power supplied by each generator in the microgrid will slightly lead the phase angle of the grid's power. This slight change in phase angle will be in correspondence with the power they deliver to the grid. An increase in the power demanded by the loads will result in an increase in the power supplied by the system's generators. In this case, the engines require more fuel to increase power yield, so the governor automatically opens a steam or gas inlet valve to supply more power to the turbine.

However, if for some reason there is not enough capacity to meet the demand for power, even for a brief period of time, then generators' RPM and the frequency drops. For large power grids having large distributed loads and a number of good-size AC generators plus other sources of energy, like for example solar or wind, makes frequency management easier because any given load is a much smaller percentage of the combined capacity. For smaller grids like the microgrid, there will be a much larger fluctuation in capacity as delays in matching power supplied are harder to manage when the loads represent a relatively larger percentage of the generated power [61].

However, in addition to the above, a good HC system operating with adequate performance levels of EE and thriftiness can be used in order to impact the exergy of the system. This can help keep short-term fluctuations in power requirements from dropping the frequency or at least helping this situation become milder [64]. This is because sometimes there are lags in the system's governor and generators' output which require a finite time to adjust to the new power requirements. Such actions can be aided by reactive and predictive homeostasis functions built in the system in order to act as an energy consumption-based frequency regulator, $E_{consump}(u,v,\alpha)+\frac{d}{dt}E_{consump}(u,v,\alpha)$ aside from the role of power enabler. Such a role is played whenever EE and thriftiness in energy consumption make it possible that more power is made available in the system for those that need more while others need less [32–37,46]. No doubt the changing frequency in the SES will influence the power flow but, at the same time, frequency is a function of the energy consumption and the rate of change of such consumption $E_{consump}(u,v,\alpha)$, therefore, employing the principle of demand response in energy management, power flow quality and energy sustainability of the energy system can also be influenced by $E_{consump}(u,v,\alpha)+\frac{d}{dt}E_{consump}(u,v,\alpha)$.

The difference in energy consumption will impact system's frequency and voltage level which will ultimately impact the quality of the power supply and overall system's dynamics in a sustainable energy system (SES).

With the droop method the power angle depends heavily on the real power R generated while the voltage depends on the reactive power Q. If real power R can be adequately controlled, then so can the power angle, and if the reactive power Q can be regulated as well, then the voltage V_1 will be controllable too [62]. The droop control method has an inherent trade-off between the active power sharing and the frequency accuracy, resulting in the frequency deviating slightly from the nominal frequency [62]. The relationships between real power and frequency and the reactive power with voltage can be expressed as:

$f=(f_0-k_p)(P-P_0)$ *and* $V_1=(V_0-k_v)(Q-Q_0)$ where k_p and k_v are the power and voltage droop gains respectively; f_0 and V_0 are the energy system's base frequency and voltage respectively, and P_0 and Q_0 are the set points for the real and reactive power of the SES at a given point in time and are subject to change as dynamic conditions evolve [62]. Reactive power regulation is used to impact voltage regulation and real power depends on frequency and phase angle, thus voltage will be controllable as well.

In the droop method, each unit uses the frequency, instead of the power angle or phase angle to control the active power flows since the units do not know the initial phase values of the other units in the standalone system. By regulating the real and reactive power flows through a power system, the voltage and frequency can be determined [62]. In the droop method, each unit uses the frequency instead of the power angle or phase angle, to control the active power flow since the units do not know the initial phase values of the other units in the standalone system [62]. Fig. 6.4 shows a microgrid architecture model with several energy sources and loads.

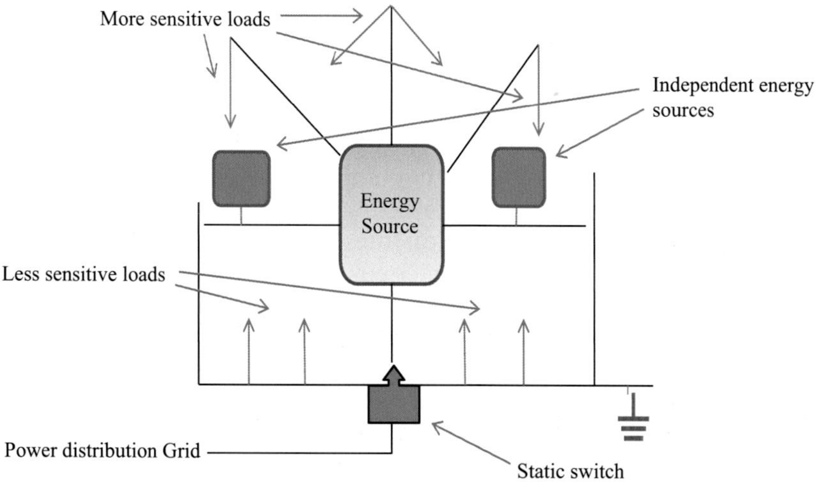

Figure 6.4. A diagram representing a particular HC model, which illustrates an example of a potential microgrid architecture with several energy sources and loads. Please notice the pivotal and very strategic role of the energy storage in the system.
Source: Own elaboration.

4 Sustainable hybrid energy systems (SHES) as living open systems

The role of exergy, exergy management and how to apply it in on-grid microgrids for buildings and condos

The potential for using sustainable hybrid energy systems (SHES) as part of electric utilities' plan to decentralize the energy grid, as well as to further personalize or customize their vast services to their ample and diverse customer base, and to incorporate more localized DG solutions like the microgrid to be installed in buildings based on renewables, in tandem with the electric power distribution networks is part of the industry transformation in Chile which is being led by industry icons like ENEL. For this purpose, proper drivers for implementing such solutions are to be identified in each case as one size does not fit all. Particularly so in a country where there are ample differences in socioeconomic strata.

There are also distinct factors and conditions in the different regions of Chile that offer geographic potentials for production with the available technology and also the social and political support and active involvement of the local community and authorities respectively. Therefore, to meet the energy requirements of both urban and rural communities, SHES stand as a viable and sensible distributed generation (DG) solution for electricity supply working in a smart grid configuration and operating with non-conventional renewable energies (NCRE) can be a viable and convenient solution to electric utilities' electrical energy distribution networks available. It can also be a much more convenient solution both socially and economically for the different regions and localities when as is the case in Chile, it is too expensive to extend AC power transmission lines and electric distribution networks thereof to connect new customers. However, the cost of hybrid energy systems based on renewable energy technologies (RETs) is generally high and there is also the problem of reliability associated with the NCRE due to their intermittent nature. Thus there arises the need to design and develop small, modular microgrid systems like the SHES based on NCRE that are cost-efficient and economically profitable as an investment for utilities to widen their service pool.

Notwithstanding their small size and limited power generation range, which can operate in the kW range for residential purposes and also for small size industrial applications, rather than in the Mega Watt (MW) range for power utility size application, there is always the back up from the mains to which the SHES is tied.

However, integrating renewables, particularly NCRE, requires a transition that for some communities may be much more difficult and complex than for others depending on local socioeconomic conditions, availability of subsidies, climate conditions due to their geographic location. The most pressing factor nowadays that is hindering such transformation towards a more widespread use of NCRE is the socioeconomic and cultural characteristics of the communities. That is where the utility arises as an enabler and promoter of green electricity and efficient, sound and responsible energy consumption in order to create sustainable conditions for the country's growth and the wellbeing of a modern society like the Chilean that is becoming more and more dependent on electricity. However, in every case, it certainly must mean an upgrade to something better than what they had before and to ensure employing the right energy policies and industry regulations that such improvements in living conditions and standards are sustainable over time [29,30].

Thus HC of SHES operated by electric utilities like ENEL that are tapping into the country's vast pool of renewables, employing predictive and reactive homeostasis control mechanisms have no doubt significant potential for contributing to the economic, social and environmental sustainability of a country like Chile. They also reduce emissions of local and global pollutants and may create local socioeconomic development opportunities for the communities to which they provide services as well [3].

4.1 HC system installed in a SHES

An example of HC system installed in a SHES in the form of an electric utility's run microgrid for a building community in Chile

The above flowchart is an example of several HC algorithms. It depicts the management of the energy flow from a set of customers with distributed generation incorporated in an apartments building community in Santiago through the use of a supervisory HC system that considers storage energy unit operated by ENEL Distribucion.

The flowchart in Fig. 6.1 represents the type of HC logic proposed in the work developed as a proposal for ENEL Distribucion in Chile, which relates back to previous papers [3,29–35] on the subject albeit without considering predictive homeostasis as this one does. The example shown in Fig. 6.5 illustrates what is expected in real life once the HC system is installed by Enel Distribucion, along with the smart microgrids operated by them as well. The diagram shows the predictive homeostasis mechanism built into the system, which allows the utility to efficiently manage electrical energy in a residential building with smart metering. The building will have its own power plant with photovoltaic generation and energy storage. The different customers that integrate the residential community of the building are taken as one single unit (the entire block) for the local utility company in charge of distributing electricity whose rates present charges for energy consumed, maximum demand reached as well as maximum demand reached during peak hours.

The control strategy is based on the principles of homeostasis so that both the generation and supply on the one hand and the energy demand of the customers that comprise the residential block respond to one another in an efficient and balanced manner, thus preserving the system's optimal efficiency and cost-effectiveness. This way it is possible to make compatible the grid-tied microgrid's supply capability with the users' demand through a system of compensation among the different types of energy users. This allows those users that consume less (those who are thrifty, and seek to economize wherever possible, or else are more energy conscious in their energy consumption) in hours where the system is facing peak hourly demand (so called peak hours) to obtain an economic benefit or reward that is provided by those who require to consume more [30,31], in order to achieve a mutually beneficial arrangement between the parties, that minimizes

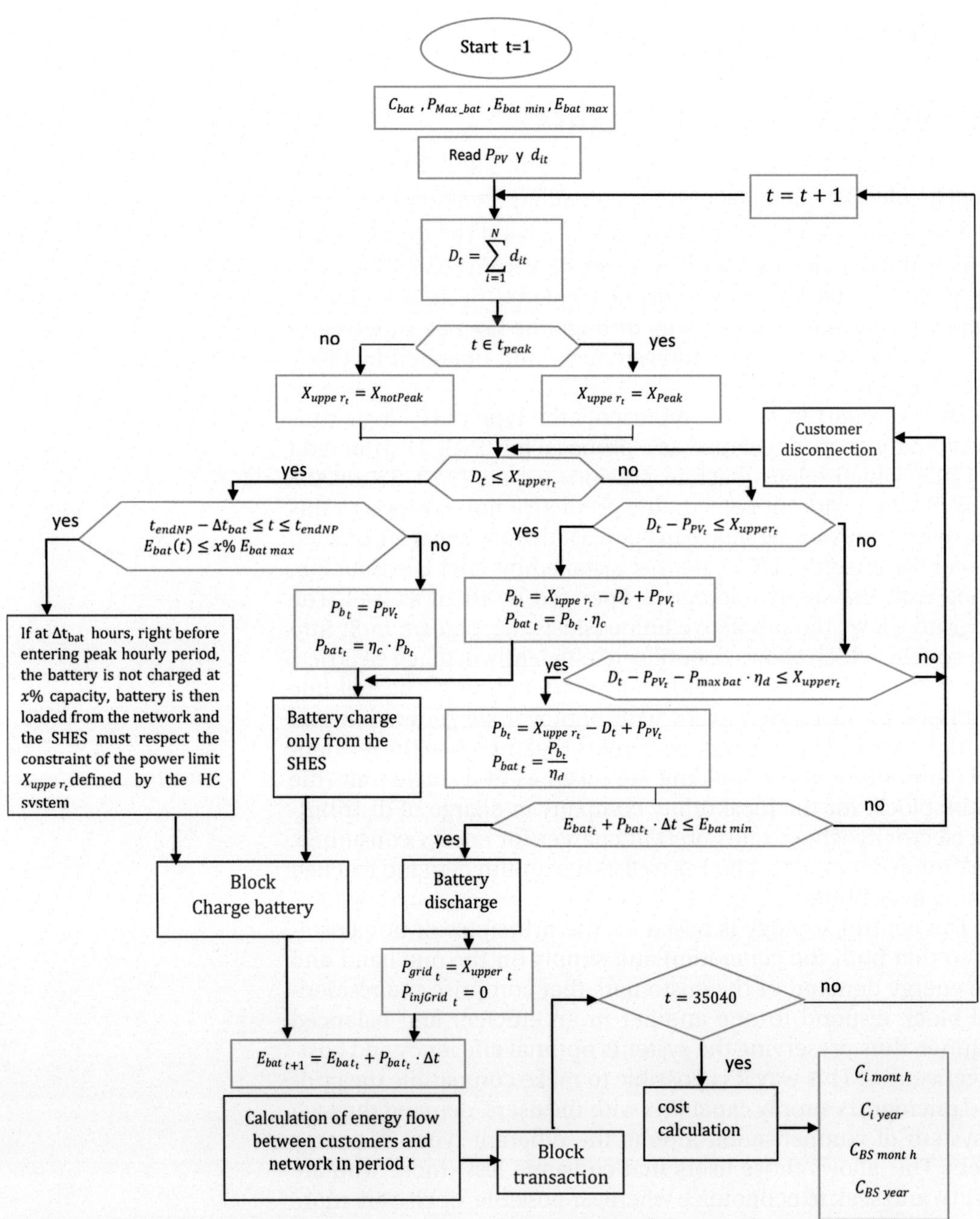

Figure 6.5. A flow diagram that illustrates the supervisory control of a grid-tied DG system that must balance electrical power supply and consumption for a set of residential customers through a particular control logic algorithm, based on the principles of energy homeostasis. Source: Own elaboration.

their social and economic costs [65]. At all times, the SES seeks to maintain a state of stable and efficient equilibrium based on self-regulation of the meta-system, something that also benefits the electric utility and the energy distribution networks as a whole.

The control cycle begins with the battery, seeking to maximize the power generated by the battery, while observing the minimum and maximum energy levels of the battery charge cycle respectively. $C_{bat}, P_{Max\_bat}, E_{bat\,min}, E_{bat\,max}$. Then there comes the reading of the photovoltaic power available in the system and the energy demand of each home P_{Pv} and d_{it} respectively. Then the system calculates the total energy demanded by the sustainable block (D_t) and the energy demand limit of the sustainable block X_{upper_t} at this point in time, depending on whether it is during peak hourly demand period or not. If the total energy demand of the sustainable block is less than the block's demand limit $D_t \leq X_{upper_t}$, (both energy and power consumption are bounded by limits being set by the HC algorithms of the SES) based on the energy being produced by the microgrid and the energy available in storage at any point in time, all the photovoltaic (PV) energy generated will then be assigned to charge the battery $P_{b_t} = P_{PV_t}$ to its maximum charge limit.

The control module charge_battery is the energy manager in charge of charging the battery and handling its load while meeting all technical constraints of the same. The PV energy that cannot be uploaded to the battery because it either has already reached its maximum charge limit or because the battery is quite near its full energy storage capacity while demand is still strong, it will be automatically consumed by the users based on their energy consumption patterns which are registered by the HC system of the microgrid controlled by ENEL. At this time the utility will inform its customers of the supply surplus. In case the PV energy minus the energy destined for the battery is greater than the total demand $P_{PV_t} - P_{b_t} \geq D_t$, the remaining power is injected to the network according to the following $P_{injRed_t} = P_{PV_t} - P_{b_t} - D_t$. With this the customer obtains the benefit that the net billing system offers to the energy users in the community. ENEL is keen on introducing new, more personalized, flexible, and cost-effective alternatives for its millions of customers in Chile, and to expand the massive use of the net billing system and others to come throughout its vast customer base.

On the other hand if $D_t \geq X_{upper_t}$, the system will proceed to evaluate whether the demand of the sustainable block minus the photovoltaic power being produced is less than X_{upper_t} that is, if the condition $D_t - P_{PV_t} < X_{upper_t}$ verifies itself. If this is the case, the remaining PV power available is injected into the battery such that $P_{b_t} = X_{upper_t} - \left(D_t - P_{PV_t}\right)$ observing battery constraints just as in the prior case. At the other end, when $D_t - P_{PV_t} \geq X_{upper_t}$, energy consumption will be drawn from the battery so long as the restriction of keeping the block's demand in the upper limit X_{upper_t} is met, that way the sustainability of the system is not compromised.

At this instance, the system will assess if discharging the battery to its maximum discharge level (as permitted by the specified control system's constraints) is sufficient to maintain the demand's upper limit X_{upper_t} in check. If this is the case, the battery is capable of meeting the block's demand so the power needed is extracted from it in order to maintain said limit whereby $P_{b_t} = X_{upper_t} + P_{PV_t} - D_t$. The system then proceeds to verify if there is enough energy already in the battery. If this is the case, then the battery discharging is done until the system's constraint is met. During this period the power supplied by the mains will be $P_{Grid_t} = X_{upper_t}$. In case the energy discharge of the battery is not enough to meet the power demand limit of the sustainable block, the system will discriminate according to the rules incorporated by design in the SES. Hence, those customers who are demanding the greater power from the microgrid will be disconnected to favor the less demanding or thriftier ones. Thus, they will only receive power supply from the utility's electric power grid, and this will go on until it becomes possible again to meet the control system's constraint that is $D_t - P_{PV_t} \geq X_{upper_t}$.

The transaction module is in charge of calculating the energy flows among the customers and the mains at all times. Thrifty customers with energy surplus can make this surplus available to those residents who consume more or else they can choose to allow this surplus to be injected in the mains, having the right in each case to an economic incentive defined by the utility which compensates their action. Of course they will be informed at all times of the economic benefit that either one represents for them in terms of the money being saved from their energy bill, as a credit earned by the energy user. All customers are responsible for the loading and unloading of the battery alike, thus avoiding the decoupling that may be produced in time.

The battery is discharge during peak demand periods since the price of the maximum demand registered by the system in such periods is high. It is necessary that the battery is at its maximum capacity before entering peak hourly price period so as to respect the power limit defined by the control system, thus avoiding the disconnection of customers. The homeostatic controller that employs predictive homeostasis will then predict or anticipate the

Δt_{bat} time in hours left before entering the peak hourly demand period, where the highest price for the electricity is charged to determine if the PV generation will be capable of charging the battery or not. If this were not the case the controller will then charge the battery from the mains observing the constraints both of the battery and that of the maximum power demand of the sustainable block $D_t \geq X_{upper_t}$. Δt_{bat} is the time needed for the battery to be recharged entirely to its maximum storage capacity and the x% will be adjusted empirically.

For simulation purposes, in order to illustrate the example shown here, once a year's worth of evaluation is completed by the system ($t = 35{,}040$, $\Delta t = 15$ min), the monthly and annual costs of each customer are computed as well as that of the entire sustainable block.

4.2 Discussion

The key concept here is understanding energy homeostasis and learning to view and manage sustainable hybrid energy systems (SHES) like living open systems, which is the very nature of homoeostatic control of EPS, which was first introduced by Fred Schweppe and his team [53–60] back in late 1970s the early 1980s, and which has been extended here, to include not only reactive homeostasis as first proposed by Schweppe and his team at MIT, focusing on HC of utilities. This time we also include predictive homeostasis (PH) mechanism operating in a meta-system which is comprised of the loads in an apartment building (the sustainable block), the utility grid and the DG solution in the form of a microgrid that constitutes the SHES. The meta-system being discussed here is all connected and operates with Net Billings, one of the recent advances in the law and regulation of the electric market in Chile. It is based on a future program being evaluated now by the utility company ENEL of Santiago, Chile, part of the ENEL Group, to implement this technology in buildings in Santiago in the future, with SHES being installed and operated by them. This joint venture initiative is circumscribed in a research project being carried out at the Universidad Tecnica Federico Santa Maria in Santiago with the support and participation of ENEL.

The HC system is part of the energy system (microgrid) which combines DG with the utility grid considering both options, with and without energy storage, in an apartment building and where all energy sources can contribute to make SHES more resilient, efficient and reliable while complementing more traditional control methods. PH involves a recursive set of mechanisms that trigger appropriate responses in the prospect of near future stimuli manifestation. In the case of sustainable energy systems, these constitute a set of corrective responses initiated in anticipation of a predictably environmental (both internal and external) challenge.

In general terms, predictive homeostasis is an anticipatory response to an expected homeostatic challenging event in the future. Seasonal migration of animals and birds in particular are examples of predictive homeostasis. Predictive responses often compromise the effectiveness of reactive homeostatic mechanisms, even to the point of risking the survival of the organism itself. Nevertheless, both predictive and reactive homeostasis (RH) must be in equilibrium and perfectly synchronized just as large rotating AC generators must operate in synchrony with the frequency of the grid. If this was not the case and both RH and PH were to operate uncoordinatedly, they may become in conflict. In such cases predictive responses may compromise the effectiveness of reactive HC mechanisms to the point of jeopardizing the sustainability of the energy system. There are many examples of such catastrophes in recent history and everything seems to indicate that things aren't getting any better and that more severe weather patterns are expected along with natural disasters. Chile is a good example of such calamities, showing just how much chaos and destruction they can bring to its population. The February 27th giant earthquake and tsunami are still vivid in the memory of Chileans and there is clear memory of how the major electric power and telecommunications networks collapsed as a result. There have also been more frequent occurrences of strong winds, large fires extending to large forest and urban areas, and volcano eruptions, all of which cause havoc in our energy and communication networks.

The HC-based energy management system must respond to changes in the environment both in the immediate as well as in the short to medium term span. It must do so in such a way that its responses conform to sustainable community-driven HC strategies [3,29–35,53–63], that are best suited for particular environments and energy consumption schemes. Such strategies involve distinct homeostasis mechanisms that are aimed at preserving the efficiency and thriftiness in the energy supply and consumption process of the SES so as to ensure its well-being under any circumstance. Every energy system can only be truly sustainable and functional as a SES if and only if its internal mechanisms operate efficiently and opportunely as it is expected to occur in any healthy living system. It does so employing all means at its disposal (internal system's design engineering as well as the energy consumption management, e.g., the loads) to attain optimal system's response to environmental conditions. This is true also in any living system that is made aware that there are enough resources available to it in the environment, so it can safely exercise restraint, thriftiness and energy efficient behavior without overdoing homeostatic regulation (HR) in a way that overstrains the system to a point where living conditions are compromised.

With the latter in mind, one can visualize that SES are moving toward the attainment of a new, safer, and more sustainable equilibrium point—one that is conditioned by the sensory communication and information processing capabilities that are engineered into the SES itself, including the loads and their energy consumption characteristics [3,29–36,45,46]. Such capabilities can be enhanced and complemented by the presence of an energy buffer operating in the grid-connected microgrid. Being a definitely environmentally circumscribed artificial system, the SES in the form of an intelligent microgrid can only continue to exist and thrive so long as it is in continuous equilibrium with the forces that are internal and external to it. For this efficient equilibrium to exist and maintain itself under different circumstances, homeostatic regulation and adaptive energy consumption control must work coordinatively and flawlessly within a highly dynamic, real time environment whatever its characteristics may be. Therefore, different options ought to be considered on how to engineer reactive and predictive homeostasis in SES design and later implementation, whether the microgrid is to operate tied to the grid or not. Such is the challenge to be analyzed for crating sustainable communities facing uncertain times. The characteristics of such communities' behavior in terms of their use of energy and the goals that such behavior entail should certainly be considered in the design of HC systems for SES like the microgrid, and assess how the responses of energy users can be influenced by expectations and anticipatory behavior therein.

5 Conclusions

Climate change and natural disasters are a very serious threat that has brought worldwide attention and policy changes, increasingly so in the last few years. It is a harsh reality which demands concrete actions now, not later. World leaders have understood this and have been meeting for 5 years now, during the World Climate Summit (WCS) to discuss the matter in an effort to explore the possible mitigation measures. In WCS 2015 [66] which took place in Paris, France was particularly relevant and important accords and measures emerged from this, which will hopefully bring about much needed changes in EPS infrastructure. Nature doesn't give man a second chance, either you are prepared or you aren't.

It is therefore imperative that we understand that traditional electric power infrastructure is not only vulnerable but also dangerous when it comes to natural disasters like strong winds, earthquakes, and floods, all of which are part of an ever more complex environmental scenario. These and other dangerous phenomena like sinkholes and rising sea tides are also affecting modern infrastructure and mankind's way of life just about everywhere. Thus, decentralizing power systems by means of DG solutions employing both renewables and traditional power sources like SHES makes sense. The concept of energy hubs [37,38] is also part of this new power infrastructure vision with a much more sensible and reliable architecture than traditional energy matrix which relies solely on long transmission and distribution lines with exposure to the elements.

Furthermore, if we can equip these systems with adequate communications capabilities, AI and data sensing devices operating in interconnected fashion, and make of this an industry standard in all of the systems comprising the power generation and distribution matrix, this would make an even better solution based on the Smart Grid technology available today. This can be done using today's smart grid technology with set-points data signals programmed in the control system itself. This would enable it to control the entire system in a hybrid, distributed fashion, wherein there is a central controller interacting with each sub-system's controller as well as with the electric utility control operator as well. That would allow for a more flexible, inexpensive, fast and more robust control system than the more expensive, complex systems such as multi-agent systems or expert systems that much of the current literature on smart grid and intelligent computing is embracing. At least for smart microgrids to supply electricity and potentially heat using CHP as a generation alternative [22] (for cases in which a natural disaster strikes leaving a whole neighborhood without power), simpler, more flexible, economical, and modular control solutions are possible [67]. This paper offers a glimpse of such new solutions, incorporating reactive and predictive homeostasis control mechanisms to attain superior energy management and power supply and consumption control. These are part of the new frontier of control and communications systems engineering that will make possible for utilities to embrace DG and incorporate it into their product-service system supply.

Electricity infrastructure is feeble and largely outdated when it comes to the electric power distribution sector in Chile, as it has not kept up with the times. Environmental challenges like natural disasters and severe weather patterns brought by phenomena like El Niño leave no room for preambles. In spite of electricity being the blood flow that powers all sectors of the country, including utilities, all of which is crucial to sustain any society's well-being and economic growth, little if anything has been done to counteract environmental and other threats (like terrorist acts, for example) that put power supply stability at risk. Therefore, a new model for SES is needed whose engineering design and HC capabilities incorporate reactive and predictive homeostasis. This new model may very well coexist and complement more traditional control and energy management methods like the droop control to better equip the system with better environmental challenge management and resilience capacity. In general terms such capacity must extend the EPS functioning way beyond the perils that such systems have had to endure in the past, for example as a result of fault line power outages and abrupt shift in weather conditions for which they were originally designed. Unless the country seriously takes these challenges and upgrades its power grid looking at more modern options like SES with vast solar photovoltaic and wind penetration in the energy generation matrix [68–71] to complement its infrastructure, the effects of these calamities will continue to besiege the population of vulnerable countries like Chile, Argentina, and others in the world, affecting thousands of users each time for even longer periods than what has occurred in the past. Developing and implementing such a change in the country's power grid infrastructure will be necessary not only for the reasons already stated but also for diversifying and modernizing the power generation and distribution model as well. The latter has been understood and is beginning to be embraced by local utilities like ENEL which are currently considering employing DG solution in this endeavor. The work presented here is part of this new emerging trend in Chile and is supported by the most important local electric power distribution company ENEL Distribucion S. A. in Santiago, Chile. In the work to come, we will show the simulation analysis of said project.

Sustainability and systems resilience can also be built through homeostasis control mechanisms engineered in EPS [3,30,31] that can promote and potentiate energy efficiency, thriftiness as exergy drivers, thus making the energy system more capable. The aim is for users to be able to respond to HC strategies designed in the system to further increase its resilience and adaptability capacity in the face of an environmental challenge like a natural disaster. Such systemic initiatives can complement well other measures that may be taken by local authorities and by the utility agents to better prepare the community for these occurrences. The benefits of the action scheme proposed here can be better realized and measured by the shared responsibility of the different parties involved, unlike the traditional scheme of the past that we have been accustomed to for so long.

Acknowledgments

The corresponding author wishes to thank FONDECYT of Chile for the Grant FONDECYT Postdoctoral Project No 3170399 which began in March, 2017, and ended officially on March 14, 2020. The author also wishes to thank ENEL Distribucion S.A. for its valuable and unrelenting support, patronage, and contribution to this scientific research project.

References

[1] W.B. Cannon, Organization for physiological homeostasis, Physiol. Rev. 9 (3) (1929) 399–431.
[2] W.B. Cannon, Stresses and strains of homeostasis, Am. J. Med. Sci. 189 (1) (1935) 13–14.
[2a] M. Gong, G. Wall, Exergy, Int. J. 1 (2001) 217–233.
[2b] A.M. Rosen, I. Dincer, Int. J. Energy Res. 27 (2003) 415–430.
[3] F.M. Cordova, F.F. Yanine, Homeostatic control of sustainable energy grid applied to natural disasters, Int. J. Comput. Commun. Control 8 (1) (2012) 50–60.
[4] https://earthsky.org/earth/powerful-earthquake-strikes-off-chiles-coast.
[5] http://www.theguardian.com/world/live/2015/sep/17/chile-earthquake-massive-83-magnitude-tremor-strikes-santiago-live-updates.
[6] http://www.britannica.com/event/Chile-earthquake-of-2010.
[7] http://www.renewableenergyworld.com/articles/2015/12/u-s-senate-expected-to-take-up-microgrid-policy-in-early-2016.html.
[8] https://www.congress.gov/bill/114th-congress/senate-bill/1243.
[9] http://www.energy.senate.gov/public/index.cfm/2015/5/sen-murkowski-introduces-17-targeted-bills-to-modernize-america-s-energy-policies.
[10] http://www.heritage.org/research/reports/2013/10/after-hurricane-sandy-time-to-learn-and-implement-the-lessons.
[11] E.S. Blake, T.B. Kimberlain, R.J. Berg, P.C. John, J.L. Beven II, Hurricane Sandy: October 22–29, 2012 (Tropical Cyclone Report). United States National Oceanic and Atmospheric Administration's National Weather Service.[Links], 2013.
[12] M. Diakakis, G. Deligiannakis, K. Katsetsiadou, E. Lekkas, Hurricane sandy mortality in the Caribbean and continental North America, Disaster Prev. Manag. 24 (1) (2015) 132–148.
[13] K.D. Sullivan, L.W. Uccellini, Service assessment: Hurricane/post-tropical cyclone Sandy, October 22–29, 2012. US Department of Commerce NOAA and NWS, Silver Spring, Maryland, 66 (2013).
[14] http://www.livescience.com/22522-hurricane-katrina-facts.html.
[15] http://www.datacenterresearch.org/data-resources/katrina/facts-for-impact/.
[16] http://microgridknowledge.com/think-microgrid-government-leaders-serious-quest-build-microgrids/.
[17] http://kresge.org/sites/default/files/Resilient-Power-report-2014.pdf.
[18] L. Che, M. Khodayar, M. Shahidehpour, Only connect: Microgrids for distribution system restoration, IEEE Power Energy M. 12 (1) (2014) 70–81.
[19] D.E. Olivares, A. Mehrizi-Sani, A.H. Etemadi, C.A. Canizares, R. Iravani, M. Kazerani, et al. Trends in microgrid control, IEEE Trans. Smart Grid 5 (4) (2014) 1905–1919.
[20] C. Chen, J. Wang, F. Qiu, D. Zhao, Resilient distribution system by microgrids formation after natural disasters, IEEE Trans. Smart Grid 7 (2) (2016) 958–966.

[21] M. Roach, M.H. CEO, Hurricane Sandy & the Emperor's New Clothes: Microgrids as a Risk Mitigation Strategy for Extreme Weather Events (2012).

[22] B. Hedman, Combined Heat and Power and Heat Recovery as Energy Efficiency Options. Presentation on behalf of US CHP Association and ICF Consulting, Washington, DC (2007).

[23] T. Olinsky-Paul, CESA Energy Storage Technology Advancement Partnership (No. SAND2017-10862C), Sandia National Lab.(SNL-NM), Albuquerque, NM, United States, (2017).

[24] Resilience for free, http://www.cleanegroup.org/ceg-resources/resource/resilience-for-free-how-solar-storage-could-protect-multifamily-affordable-housing-from-power-outages-at-little-or-no-net-cost/, 2015.

[25] A. Haines, K.R. Smith, D. Anderson, P.R. Epstein, A.J. McMichael, I. Roberts, et al. Policies for accelerating access to clean energy, improving health, advancing development, and mitigating climate change, Lancet 370 (9594) (2007) 1264–1281.

[26] K.B. Jones, S.J. Bartell, D. Nugent, J. Hart, A. Shrestha, The urban microgrid: Smart legal and regulatory policies to support electric grid resiliency and climate mitigation, Fordham Urb. Law J. 41 (2013) 1695.

[27] A. Khodaei, Resiliency-oriented microgrid optimal scheduling, IEEE Trans. Smart Grid 5 (4) (2014) 1584–1591.

[28] B. Warshay, Upgrading the Grid, Foreign Affairs 94 (2) (2015) 125–131.

[29] F.F. Yanine, E.E. Sauma, Review of grid-tie micro-generation systems without energy storage: Towards a new approach to sustainable hybrid energy systems linked to energy efficiency, Renew. Sustain. Energy Rev. 26 (2013) 60–95.

[30] F.F. Yanine, F.I. Caballero, E.E. Sauma, F.M. Córdova, Homeostatic control, smart metering and efficient energy supply and consumption criteria: A means to building more sustainable hybrid micro-generation systems, Renew. Sustain. Energy Rev. 38 (2014) 235–258.

[31] F.F. Yanine, F.I. Caballero, E.E. Sauma, F.M. Córdova, Building sustainable energy systems: Homeostatic control of grid-connected microgrids, as a means to reconcile power supply and energy demand response management, Renew. Sustain. Energy Rev. 40 (2014) 1168–1191.

[32] F.F. Yanine, E.E. Sauma, F.M. Cordova, An exergy and homeostatic control approach to sustainable grid-connected microgrids without energy storage, Applied Mechanics and Materials, 472, Elsevier, 2014, pp. 1027–1031.

[33] F. Caballero, E. Sauma, F. Yanine, Business optimal design of a grid-connected hybrid PV (photovoltaic)-wind energy system without energy storage for an Easter Island's block, Energy 61 (2013) 248–261.

[34] F.F. Yanine, F.M. Córdova, L. Valenzuela, Sustainable hybrid energy systems: An energy and exergy management approach with homeostatic control of microgrids, Procedia Comput. Sci. 55 (2015) 642–649.

[35] F. Yanine, F.M. Córdova, Homeostatic control in grid-connected micro-generation power systems: A means to adapt to changing scenarios while preserving energy sustainability, in: Renewable and Sustainable Energy Conference (IRSEC), 2013 International, IEEE, 2013, pp. 525–530.

[36] W. El-Khattam, M.M.A. Salama, Distributed generation technologies, definitions and benefits, Electr. Power Syst. Res. 71 (2) (2004) 119–128.

[37] M. Geidl, G. Koeppel, P. Favre-Perrod, B. Klockl, G. Andersson, K. Frohlich, Energy hubs for the future, IEEE Power Energy M. 5 (1) (2007) 24.

[38] M.C. Bozchalui, S.A. Hashmi, H. Hassen, C.A. Cañizares, K. Bhattacharya, Optimal operation of residential energy hubs in smart grids, IEEE Trans. Smart Grid 3 (4) (2012) 1755–1766.

[39] M. Amin, North America's electricity infrastructure: Are we ready for more perfect storms?, IEEE Security Privacy (5) (2003) 19–25.
[40] M. Amin, Toward self-healing energy infrastructure systems, IEEE Comput. Appl. Power 14 (1) (2001) 20–28.
[41] S.M. Amin, B.F. Wollenberg, Toward a smart grid: power delivery for the 21st century, IEEE Power Energy M. 3 (5) (2005) 34–41.
[42] M. Amin, P.F. Schewe, Preventing blackouts, Sci. Am. 296 (5) (2007) 60–67.
[43] SCHOTT White Paper on Solar Thermal Power Plant Technology, http://www.schott.com/newsfiles/com/20070531190119_SCHOTT_US_White_Paper_07.02.06.pdf, 2006.
[44] A.A. Sallam, O.P. Malik, Electric Distribution Systems, Vol. 68, John Wiley & Sons, 2011.
[45] G. Walker, S. Hunter, P. Devine-Wright, B. Evans, H. Fay, Harnessing community energies: explaining and evaluating community-based localism in renewable energy policy in the UK, Global Environ. Polit. 7 (2) (2007) 64–82.
[46] J.J. Ding, J.S. Buckeridge, Design considerations for a sustainable hybrid energy system, Trans. Institut. Prof. Eng. NZ 27 (1) (2000) 1.
[47] https://energypedia.info/wiki/Net-Metering_/_Billing_in_Chile.
[48] A. Afzal, M. Mohibullah, V. Kumar Sharma, Optimal hybrid renewable energy systems for energy security: a comparative study, Int. J. Sustain. Energy 29 (1) (2010) 48–58.
[49] N. Dawar, T. Frost, Competing with giants: Survival strategies for local companies in emerging markets, Harvard Bus. Rev. 77 (1999) 119–132.
[50] Power transactions and trends Global power and utilities mergers and acquisitions review (Q2 2014). http://www.ey.com/Publication/vwLUAssets/EY-power-transactions-and-trends-Q2-2014/$FILE/EY-power-transactions-and-trends-Q2-2014.pdf.
[51] G. Fettweis, E. Zimmermann. ICT energy consumption-trends and challenges, in: Proceedings of the 11th International Symposium on Wireless Personal Multimedia Communications, Vol. 2, No. 4, p. 6, 2008.
[52] O. Zinaman, M. Miller, A. Adil, D. Arent, J. Cochran, R. Vora, et al., Power Systems of the Future: A 21st Century Power Partnership Thought Leadership Report (No. NREL/TP-6A20-62611). National Renewable Energy Laboratory (NREL), Golden, CO, 2015.
[53] MIT Energy Laboratory; MIT Homeostatic Control Study Group, New electric utility management and control systems: proceedings of conference, held in Boxborough, Massachusetts, May 30-June 1, MIT Energy Laboratory, 1979.
[54] F.C. Schweppe, R.D. Tabors, J.L. Kirtley Jr., S.R. Law, P.F. Levy, H. Outhred, et al., Homeostatic Control of Power Systems, Fourth Energy Monitoring and Control System Conference, Norfolk, VA November, 1979.
[55] F.C. Schweppe, R.D. Tabors, J.L. Kirtley Jr., H.R. Outhred, F.H. Pickel, A.J. Cox, Homeostatic utility control, IEEE Trans. Power Apparatus Syst. (3) (1980) 1151–1163.
[56] F.C. Schweppe, R.D. Tabors, J.L. Kirtley, Homeostatic control: the utility/customer marketplace for electric power. Massachusetts Institute of Technology, Energy Laboratory, 1981.
[57] T.L. Sterling, R.D. Williams, J.L. Kirtley Jr., Control and Monitoring System Communications for Effective Energy Use, IEEE Power Engineering Society 1981; Summer Meeting, Portland, OR, July 1981, paper no. 81 SM 307-8.
[58] F.C. Schweppe, R.D. Tabors, J.L. Kirtley, Power/energy: Homeostatic control for electric power usage: A new scheme for putting the customer in the control loop would exploit microprocessors to deliver energy more efficiently, IEEE Spect. 19 (7) (1982) 44–48.

[59] F.C. Schweppe, R.D. Tabors, J.L. Kirtley Jr., Homeostatic control for electric power usage, IEEE Spect. (July, 1982) 44–48.
[60] R.D. Tabors, F.C. Schweppe, J.L. Kirtley Jr, Homeostatic Control: The Utility/Customer Marketplace for Electric Power, Proc Local Heat and Power Generation: A New Opportunity for British Industry, Inderscience Enterprises, St. Helier, Jersey, UK, 1983, pp. 66–88.
[61] C. Zhao, U. Topcu, N. Li, S. Low, Design and stability of load-side primary frequency control in power systems, IEEE Trans. Automatic Control 59 (5) (2014) 1177–1189.
[62] M. Beckerman, Homeostatic control and the smart grid: Applying lessons from biology, Optimization and Security Challenges in Smart Power Grids, Springer Berlin Heidelberg, 2013, pp. 39–52.
[63] S.D. Ramchurn, P. Vytelingum, A. Rogers, N.R. Jennings, Agent-based homeostatic control for green energy in the smart grid, ACM Trans. Intell. Syst. Technol. 2 (4) (2011) 35.
[64] A.M. Bollman, An experimental study of frequency droop control in a low-inertia microgrid, Doctoral dissertation, University of Illinois at Urbana-Champaign, 2009.
[65] R.H. Coase, The problem of social cost, Classic Papers in Natural Resource EconomicsPalgrave Macmillan UK, 1960, pp. 87–137.
[66] http://www.wclimate.com/world-climate-summit-2015/.
[67] A. Hartmanns, H. Hermanns, Modelling and decentralised runtime control of self-stabilising power micro grids, Leveraging Applications of Formal Methods, Verification and Validation. Technologies for Mastering Change, Springer Berlin Heidelberg, 2012, pp. 420–439.
[68] G.A. Montoya, C. Velásquez-Villada, Y. Donoso, Energy optimization in mobile wireless sensor networks with mobile targets achieving efficient coverage for critical applications, Int. J. Comput. Commun. Control 8 (2) (2013) 247–254.
[69] M. Ilic, J.W. Black, M. Prica, Distributed electric power systems of the future: Institutional and technological drivers for near-optimal performance, Electr. Power Syst. Res. 77 (9) (2007) 1160–1177.
[70] J.P. Lopes, N. Hatziargyriou, J. Mutale, P. Djapic, N. Jenkins, Integrating distributed generation into electric power systems: A review of drivers, challenges and opportunities, Electr. Power Syst. Res. 77 (9) (2007) 1189–1203.
[71] A. Rearte-Jorquera, A. Sánchez-Squella, H. Pulgar-Painemal, A. Barrueto-Guzmán, Impact of residential photovoltaic generation in smart grid operation: real example, Procedia Comput. Sci. 55 (2015) 1390–1399.

7

Smart energy systems and the need to incorporate homeostatically controlled microgrids to the electric power distribution industry: an electric utilities' perspective

Fernando Yanine[a], Antonio Sanchez-Squella[b], Aldo Barrueto[b], Sarat Kumar Sahoo[c], Felisa Cordova[a]

[a]Faculty of Engineering, Universidad Finis Terrae, Santiago, Chile; [b]Department of Electrical Engineering, Universidad Tecnica Federico Santa María (UTFSM), Santiago, Chile; [c]A Constituent College of Biju Patnaik Technological University, Parala Maharaja Engineering College, Department of Electrical Engineering, Govt. of Odisha, Berhampur, Odisha, India

Chapter outline

Low Carbon Energy Technologies in Sustainable Energy Systems. http://dx.doi.org/10.1016/B978-0-12-822897-5.00007-9

1 Smart energy systems, energy sustainability, and grid flexibility

The concept of smart energy systems, energy sustainability, and grid flexibility in the smart grid agenda of electric utilities like ENEL Distribucion in Chile

1.1 Toward a new electric utilities' perspective

For no one is a secret nowadays that electric power generation and distribution systems are being faced with a number of challenges and concerns, which emanate not so much from a shortage of energy supply but from environmental, infrastructural, operational, and legal issues. They are required by law—as all public utilities are—to respond to such challenges and threats as, for example, transmission and distribution lines failure, transformers break-down and power lines failure due to extreme or harsh weather conditions such as heavy rains, hail and snow fall, violent and unanticipated winds or sudden natural disasters like earthquakes floods and landslides, as we have seen in Chile, North America, and in many other places recent years. Torrential rains, fires, and earthquakes are to be dealt with very rapidly and effectively so as to preserve stability and continuity of operations at any time no matter what, regardless of what may occur in the surroundings. Public utilities, especially electric utilities like ENEL, know this quite well, and they are doing something about it. This in fact is the true measure of what sustainable hybrid energy systems (SHES) tied to the grid are all about, and homeostaticity in energy systems seeks just that: to enable distributed energy systems (DES) tied to the grid and operated by electric utilities like ENEL, to operate rapidly and effectively in order to restore stability and continuity of operations. Energy homeostaticity is, in and of itself, an engineering concept that points to the system's capacity to restore efficient, stable equilibrium between supply and demand. The energy system is to be proactive thanks to its predictive homeostasis function, and able to react and respond effectively and rapidly to stimuli under any circumstance, based on the built-in intelligence, and installed capacity engineered in the energy system which allows the system to manage its generation and supply requirements. They do this by attaining a state of optimally efficient equilibrium (homeostaticity) between energy supply and energy expenditure in electric power systems (EPS) operation. To accomplish they ought to imitate homeostasis mechanisms present in all living organisms. Ever since Cannon (1929, 1935) [1,2] first introduced the concept, attention on homeostasis and its applications have been the sole patrimony of medicine and biology to find cures for diseases like diabetes and obesity. Nevertheless, homeostasis is rather an engineering concept in and of itself—even more so than in the natural sciences—and its application in the design and engineering of SHES is a reality with the outlook of incorporating DES to the utility grid. Thus homeostasis mechanisms are present in all living organisms, and as such are also applicable to EPS engineering in order to enable and maintain a sustainable performance when they are linked to energy efficiency (EE) and thriftiness. In doing so, both reactive and predictive homeostasis play a substantive role in the engineering of such mechanisms [3,4]. Reactive homeostasis (RH) is an immediate response of the SES to a homeostatic challenge such as energy deprivation, energy shortage or imbalance. RH, on the other hand, entails feedback mechanisms that allow for reactive compensation, reestablishing homeostasis or efficient equilibrium in the system. Predictive homeostasis (PH), on the other hand, is a proactive mechanism that anticipates events that are likely to occur, sending the right signals to the central controller, enabling SES to respond early and proactively to environmental challenges and systems' concerns [3,4].

Therefore, based on the above arguments, it is reasonable to expect that government authorities as well as legislators and industry pundits do something about it with the degree of responsibility shown elsewhere in the world, as for example in North America [5–7]. It is time that those responsible for these issues affecting electricity transmission and distribution, like the Superintendencia de Electricidad y Combustible, SEC in Chile or the Agencia Nacional de Energía Eléctrica in Brazil take action, instead of just worrying about the problem as we wait for the next outage; in regards to having critical outposts such as hospitals, airports, emergency healthcare units, residential areas and transport systems being equipped with a fully functional, powerful and ready-to-go microgrid that can be incorporated in the electricity distribution grid [8–10]. We know that we need a new solution to an old problem: the vulnerability and feebleness of our current EPS infrastructure, especially in the distribution sector. That solution comes in the form of a smart distributed generation (SDG) solution like the microgrid. One that is connected to the grid, and that includes both solar photovoltaic and thermal energy sources, as well as wind generation where available and possibly energy storage for critical processes, if necessary [3]. This could greatly minimize the impact on such critical services and other equally important ones as well, in case urban areas like Santiago de Chile were to suffer another power outage as in the past [11,12]. Unfortunately, however, at present there are no incentives in place in Latin America to encourage electric utilities like ENEL to pursue microgrids. Moreover, no guarantee exists that industry regulators like the SEC in Chile, will allow the utility to recover the capital costs of the microgrid through rates as being discussed in North America [13,14]. In Chile, in particular, the matter is not even being discussed in terms of cost recovery options for investors, arguing that customers are already paying enough in their rate structure to cover for energy security, and therefore the utility should provide such security in spite of the high costs involved. A dismal outlook if one is to review the huge losses and hefty fines that they have had to face in recent years due to environmental and natural disasters. Of course, one would expect that specific critical customers in the private sector such as private hospitals and clinics and private airports should pay an additional rate fee for getting such energy security, while public service customers like the electric underground mass transport system (subway) should be subsidized by the government, and in part also absorbed by the utility. Today, thanks to Net Billing Law 20,571 for Distributed Generation of electricity in Chile, there is an alternative [15]. The law grants the right to customers of electric power utilities like ENEL Distribution, to generate their own electrical energy, self-consume and inject their surplus into the network. Thanks to this and to lowering costs of solar and wind energy generation technologies, the installation of renewable energy sources by independent clients or residential communities connected to the power distribution grid is growing, together with the introduction of some Smart Grid initiatives, as for example: ENEL's smart metering unit offering several services [16,17]. However, as it is usually the case, legislation does not move at the same pace as the market sector, and much less so as than the rate at which society's needs grow. Therefore, this new scenario is presenting new and complex challenges to utility operators like ENEL, which see private enterprises encroaching into an until very recently protected market turf. Such challenges are two-fold. They come mainly from a regulatory but also from a competition viewpoint. Hence, in this article we present a hypothetical case being considered by ENEL of a community belonging to an average residential building, as the subject of study in upper-side neighborhood of Santiago, Chile. The community is considering the installation of a photovoltaic (PV) energy generation microgrid, with and without energy storage unit to supply electricity to 60 apartments of various sizes and consumption (referred to as a "sustainable block"). Under this scenario, a set of strategies for the coordination and supervisory control through energy homeostasis is considered [18–22], adapted for specific needs and consumption characteristics of the customers as well as of the power infrastructure of the distribution network. These are to be applied with the objective of efficiently managing the supply and consumption of energy and power in the residential community maintaining systems homeostaticity. The proposed supervisory control is designed based on homeostatic control of EPS and simulations results can be seen in [9,10,19,20] under different scenarios and with various operational options.

The present work addresses these and other important concepts, particularly homeostasis of energy systems as well as the design and engineering of what we have termed **homeostaticity** in SHES like grid-tied microgrids which is a current concept being considered by utilities like ENEL to complement and diversify their electric power distribution resources and services. In doing so, we employ engineering concepts and empirical arguments from previous work, which are further elaborated herein, plus new material that is being presented as well, in order to advance the research in the field of homeostatic control applied to electric power systems. Hence, section one serves as introduction where the main issues and arguments are laid out, emphasizing the need for SES incorporated to the electricity distribution grid in light of climate change and the need for energy resilience. This section also underlines the hurdles of electric power systems (EPS) decentralization and the roadblocks for adopting SHES like the microgrid concept. Section 2 elaborates on the current shift in microgrid trends from an alternative energy generation solution to a more active power industry player. A reality that although timidly moving forward is no doubt a much needed shift in the current energy matrix of countries like Chile and others especially hit by environmental and natural disasters. It also points to the need to incorporate homeostasis-based control systems in the design of SHES tied to the grid for ensuring optimal power and energy management control of such systems. Section 3 shows brief experimental results. Section 4 offers a discussion. Conclusions come afterwards.

1.2 Homeostaticity of energy systems

How homeostaticity of energy systems works and why electric utilities need it

Today's electrical energy generation and distribution systems are being faced with a number of challenges and concerns which emanate from both environmental as well as operational issues. Therefore, they are required to respond to such challenges very rapidly and effectively so as to preserve stability and continuity of operations. This is the true measure of what sustainable energy systems (SES) are all about and homeostaticity of energy systems seeks just that: to bring about a rapid, highly effective and efficient state of equilibrium between energy supply and energy expenditure in electric power systems (EPS) [3,4]. To accomplish so they ought to emulate homeostasis mechanisms that are present in all living organisms. Homeostasis and its applications have been the sole patrimony of medicine and biology for several decades in order to find cures for diseases like diabetes and obesity. Nevertheless, homeostasis is rather an engineering concept in its very essence—even more so than in the natural sciences—and its application in designing and engineering SHES is what we term homeostaticity. In this chapter we present the groundwork behind the theory and offer a prescriptive model for the operation of SHES which is supported by the theoretical and empirical results. The work presented explains how the engineering of homeostaticity in SES is done and how reactive and predictive homeostasis plays a key role in this system dynamics. Reactive homeostasis (RH) is an immediate response of the system to a homeostatic challenge such as energy deprivation, shortage or an energy imbalance. RH entails a feedback mechanism that allows for reactive compensation, reestablishing homeostasis or efficient equilibrium in the system. Predictive homeostasis (PH), on the other hand, anticipates the events that are likely to occur, enabling SES to respond early and proactively to environmental challenges and concerns by foreseeing when these are most likely to occur, adjusting their energy management to maintain sustainability. Environmental challenges like natural disasters and hazardous climatic events are becoming more severe and recurrent in many parts of the world, and they are here to stay, affecting millions. Nowhere is the matter being taken more seriously than in the United States where the US Senate passed a bill that supports grid-connected hybrid microgrids to tackle, among others, the issue of electric power grid readiness and resilience [5,6]. Such initiative is one of several pieces of legislation that are being studied by industry stakeholders, legislators, and also local and federal authorities in North America, to promote a range of technologies and policies that can make the grid more reliable and cyber-secure in the US and Canada. All these steps fall, in one way or another, on the path set forth by President Obama in 2013, when he introduced the Energy Independence Roadmap for the country [23]. However, the pace towards more concrete and expedient changes has somehow sped up even more, following the catastrophic natural disasters that the country has had to endure in recent years [24–27]. Among such initiatives are grants, microgrid technologies' demonstration projects and a variety of studies to determine the costs involved and to define the precise scope of action in order to shape federal microgrid policy in the immediate future [28–34]. The Energy Policy Modernization Act of 2015 [35,36] was introduced in the US Senate on September 2015. This bill amends the Energy Conservation and Production Act, the Energy Policy and Conservation Act (EPCA), and the Energy Independence and Security Act of 2007 with respect to energy efficiency in buildings and appliances. The EPCA is amended regarding the Strategic Petroleum Reserve as well. All of the above has heightened the federal role of supporting microgrids integration to the current EPS' infrastructure, something which has been largely a state endeavor to date [36–38]. In North America, for example, large-scale power outages spanning extended urban areas are not new. Still fresh in people's memory is Hurricane Sandy, known as "Superstorm Sandy." This natural event was the deadliest and most destructive hurricane of the 2012 Atlantic hurricane season, and the second-costliest hurricane in United States history [24–26], with damages estimated to be over $75 billion (2012 USD), a total surpassed only by Hurricane Katrina [27]. This monstrous calamity caused unprecedented infrastructure damages including major power outages. Yet there were other power disruptions as predecessors of Sandy, among them the Northeast Blackout of 2003, Hurricanes Katrina and Rita in 2005 and Hurricane Irene and the Northeast's freak Halloween snowstorm in 2011[4,39]. After each of these events, more consensuses were built among public opinion and industry leaders. Both local authorities and power industry experts were in accord that something new had to be done fast to strengthen the power distribution grid against such recurrent catastrophes. The problem lies in where to start and how much to spend, this in light of other equally pressing needs that demand attention and resources, aside from the regulatory issue which has proven quite stiff [4]. Nevertheless, on September 2014 the Clean Energy Group presented its report: "Resilient Power: Evolution of a New Clean Energy Strategy to Meet Severe Weather Threat" [40], which marked a turning point on this issue and provided a roadmap for states like New Jersey, New York and Connecticut to begin a new era of development in the fight against severe environmental challenges. Nevertheless, much work needs yet to be done going forward in North America and also in South America in order to understand the dimensions of the challenges we are facing and how the influence of extreme weather, natural disasters and all of what climate change has in store for us will have an impact on the resilience of power systems. We need to seriously study what are the possible mitigation strategies at hand and what are the legislative changes needed to enable such industry change therein [3,4,41,42]. Hence in this context, the chapter presents a prescriptive energy management and homeostatic control model for incorporating microgrids in residential and commercial buildings serviced by ENEL Distribucion, part of ENEL, the largest electric utility in Chile. The work presented is part of an ongoing research program funded by CONICYT of Chile under the auspice of ENEL.

1.3 Climate change and the current energy transition

Building the case for homeostaticity of sustainable hybrid energy systems (SHES)

Like the United States, Chile is no stranger to these scenarios either, and has had its share of disasters too. The country is "sitting on a hot stove" so to speak, with earthquakes, volcano eruptions, and rain floods becoming increasingly present in the collective consciousness of its people. Such events are simply not uncommon but are becoming prevalent not only in Chile but in many parts of the world, with climate change and harsher weather on the rise. The difference is that in today's 21st century world, much of the fragile living systems and economic sustainability depend on modern utilities' infrastructure of which roads, water, electric power transmission, and distribution networks, and telecommunications are a vital part, yet increasingly vulnerable when faced with such phenomena [3,4,8,10]. On September 17, 2015 a powerful 8.3-magnitude earthquake struck off Chile's coast causing havoc and chaos in an otherwise tranquil Wednesday afternoon [11,12,43]. Unlike its predecessor of 2010, the natural disaster triggered an immediate tsunami alert and coastal evacuations were readily executed yet utility infrastructure was compromised, particularly electricity. The tremendous earthquake that struck Chile in 2010 was much worse and found the country largely unprepared. It occurred on February 27, 2010 at 3:34 a.m., off the coast of south-central Chile, taking everyone by surprise. The 8.8-magnitude earthquake had its epicenter some 200 miles (325 km) southwest of the country's capital, Santiago, causing widespread damage on land and initiating a tsunami that devastated some coastal areas of the country. Together, the earthquake and tsunami were responsible for more than 500 deaths and caused major damage to infrastructure [12]. Yet, despite these and other natural disasters, the country remains largely unprepared against massive telecomm and electric power systems brake-down [4]. The problem lies in the high percentage of centralized electric power and communication systems—a model than once proved efficient and secure but which no longer holds—as well as the lack of adequate technologies and back-up/emergency power systems for disaster recovery, something that even extends to the armed forces today. There is no energy sustainability roadmap for the country whatsoever and environmental policy is also weak and short sighted [4]. Nowhere is this more evident than when one analyzes the flaws that are built into the very fabric of our presently centralized power systems, which were on full display in the aftermath of the February 27, 2010 earthquake in Chile [12]. It is never more evident that hugely centralized power generation and distribution systems are extremely vulnerable and ineffective to disruptions from natural disasters, human error or other calamities than in a situation like this. The large power and telecommunication networks that once proved very efficient and secure, are now at the center of discussion fueling the need for decentralization and the rapid growth of distributed generation (DG). Hence it makes sense to follow other nations example seeking more decentralized, diversified and distributed generation (DG)-oriented energy matrix, a solution that is notoriously much better suited to withstand these disasters [44–51].

1.4 Electric power systems' decentralization

The hurdles and roadblocks of electric power systems' decentralization agenda: is it time for electric utilities to adopt SHES?

The saying is clear: a chain is only as strong as its weakest link, where the weakest link, figuratively speaking, applies to a system's characteristic or technical feature that makes it quite vulnerable in terms of its design, rather than the link of an actual chain [3,4]. Due to its geography, utility infrastructure design and operational conditions, Chile is a country that is quite susceptible to be struck by natural disasters including landslides, floods and earthquakes which can seriously impair utilities' infrastructure such as electricity, water and gas, let along roads and transportation [3,4,8]. Such events can cause major damage, producing havoc and mayhem all around, compromising the operation of key infrastructure like the power grid. Therefore, there is a clear need—as it has been already understood and acted upon in North America—to develop better, more resilient and robust approaches to enable today's EPS infrastructure to successfully withstand and overcome such adverse conditions [28–36].

The weakest link in the case of Chile's electric power distribution system is its inability to adequately sort out these events, as it was designed for normal conditions, without the level of stress and severity being imposed on the system by such scenarios [3,4]. For this very reason, DG solutions ought to be designed around the idea of flexibility, resourcefulness and energy independence, all common features of distributed control systems. These solutions may take several forms, sometimes with autonomous control coexisting with other forms of control like the traditional centralized control, but they all point to the same goal [18,22]. This way, if a sudden power failure were to occur, like a distribution line being brought down or a power transformer being lost as a result of large violent lightning storm or wind gusts, the result would be widespread shutdown [3,4,8,10]. A utility service supply disruption would impact an entire region, with long periods of limited or no electricity or water for the population until the damaged is repaired and service is brought back up again.

While the nuclear energy issue is still a double edged knife, with its pros and cons in today's electricity generation market, it is off track when it comes to energy matrix modernization and rather antagonistic when it comes to DG and renewables-based sustainable energy world trend. Nuclear energy plants can have disastrous implications to humanity should an accident or negligent act occur again (like the disasters as a result of the Fukushima nuclear power plant accident in Japan or the Chernobyl nuclear power plant meltdown in the old Soviet Union). Although still relevant to energy sustainability and security, the nuclear power issue and its future standing in today's world energy matrix is a case of profound implications on its own right, and would therefore require an entire chapter to discuss it. Yet if we are to focus too much on power generation technologies like nuclear, fossil fuels or hydroelectricity generation, we may be missing the larger picture or at least not giving it its proper place in the scale of concern it deserves [4].

Although economically efficient, traditional centralized EPS (including nuclear energy) are not only vulnerable in regards to natural disasters and other environmental challenges that may threaten our energy supply, presenting very little flexibility and no diversification of energy sources [3,4]. There is also the vulnerability of EPS infrastructure and how its collateral damage in case of collapse can manifest itself in a variety of forms, like the still huge concentration on fossil, non-renewable fuels, the need for safe and steady fuel provisioning, and large hydroelectric projects which require building large dams, inundating vast extensions of fertile land [10].

Centralized electric power generation and distribution systems as well as large telecommunication networks have, on the one hand, large economies of scale and are very efficient, especially when it comes to serving large interconnected metropolitan areas, as in North America for example, but that comes at a cost. Their major drawback and weakness however are quite evident and alarming when large power black-outs occur (Chile has had several in the last few years) which leave large populated areas in complete and utter darkness, sometimes for several hours, causing widespread chaos, mayhem and rampant looting all around [4,10]. Their sheer size and highly centralized architecture makes them extremely vulnerable to natural disasters and major accidents due in large part to human error. In this way all the economic gains as a result of high efficiencies, power quality, and stability achieved by creating these huge electric power grids are all of a sudden lost when a disaster like Hurricane Sandy or a major earthquake strikes. Centralized EPS are concentrated usually on a few, very large power plants, operating on thermal and hydroelectricity generation for the most part, and distributing power in a radial-type distribution scheme, with each substation supplying electric power to radially connected nearby communities [52,53]. They provide service across a wide range of consumers over vast distances that span hundreds and even thousands of miles, all of which increases the risk of disruption dramatically (generation and distribution power topologies) [4].

Hence, the sheer forces involved in just about any natural disaster (whether it is a storm bringing strong winds and snow, flood waters, violent quakes or volcanic eruptions) are no match when it comes to our presently centralized power systems, especially in the case of the two most vulnerable parts of any power system: transmission and distribution. As an example, just one afternoon of strong winds, although rare in the Santiago metropolitan area, can knock dozens of trees and blow away roof tops, disrupting electricity distribution to several areas at once, with fallen trees over power lines, damaged transformers, and other similar havoc that can deprive whole metropolitan areas of power for several hours [3,4]. All of these are strong arguments in favor of decentralization of power systems and the need for more rapid advancements in DG penetration in the form of SES [54–59]. Hopefully, adequate legislation initiatives will be more forthcoming in the years to come, bringing changes that can make possible the transition to a more secure, robust, resilient, and better prepared EPS come to fruition [4] (Fig. 7.1).

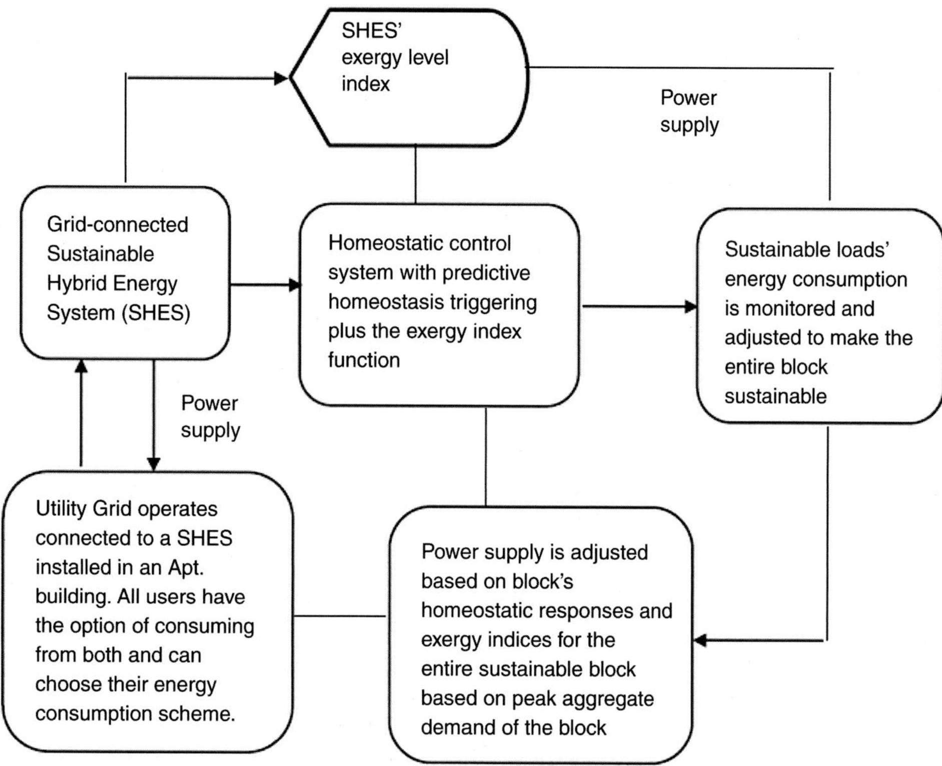

Figure 7.1. A particular example of the homeostatic control (HC) model for SHES. Here both energy and exergy management are built in the SHES to enable resilience and sustainability [10,18,19,21,22]. Source: Own elaboration.

2 Electric power distribution's decentralization agenda

The electric utilities' perspective regarding electric power distribution's decentralization agenda and the much needed role of the state and local government in enabling and supporting appropriate legislature, technology innovation, green energy integration, and the concept of energy hubs

2.1 Microgrid trends

The shift in microgrid trends from an alternative energy generation solution to a more active power industry player

While DG solutions like the microgrid or diesel generator sets were first thought as an alternative solution for remote and isolated areas only, as back-up in case of a power outage, or because grid power was unreliable or simply non-existent, like in many parts of Canada and other places in the world where no transmission lines exist, that is no longer the case [4,8–10]. Yet in many places in Latin America and elsewhere in the world there are still a number of rural and semi-rural areas with weak or no connection to the main grid. Thus, microgrids are becoming increasingly more relevant and viable as a solution to both urban and rural areas of all sizes and configurations and are now being considered by electric utilities, just like in North America. Microgrids are first and foremost local DG solutions that comprise a number of feeders (one or more), servicing a clusters of loads not necessarily grouped together. Some of these loads are more sensitive while others are considered less or non-sensitive, therefore the microgrid system's design and configuration, including the choice of control system, is in part dependent on the role of the microgrid itself, the energy sources employed, and the nature and size of the loads to be serviced [4,10]. In order to operate autonomously or semi-autonomously so as to reliably supply electricity and heat to a local community whether urban or rural for example, the microgrid may or may not have energy storage systems, depending on budget and system's operation necessities, and may operate connected to the grid if need be and conditions allow so or operate as a stand-alone system [8–10,18–22]. The degree of resiliency and robustness of the microgrid itself is given by the choice of engineering design and configuration characteristics, both of which depend on technical and budget constraints. Such aspects will influence how much the microgrid, and any other form of DG for that matter, will be prepared to act as a potential solution to endure adverse environmental conditions, while continuing to provide service following a natural disaster or harsh climatic event. Also the question of how microgrids can be designed in a way that can be incrementally integrated to the current EPS will depend on adequate legislation like the net billing law passed in October 2014 in Chile [15] as well as finding the right business and operational architectures that can accommodate these new market solutions. Thus, in light of the current necessities and changing industry trends, alliances involving communities, power utilities and independent local operators are likely to emerge [48–53], just like it has occurred in North America, with notable examples in the United States and Canada as well as in the United Kingdom, and other parts of Europe [3,4].

The interesting Smart Grid concept of energy hubs [54,56] built around the current EPS infrastructure is an idea that is being explored in Chile at present, where such localized energy hubs could be built around residential and commercial clusters incrementally, provided that adequate legislation and industry incentives are forthcoming in order to transition to a new EPS reality. This is needed in order to break away with the old unfit paradigm of today's power long distance transmission and distribution networks [10,55,57]. SHES not only will keep the lights on and basic services running for the residential and critical facilities' loads they serve, but can also act as a power source that can aid the grid in times of trouble. Microgrids constitute a fundamental resource as well as an electric power industry's paradigm change, as demonstrated in New York city, for example, during Hurricane Sandy, and it is at the forefront of the new energy independence (from the grid-only scenario) trend toward localized energy production and provisioning in the United States, Canada and several places in Europe as well. In fact, North America, along with several European countries and Australia, are now looking at microgrids not just as a means toward more energy independence and more resilient and robust EPS, but also as an industry game changer in the face of very pressing energy needs like we see in Chile for the years ahead. Important technological advances in power electronics and renewable energies, the falling prices of certain energy generation technologies like solar photovoltaic and some wind turbines are significant drivers. These, along with a rise in very harsh weather patterns and natural disasters like earthquakes and landslides, have accelerated this change of thinking on the part of industry agents, local authorities, and legislators alike [3,4].

2.2 Homeostaticity in electric utility-operated microgrids

How to incorporate homeostaticity in electric utility-operated microgrids: making the case for engineering resiliency and grid flexibility in the electric power distribution infrastructure

Since Cannon first formulated the concept of homeostasis over 80 years ago [1,2], attention has largely been focused on the corrective responses initiated after the steady state of the organism is perturbed. However, the concept of homeostasis should be extended not only to include reactive homeostasis but also the precise homeostatic control mechanisms that can be designed to enable a sustainable energy system to predict when environmental challenges are approaching or are most likely to occur [3,4]. Sustainable energy systems (SES) encompass both reactive and predictive homeostasis mechanisms operating recursively and in coordination with one another in the face of environmental challenges. Hence, the ability of the energy system to respond to such challenges rapidly and effectively so as to attain equilibrium between power supply and energy demand to preserve stability and continuity of operations is the true measure of what sustainable energy systems (SES) are all about. This set of mechanisms is what we term homeostaticity in electric power systems' operation. Reactive homeostasis (RH) in SES, as the name suggests, is a feedback-enabled mechanism driven by energy generation and supply versus consumption or expenditure of energy [3,4,10,18–22]. This can be engineered in microgrids built around the concept of SHES by employing advanced sensors, control limit actuators (e.g., set-point fired responses) and artificial intelligence (AI) algorithms that allow the system to make timely decisions in order to respond to changes in a predetermined set of systems control variables [3,4]. Thus, SES take actions to counteract or fend off adverse conditions and noise that may affect the system's normal operation [3,4,10].

On the other hand, predictive homeostasis (PH) mechanisms generate responses well in advance of potential or possible challenges, once the system has reached a threshold signaling a predetermined degree of likelihood that an event will occur. Hence, there is a set of precise SES responses that come about in anticipation of predictable environmental challenges. Such PH responses enable the energy system to immediately prepare itself, taking the necessary precautions and actions to adapt and even reconfigure itself if necessary, in order to respond to the challenge ahead of time. Such actions may come in several forms and will depend on the resources and intelligence built into the system, but they are all geared toward making the SES more secured and able to withstand the upcoming challenge by activating its readiness control mechanisms [3,4,10].

Actions may come differently in magnitude and timeliness; some may be big and come immediately to adjust parts of the SES operation while others may come in the form of smaller changes in the system, largely as a result of stage-by-stage preparedness protocol building over time. The decision of which changes will occur first, where and how big they will be will be determined by both RH and PH control mechanisms engineered in the SES. Some may come very soon while others may come a longer time in advance of a probable environmental challenge. However, as we all know, systems are prone to internal conflicts when control criteria superimpose on one another generating the wrong response [3,4,10]. Reality is not always adequately interpreted and possible challenges may not always be correctly anticipated or foreseen in their full magnitude and scope. Also misread signals and misfires may risk a wrong or inadequate response as systems sometimes experience false alarms and misfires [4]. These may occur in part as a result of possible conflicts over certain homeostatic control (HC) variables that may share common goals and values but different scope of action and logic sequence, depending on the scenario being faced. Such HC variables may involve PH and RH control logic sequences which, if inadequately engineered in the SES, may result in inadvertent clash that can hurt system's performance [3,4]. Therefore, careful engineering of such capabilities in SES must account for such conflict of interest and changing scenarios must also be accounted for when establishing set-points [9,10]. Adequate measures must be engineered in the system's design to prevent PH responses interfering with RH mechanisms. If these potential conflicts were not accounted for and swiftly overcome, should they arise in the course of events, undesirable conditions may emerge which can compromise the effectiveness of SES readiness mechanisms, risking the very sustainability, and efficiency of the system itself [4].

2.3 Homeostaticity of SHES

Homeostasis-based power and energy management system for microgrids tied to the grid

In the case of PH the system responses will come as a result of information being processed by the system as the stimulus approaches and is detected by the sensing devices. Here there are both RH and PH sensors and HC software that account for an ample array of control mechanisms ready to act whenever conditions arise [4]. Thus homeostaticity of SHES requires a careful equilibrium of such control mechanisms and the coordination of internal and external decision variables—all of this as part of the particular HC strategy designed in the SHES—which will stand guard against a variety of adverse conditions and possible challenges [3,4,18–22]. Thus, the SHES will control the use of its energy resources including the grid and the use of alternative energy sources like energy storage if the grid is off. It will do so recursively and permanently in order to generate and supply enough energy to meet the loads demand, while at the same time, signaling to consumers how much energy is the SHES capable of supplying [3,4,18–22]. The question of if, when and how much energy should go into the energy storage unit will be determined by the HC system in charge of the energy management. Moreover, this will be based on the situational awareness and degree of criticality being experienced by the energy system itself. The HC system will therefore decide when and how much energy to store, based on supply surplus and the energy being demanded by the loads. Some loads will be more sensitive than others, some will be intensive in power for a period of time, while others will consume energy on a moderate and regular basis. Therefore, at certain point some loads will occupy a higher hierarchy than others while some may be spared or serviced partially, as conditions change [3,4]. Such control mechanisms will involve both PH and RH operating in unison, determining a generalized state of energy equilibrium between supply and demand, as dynamic scenarios unfold [3,4].

3 Homeostaticity in energy systems

The potential incentive of tariff differentiation, the frequency footprint concept, and why a possible reward-based system for green energy integration might be a good idea, when considering the novel concept of green tariffs analysis with energy sharing innovations

Next, we show the mathematical equations that represent the HC model incorporating energy homeostaticity capabilities by means of PH and RH, both of which are to be engineered in the energy system. The expression representing the attainment of energy equilibrium in the SHES (grid-tied microgrid) installed in a residential or commercial building is given in terms of the total power supply by the metasystem (the microgrid plus the grid acting concomitantly), and the loads [3,4]. Homeostaticity in the energy system is achieved through specially designed homeostasis regulation mechanisms that are engineered in the energy management and control system, as discussed earlier, so that:

$$E_{equilb} = P_{supply}(x)PH(u)RH(v)S(\alpha) = E_{consump}(u,v,\alpha) + \frac{d}{dt}E_{consump}(u,v,\alpha)\, and$$

$$where\, P_{supply} = Real\, Power + Reactive\, Power = (P+Q) - Losses$$

Where x represents the internal state of the energy systems at time $t = 0$ and the Energy equilibrium E_{equilb} is dependent upon several factors operating in the SHES [14]. Both u and v represent the specific predictive and reactive homeostasis variables respectively, which are designed in the HC model. These are designed in the software of the energy management and control system supported by appropriate hardware such as power electronics and the like. Later these variables undergo various simulation stages, based on extensive data modeling, taking into account local factors inherent to the community and/or the locality so as to incorporate as much accuracy in the system's response as possible [4]. $S(\alpha)$ represents the conditioning function of the SHES and operates to alert and condition the system's adaptive capacity recursively in order to respond to a wide range of stimuli. Its actions are based on the situational awareness and degree of criticality being experienced by the metasystem (microgrid plus grid) and the solution incorporates AI algorithms and intelligent control.

Thus, $S(\alpha)$ is a function of the adverse conditions and environmental challenges being sensed by the energy system at any point and also foreseen in the near future. These are represented by the awareness and criticality variable α. Such variable may also be influenced by transient system conditions such as a sudden increase in reactive power demand in a reactive power constrained line, which is generally due to a contingency in transmission network causing an increase of the load burden of the adjacent line(s) in order to maintain the constant system loads, for example. These three variables: u, v, α are the equivalent of metabolic variables present in all living organisms, like the physiological and endocrine systems variables for example, which are present in humans and animals and affect the system's energy expenditure and storage. The expression $\frac{d}{dt}E_{consump}(u,v,\alpha)$ stands for the rate of change of energy consumption of users, let us say a sustainable block [8–10,18–21] somewhere, and is a direct indicator of thriftiness and energy efficiency present in the metasystem. It is linked to powerful sustainability performance indicators of SES introduced previously in the literature [10], such as the homeostatic index H_i and the Grid_Frac [10]. The homeostatic index H_i is a powerful new concept previously introduced [10] which measures how much electricity is being consumed from the grid per home as a percentage of the total electricity (renewable plus non-renewable) being consumed by the entire sustainable block [9,10]. This is being monitored and recorded in real time and shown to the consumer as a monthly reading and/or on a daily and hourly basis, as preferred [9]. Thus, homeostatic index H_i shows how thrifty and energy efficient each home is with respect to the power supplied by the microgrid [10]. This is important, not only because of what these two aspects of energy consumption represent but also because they drive the exergy level being built in the SHES.

3.1 Exergy and energy efficiency

How to engineer sustainability and resiliency in sustainable hybrid energy systems (SHES)

Exergy is the maximum useful work, which can be extracted from a system as it reversibly comes into equilibrium with its environment. Exergy also expresses the quality of a particular energy source and also quantifies the useful work done by a certain amount of energy employed [19,21,22] in any given process where there is energy intake and expenditure. Finally, we have that exergy is also defined as a measure of the actual potential of a system to do work with reference to a given environment. By means of the right homeostatic control (HC) strategies [9,10] and the given system conditions, one can elicit higher degrees of thriftiness and energy efficiency increasing the level exergy in the energy system. We should not forget that the energy system represents the microgrid tied to the grid both of which are connected to a number of energy consumers, all three acting concomitantly. A value of homeostatic index H_i below 1 is considered acceptable yet ideally values closer to 0.50 or below are a truer indicator of a high degree of thriftiness and energy efficiency for the home [10]. H_i = A measure of the energy efficiency and thriftiness of the energy consumers in a sustainable block (Fig. 7.2).

On the other hand, Grid_Frac is an indicator of the fraction of total electricity drawn from the grid per each home. It is also a measure of EE and thriftiness just like the homeostatic index H_i whereas Grid_Frac shows the fraction (in percentage) of the total electricity consumption drawn from the grid by each home in the sustainable block [10]. Below is a diagram which illustrates the concepts presented here and incorporates an exergy index function which is also related to the quality of the energy being produced by the SHES and with the amount of thriftiness and EE being exercised by consumers which directly impacts $E_{consump}(u,v,\alpha)+\frac{d}{dt}E_{consump}(u,v,\alpha)$ determining how much energy is being made available in the energy system by the consumers of the sustainable block [10] (Fig. 7.3).

3.2 Role of the electric tariff differentiation

The role of the electric tariff differentiation factor and how it can be implemented upon incorporating thriftiness-based reward systems to residential consumers

In the model depicted earlier, there is a particular control and energy management architecture involving independent energy

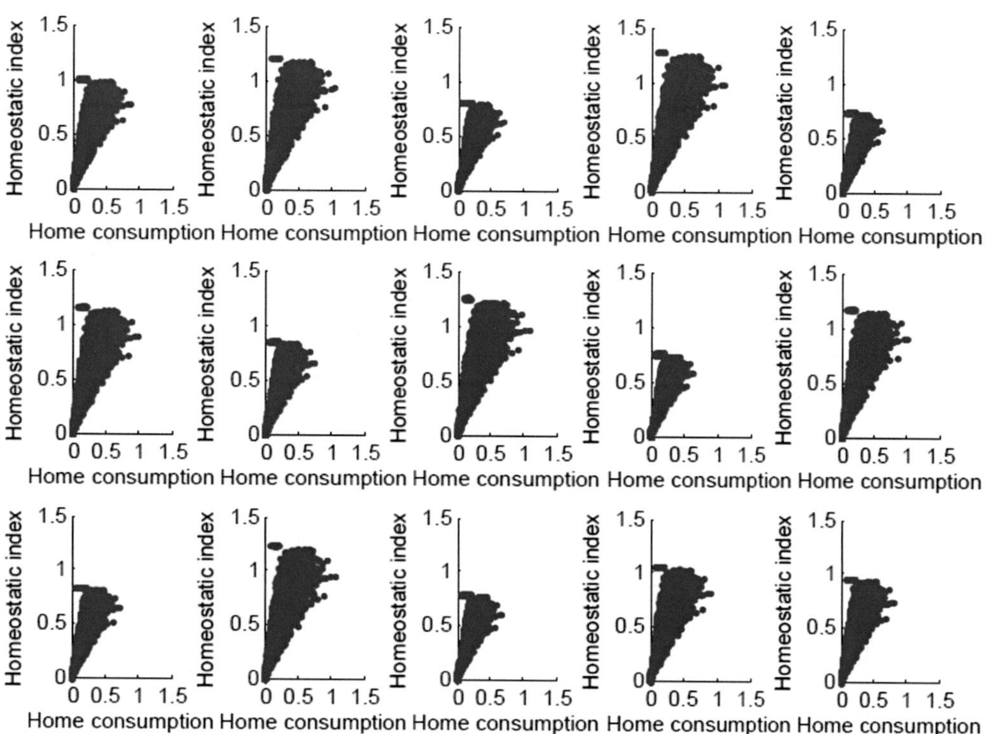

Figure 7.2. The simulation of the homeostatic index function H_i for a sustainable block of 15 homes in a residential neighborhood, being supplied by a grid-tied micogrid and by the electric power distribution grid [17].

sources and the grid. The sustainability of the system is in part safeguarded by AI algorithms that comprised the HC system which make up the autonomous mission control of the microgrid [9,10]. There is also an Exergy index function X_i that, like the homeostatic index, is a measure of the quality and efficiency of the energy being generated and utilized by the microgrid including the energy consumers in the sustainable block (the loads) [9,10]. The power supply by the SES is adjusted based on the block's homeostasis regulation (HR) and exergy indices for the whole sustainable block, based on aggregate demand of energy in the SHES [9,10].

However, in addition to the above, a good HC system operating within adequate performance levels of EE and thriftiness can greatly impact the exergy of the system and with this, effective and timely real and reactive power levels can be ensured to maintain system's stability through equilibrium between supply and demand. This can help keep short-term fluctuations in power requirements, both active and reactive power, from dropping the

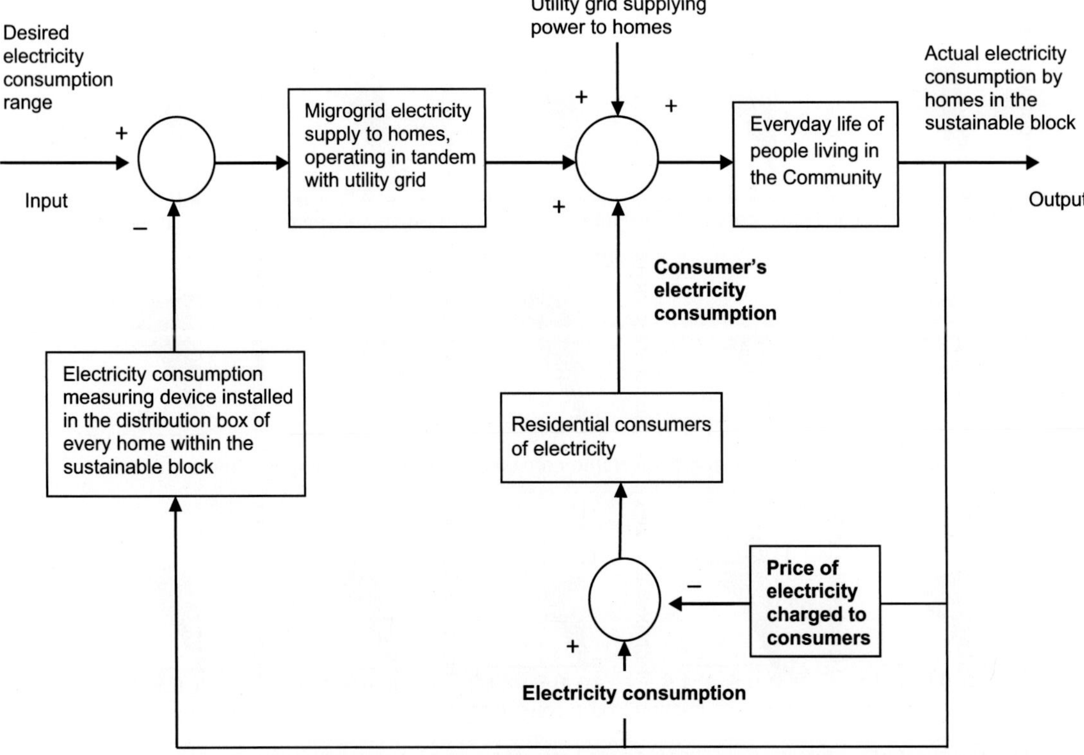

Figure 7.3. Flow diagram depicting the drivers of homeostatic control strategies to ensure that efficient electricity supply and demand equilibrium is achieved by the SHES based on energy and power consumption ranges per home consumers [10]. Source: Own elaboration.

frequency or at least helping this situation become milder. This is because sometimes there are lags in the system's governor and generators' output which require a finite time to adjust to the new power requirements. Such actions can be aided by reactive and predictive homeostasis functions built into the system in order to act as an energy consumption-based frequency regulator, $E_{consump}(u,v,\alpha)+\frac{d}{dt}E_{consump}(u,v,\alpha)$ aside from the role of power enabler. Such a role is played whenever EE and thriftiness in energy consumption make it possible for more power to be available in the system for those that need more while others need less [14,16,17].

No doubt the changing frequency in the SHES will influence the power flow but, at the same time, frequency is a function of the energy consumption and also the rate of change of such energy consumption, $E_{consump}(u,v,\alpha)$ therefore power flow and energy sustainability can also be influenced by

$E_{consump}(u,v,\alpha)+\frac{d}{dt}E_{consump}(u,v,\alpha)$. The difference in energy consumption will impact system's frequency and voltage level, which will ultimately impact the power supply quality and overall system's dynamics in a SHES. We will talk about this further as we discuss next the reactive power role and how the prescribed energy and power management model can help reactive power management and thus voltage control.

With the droop method [60], however, the power angle depends heavily on the real power R generated while the voltage depends on the reactive power Q; AC systems like the grid supply or consume both: real power and reactive power, and when DG is operating tied to the grid, it can be difficult to maintain voltage levels. Real power accomplishes useful work on the loads that demand it while reactive power supports the voltage that must be controlled for system reliability and efficiency.

In the case where the DG systems is tied to the grid, like the utility-operated microgrid being discussed here, it is hard to control voltage unlike when there is grid-only supply. This turns particularly true when sharp increases in power demand occur, as in early hours and towards the evening hours of the day. With homeostatic control of SHES it is easier to manager real and reactive power balance and to control reactive power in the grid-tied microgrid so as to ensure its effect on the reliability and security of EPS. This is so because HC enables reactive power management more easily by controlling the system's optimal equilibrium between supply and demand. Reactive power affects voltages throughout the system, and we all know how important and also difficult it is to keep voltage control steady in electrical power systems (EPS) in order for proper operation of electrical power equipment to prevent damage such as overheating of generators and motors, to reduce transmission losses and to maintain the ability of the EPS to withstand and prevent voltage sagging or simply power failure.

In general terms, when we decrease reactive power, we cause the voltage to fall while increasing it causes voltage to rise. A voltage sag or voltage collapse may occur when the system is being demanded beyond its capacity. This occurs when, for example, the grid-tied microgrid may try to supply too many loads at once, within a certain range of consumption, or it is being demanded too much power by certain loads at certain times, all of which plays against the homeostasis aquarium between supply and demand. This invariably will affect the voltage and the microgrid may not be able to support it. Regarding the latter and the importance of reactive power and how it is useful to maintain system voltage stability, homeostatic control of SHES seeks to do just that. If real power R can be adequately controlled, so can the power angle, and if the reactive power Q can be better regulated as well, then the voltage V will be controllable too [4,60]. The droop control method has an inherent trade-off between the active power sharing and the frequency accuracy, resulting in the frequency deviating slightly from the nominal frequency [4,60].

4 Energy homeostasis and homeostatic control strategies

Energy homeostasis and homeostatic control strategies developed to enhance and accommodate different communities' lifestyles and energy consumption needs for the integration and expansion of green energy systems tied to the grid

Next the chapter addresses the subject of energy homeostasis and homeostatic control strategies to enhance and accommodate different communities' lifestyles and energy consumption needs, all of which present different and sometimes quite diverse scenarios to the integration and expansion of green energy systems tied to the grid as the model being analyzed here shows. These are considerations that ENEL or any other electric utility must take into account in their plans to pursue residential green energy with DG systems (Figs. 7.4 and 7.5).

4.1 Discussion

We extend the concept of homoeostatic control first introduced by Fred Schweppe and his team [61–64] back in the early 1980s, to include not only reactive homeostasis but also predictive homeostasis (PH), both of which can contribute to make SES more resilient, efficient, and reliable while complementing more traditional control methods. PH involves a recursive set of mechanisms that triggers appropriate system responses in the prospect of near future stimuli manifestation [4]. In the case of sustainable energy systems, this constitutes a set of corrective responses initiated in anticipation of a predictably environmental (both internal and external) challenge. In general terms, predictive homeostasis is an anticipatory response to an expected homeostatic challenging event in the future. Seasonal migration of animals and birds in particular are examples of predictive homeostasis. Predictive responses often compromise the effectiveness of reactive homeostatic mechanisms, even to the point of risking the survival of the organism itself [4]. Nevertheless, both predictive and reactive homeostasis (RH) must be in equilibrium and perfectly synchronized just as large rotating AC generators must operate in synchrony with the frequency of the grid. If this was not the case and both RH and PH were to operate uncoordinatedly, they may become in conflict with each other. In such cases predictive responses may compromise the effectiveness of reactive homeostatic control mechanisms to the point of jeopardizing the sustainability

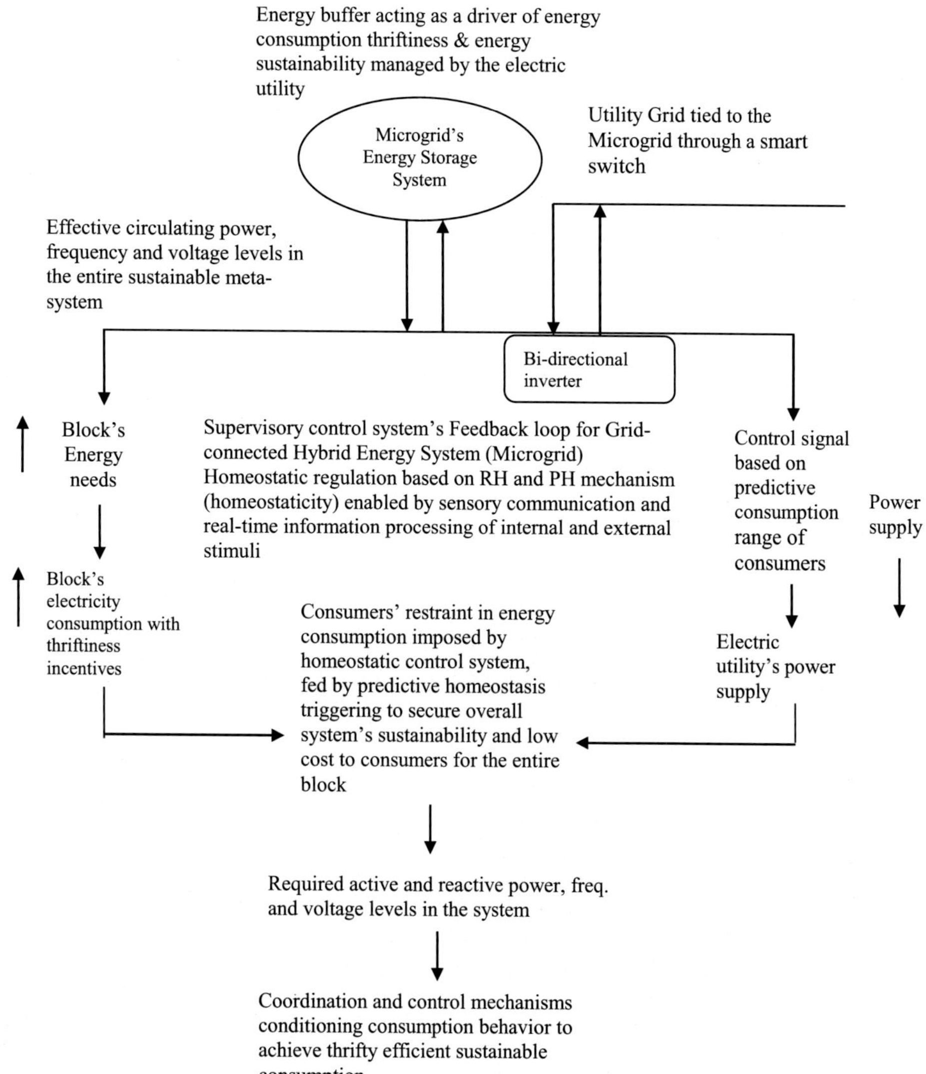

Figure 7.4. The figure depicts how energy homeostaticity mechanisms can operate for green energy systems, where grid-connected microgrid are operated by the utility company, in this particular case, with an energy buffer [4,10].
Source: Own elaboration.

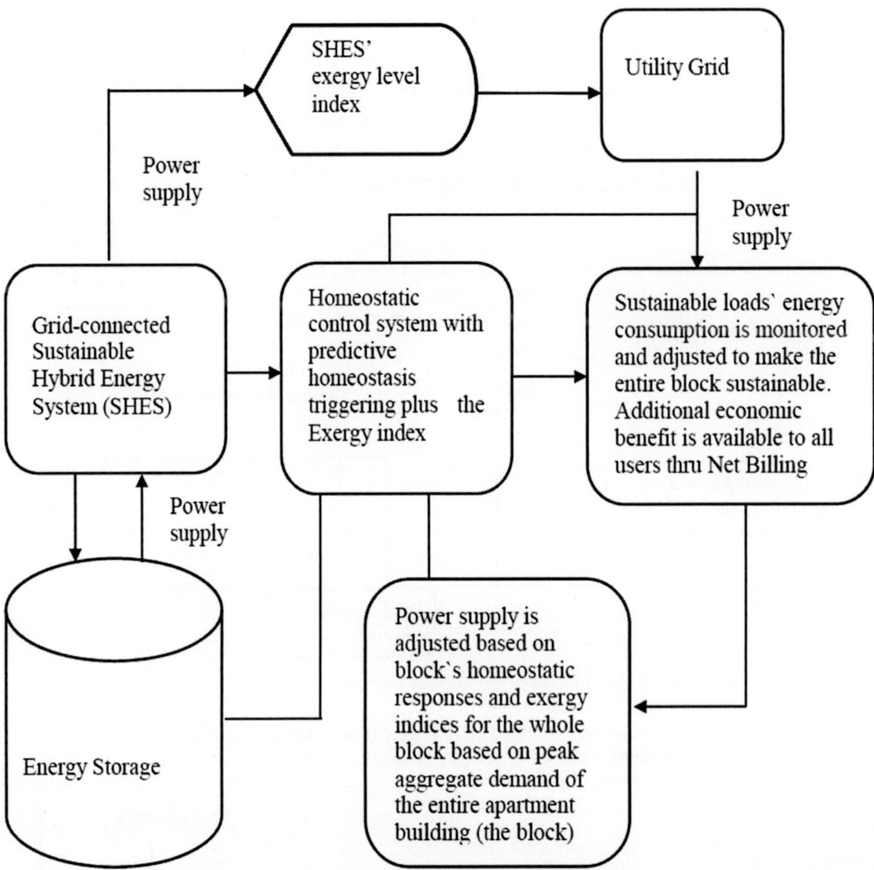

Figure 7.5. An energy homeostasis diagram for a SHES, wherein predictive and reactive homeostasis mechanisms operate, as part of the HC of the SHES, this time with energy storage [4,10]. Source: Own elaboration.

of the energy system [4]. There are many examples of such catastrophes in recent history and everything seems to indicate that things are not getting any better and that more severe weather patterns are expected along with natural disasters. Chile, in particular, bears a good example of such calamities, showing just how much chaos and destruction they can bring to the population. The February 27, 2010s giant earthquake and tsunami are still vivid in the memory of Chileans and there is clear memory of how the major electric power and telecommunications networks collapsed as a result. There have also been more frequent occurrences of strong winds, and large fires extending to forest and urban areas, and volcano eruptions, all of which cause havoc in our energy and communication networks [4].

5 Conclusions

Climate change and natural disasters are a very serious threat that has brought worldwide attention and demands concrete actions now. World leaders have understood this and have been meeting to discuss this issue and to explore mitigation measures for several years. Nature does not give man a second chance, either you are prepared or you aren't. It is therefore imperative that we understand that traditional electric power infrastructure is not only vulnerable but also dangerous and ineffective when it comes to natural disasters or harsh climate like strong winds, earthquakes and floods, all of which are part of an ever more complex environmental scenario. These and other dangerous phenomena like sinkholes and rising sea tides are also affecting modern infrastructure and mankind's way of life just about everywhere. Thus decentralizing power systems by means of DG solutions employing both renewables and traditional power sources like a SES makes sense. The concept of energy hubs is also part of this new power infrastructure vision with a much more sensible and reliable architecture than traditional energy matrix that relies on long transmission and distribution lines. Furthermore, if we can equip these systems with adequate communications capabilities and data sensing devices operating in interconnected fashion, in all systems comprising the power generation and distribution chain, this would make a better solution. It can be done using simple microprocessor-based technology with set-points data signals programmed in the microprocessor itself for the control of the entire system in a hybrid, distributed fashion, where there is a central controller interacting with each sub-system's controller as well as with the electric utility control operator as well. That would allow for a more flexible, inexpensive, fast and robust control system than the more expensive, complex systems such as multi-agent systems or expert systems that much of the current literature on smart grid and intelligent computing is embracing. At least for smart microgrids to supply electricity and potentially heat using CHP as a generation alternative [33,64–68] for cases in which a natural disaster strikes, leaving a whole neighborhood without power, a simpler, more flexible, economical and modular control solution is possible [8,10]. This chapter offers a glimpse of such new solutions, as there are others which point to the same necessity in the current Smart Grid era of giant-leap type transformations in the electric utilities' industry sector [69–71] incorporating reactive and predictive homeostasis control mechanisms, which are part of the new frontier of control and communications systems engineering.

Electricity infrastructure is feeble and largely outdated when it comes to the electric power distribution sector in Chile, as it has not kept up with the times. Environmental challenges like natural disasters and severe weather patterns brought by phenomena like El Niño leave no room for preambles. In spite of electricity being the blood flow that powers all sectors of the country, little if anything has been done to counteract environmental and other threats (like terrorist acts, for example) that put power supply stability at risk. Nevertheless, at the present time, there are electric utilities in the world exploring new solutions to a grave and growing problem. Enel Distribucion in Chile is one of them. This chapter presents a new control model to incorporate SHES in residential, commercial and other institutions like schools, hospitals and airports, to serve as a localized energy solution that can complement the power grid. Its engineering design and homeostatic control capabilities incorporate reactive and predictive homeostasis which allow for homeostaticity to occur in energy systems. We also believe this crusade should be led by electric utilities like ENEL with the aid of appropriate legislation by the government. This new model may very well coexist and complement more traditional control and energy management methods like droop control to better equip the system with better environmental challenge management and resilience capacity.

Acknowledgments

Dr. Fernando Yanine is grateful to CONICYT of Chile for his scholarship endowment of his Postdoctoral FONDECYT N° 3170399 “BUILDING SUSTAINABLE ENERGY SYSTEMS TIED TO GRID: A RESEARCH PROJECT PARTNERSHIP WITH CHILECTRA TO INSTALL HOMEOSTATICALLY CONTROLLED MICROGRIDS IN APARTMENT BUILDINGS” and to ENEL Distribucion S.A. for its auspice, sponsorship, and support. Thanks are also given to Universidad Finis Terrae for its ongoing support in this research initiative and to the Department of Electrical Engineering of UTFSM.

Dr. Antonio Sanchez-Squella is supported by a research grant from FONDECYT #11150911 of Chile.

References

[1] W.B. Cannon, Organization for physiological homeostasis, Physiol. Rev. 9 (3) (1929) 399–431.

[2] W.B. Cannon, Stress and strains of homeostasis, Am. J. Med. Sci. 189 (1935) 1–14.

[3] F. Yanine, A. Sanchez-Squella, A. Barrueto, F. Cordova, S.K. Sahoo, Engineering sustainable energy systems: how reactive and predictive homeostatic control can prepare electric power systems for environmental challenges, Procedia Comput. Sci. 122 (2017) 439–446.

[4] F. Yanine, A. Sanchez-Squella, A. Barrueto, J. Tosso, F.M. Cordova, H.C. Rother, Reviewing homeostasis of sustainable energy systems: How reactive and

predictive homeostasis can enable electric utilities to operate distributed generation as part of their power supply services, Renew. Sustain. Energy Rev. 81 (2018) 2879–2892.

[5] https://www.congress.gov/bill/114th-congress/senate-bill/2012.

[6] https://microgridknowledge.com/think-microgrid-government-leaders-serious-quest-build-microgrids/.

[7] https://www.cleanegroup.org/ceg-resources/resource/resilient-power-evolution-of-a-new-clean-energy-strategy-to-meet-severe-weather-threats/.

[8] F.M. Cordova, F.F. Yanine, Homeostatic control of sustainable energy grid applied to natural disasters, Int. J. Comput. Commun. Control 8 (1) (2012) 50–60.

[9] F.F. Yanine, F.I. Caballero, E.E. Sauma, F.M. Córdova, Homeostatic control, smart metering and efficient energy supply and consumption criteria: A means to building more sustainable hybrid micro-generation systems, Renew. Sustain. Energy Rev. 38 (2014) 235–258.

[10] F.F. Yanine, F.I. Caballero, E.E. Sauma, F.M. Córdova, Building sustainable energy systems: Homeostatic control of grid-connected microgrids, as a means to reconcile power supply and energy demand response management, Renew. Sustain. Energy Rev. 40 (2014) 1168–1191.

[11] https://earthsky.org/earth/powerful-earthquake-strikes-off-chiles-coast.

[12] http://www.britannica.com/event/Chile-earthquake-of-2010.

[13] https://microgridknowledge.com/hybrid-microgrids/.

[14] http://www.iea.org/Textbase/npsum/US2014sum.pdf.

[15] https://energypedia.info/wiki/Net-Metering_/_Billing_in_Chile.

[16] https://www.enel.com/media/press/d/2016/06/enel-presents-enel-open-meter-the-new-electronic-meter.

[17] https://www.eneldistribucion.cl/medicion-inteligente.

[18] F.F. Yanine, E.E. Sauma, Review of grid-tie micro-generation systems without energy storage: Towards a new approach to sustainable hybrid energy systems linked to energy efficiency, Renew. Sustain. Energy Rev. 26 (2013) 60–95.

[19] F.F. Yanine, E.E. Sauma, F.M. Cordova, An exergy and homeostatic control approach to sustainable grid-connected microgrids without energy storage, Applied Mechanics and Materials, 472, Trans Tech Publications, 2014, pp. 1027–1031.

[20] F. Caballero, E. Sauma, F. Yanine, Business optimal design of a grid-connected hybrid PV (photovoltaic)-wind energy system without energy storage for an Easter Island's block, Energy 61 (2013) 248–261.

[21] F.F. Yanine, F.M. Córdova, L. Valenzuela, Sustainable hybrid energy systems: an energy and exergy management approach with homeostatic control of microgrids, Proc. Comput. Sci. 55 (2015) 642–649.

[22] F. Yanine, F.M. Córdova, March Homeostatic control in grid-connected microgeneration power systems: A means to adapt to changing scenarios while preserving energy sustainability, in: Renewable and Sustainable Energy Conference (IRSEC), 2013 International, IEEE, 2013, pp. 525–530.

[23] https://obamawhitehouse.archives.gov/sites/default/files/image/president27sclimateactionplan.pdf.

[24] E.S. Blake, T.B. Kimberlain, R.J. Berg, J.P. Cangialosi, I.I. Beven, L. John, National Hurricane Center (February 12, 2013). Hurricane Sandy: October, 22, p.29. (PDF) (Tropical Cyclone Report). United States National Oceanic and Atmospheric Administration's National Weather Service, 2013. Archived from the original on February 17, 2013. Retrieved February 1, 2018.

[25] M. Diakakis, G. Deligiannakis, K. Katsetsiadou, E. Lekkas, Hurricane sandy mortality in the Caribbean and continental North America, Disaster Prev. Manage. 24 (1) (2015) 132–148.

[26] Hurricane/Post-Tropical Cyclone Sandy, October 22-29, 2012 (PDF) (Service Assessment). United States National Oceanic and Atmospheric Administration's National Weather Service. May 2013, p. 10. Archived from the original on June 2, 2013. Retrieved Feb 2, 2018.
[27] http://www.livescience.com/22522-hurricane-katrina-facts.html.
[28] L. Che, M. Khodayar, M. Shahidehpour, Only connect: Microgrids for distribution system restoration, IEEE Power Energy M. 12 (1) (2014) 70–81.
[29] L. Che, M. Shahidehpour, DC microgrids: Economic operation and enhancement of resilience by hierarchical control, IEEE Trans. Smart Grid 5 (5) (2014) 2517–2526.
[30] Z. Wang, J. Wang, Self-healing resilient distribution systems based on sectionalization into microgrids, IEEE Trans. Power Syst. 30 (6) (2015) 3139–3149.
[31] C. Chen, J. Wang, F. Qiu, D. Zhao, Resilient distribution system by microgrids formation after natural disasters, IEEE Trans. Smart Grid 7 (2) (2016) 958–966.
[32] M. Roach, Hurricane Sandy & the Emperor's New Clothes: Microgrids as a Risk Mitigation Strategy for Extreme Weather Events. White Paper, MicroGrid Horizons, 2012.
[33] B. Hedman, Combined Heat and Power and Heat Recovery as Energy Efficiency Options. Presentation on behalf of US CHP Association and ICF Consulting, Washington, DC, 2007.
[34] T. Olinsky-Paul, Using State RPSs to Promote Resilient Power at Critical Infrastructure Facilities. Clean Energy States Alliance, May, 2013.
[35] L. Steg, G. Perlaviciute, E. van der Werff, Understanding the human dimensions of a sustainable energy transition, Front. Psychol. 6 (2015) 805.
[36] http://www.cleanegroup.org/assets/2015/Resilient-States.pdf.
[37] R.K. Dixon, E. McGowan, G. Onysko, R.M. Scheer, US energy conservation and efficiency policies: Challenges and opportunities, Energy Policy 38 (11) (2010) 6398–6408.
[38] R.P. Rogers, The effect of the Energy Policy and Conservation Act (EPCA) regulation on petroleum product prices, 1976-1981, Energy J. 24 (2003) 63–93.
[39] S. Grantham, Power to the People: CL&P's Crisis Communication Response Following the 2011 October Nor'easter.
[40] https://www.cleanegroup.org/ceg-resources/resource/resilient-power-evolution-of-a-new-clean-energy-strategy-to-meet-severe-weather-threats/.
[41] M. Panteli, P. Mancarella, Influence of extreme weather and climate change on the resilience of power systems: Impacts and possible mitigation strategies, Electr. Power Syst. Res. 127 (2015) 259–270.
[42] R.J. Campbell, Weather-related power outages and electric system resiliency. Washington, DC: Congressional Research Service, Library of Congress, 2012.
[43] https://edition.cnn.com/2015/09/16/americas/chile-earthquake/index.html.
[44] D.E. Olivares, A. Mehrizi-Sani, A.H. Etemadi, C.A. Cañizares, R. Iravani, M. Kazerani, et al. Trends in microgrid control, IEEE Trans. Smart Grid 5 (4) (2014) 1905–1919.
[45] K.B. Jones, S.J. Bartell, D. Nugent, J. Hart, A. Shrestha, The urban microgrid: Smart legal and regulatory policies to support electric grid resiliency and climate mitigation, Fordham Urb. LJ 41 (2013) 1695.
[46] A. Khodaei, Resiliency-oriented microgrid optimal scheduling, IEEE Trans. Smart Grid 5 (4) (2014) 1584–1591.
[47] B. Warshay, Upgrading the grid: how to modernize America's electrical infrastructure, Foreign Aff. 94 (2015) 125.
[48] A.A. Sallam, O.P. Malik, Electric Distribution Systems, Vol. 68, John Wiley & Sons, 2011.

[49] G. Walker, S. Hunter, P. Devine-Wright, B. Evans, H. Fay, Harnessing community energies: explaining and evaluating community-based localism in renewable energy policy in the UK, Global Environ. Polit. 7 (2) (2007) 64–82.
[50] J.J. Ding, J.S. Buckeridge, Design considerations for a sustainable hybrid energy system, IPENZ Trans. 27 (1) (2000) 1.
[51] W. El-Khattam, M.M. Salama, Distributed generation technologies, definitions and benefits, Electr. Power Syst. Res. 71 (2) (2004) 119–128.
[52] G. Walker, What are the barriers and incentives for community-owned means of energy production and use?, Energy Policy 36 (12) (2008) 4401–4405.
[53] S.L. Wearing, M. Wearing, M. McDonald, Understanding local power and interactional processes in sustainable tourism: Exploring village–tour operator relations on the Kokoda Track, Papua New Guinea, J. Sustain. Tourism 18 (1) (2010) 61–76.
[54] M. Geidl, G. Koeppel, P. Favre-Perrod, B. Klockl, G. Andersson, K. Frohlich, Energy hubs for the future, IEEE Power Energy M. 5 (1) (2007) 24–30.
[55] A.Q. Huang, M.L. Crow, G.T. Heydt, J.P. Zheng, S.J. Dale, The future renewable electric energy delivery and management (FREEDM) system: the energy internet, Proc. IEEE 99 (1) (2011) 133–148.
[56] M.C. Bozchalui, S.A. Hashmi, H. Hassen, C.A. Cañizares, K. Bhattacharya, Optimal operation of residential energy hubs in smart grids, IEEE Trans. Smart Grid 3 (4) (2012) 1755–1766.
[57] G. Fettweis, E. Zimmermann, ICT energy consumption-trends and challenges, in: Proceedings of the 11th international symposium on wireless personal multimedia communications, Vol. 2, No. 4, p. 6, Lapland, 2008.
[58] O. Zinaman, M. Miller, A. Adil, D. Arent, J. Cochran, R. Vora, et al., Power systems of the future: a 21st century power partnership thought leadership report (No. NREL/TP-6A20-62611). NREL (National Renewable Energy Laboratory (NREL), Golden, CO (United States)), 2015.
[59] A. Hartmanns, H. Hermanns, Modelling and decentralised runtime control of self-stabilising power micro grids, in: International Symposium on Leveraging Applications of Formal Methods, Verification and Validation, Springer, Berlin, Heidelberg, pp. 420–439, 2012.
[60] A.M. Bollman, An experimental study of frequency droop control in a low-inertia microgrid, 2010.
[61] F.C. Schweppe, R.D. Tabors, J.L. Kirtley, H.R. Outhred, F.H. Pickel, A.J. Cox, Homeostatic utility control, IEEE Trans. Power Apparatus Syst. (3) (1980) 1151–1163.
[62] F.C. Schweppe, R.D. Tabors, J.L. Kirtley, Homeostatic control: the utility customer marketplace for electric power, 1981.
[63] Schweppe, C. Fred, D. Richard, Tabors, L. James, Kirtley, Power/energy: Homeostatic control for electric power usage: A new scheme for putting the customer in the control loop would exploit microprocessors to deliver energy more efficiently, IEEE Spectr. 19 (7) (1982) 44–48.
[64] M. Amin, North America's electricity infrastructure: Are we ready for more perfect storms?, IEEE Security Privacy 99 (5) (2003) 19–25.
[65] M. Amin, Toward self-healing energy infrastructure systems, IEEE Comput. Appl. Power 14 (1) (2001) 20–28.
[66] S.M. Amin, B.F. Wollenberg, Toward a smart grid: power delivery for the 21st century, IEEE Power Energy M. 3 (5) (2005) 34–41.
[67] M. Amin, P.F. Schewe, Preventing blackouts, Sci. Am. 296 (5) (2007) 60–67.
[68] A. Afzal, M. Mohibullah, V. Kumar Sharma, Optimal hybrid renewable energy systems for energy security: a comparative study, Int. J. Sustain. Energy 29 (1) (2010) 48–58.

[69] A. Tricoire, Uncertainty, vision, and the vitality of the emerging smart grid, Energy Res. Soc. Sci. 9 (2015) 21–34.
[70] A. Rearte-Jorquera, A. Sánchez-Squella, H. Pulgar-Painemal, A. Barrueto-Guzmán, Impact of residential photovoltaic generation in smart grid operation: real example, Proc. Comput. Sci. 55 (2015) 1390–1399.
[71] C. Uckun, Dynamic electricity pricing for smart homes. The University of Chicago, 2012.

8

Grid-tied distributed generation with energy storage to advance renewables in the residential sector: tariffs analysis with energy sharing innovations

Fernando Yanine[a], Antonio Sanchez-Squella[b], Aldo Barrueto[b], Sarat Kumar Sahoo[c], Dhruv Shah[d], Antonio Parejo[e], Felisa Cordova[a], Hans Rother[f]

[a]Faculty of Engineering, Universidad Finis Terrae, Santiago, Chile; [b]Department of Electrical Engineering, Universidad Técnica Federico Santa María (UTFSM), Santiago, Chile; [c]A Constituent College of Biju Patnaik Technological University, Parala Maharaja Engineering College, Department of Electrical Engineering, Govt. of Odisha, Berhampur, Odisha, India; [d]Mukesh Patel School of Technology Management & Engineering (MPSTME), Mumbai Campus, India; [e]Department of Electronic Technology, Escuela Politécnica Superior, University of Seville, Seville, Spain; [f]Enel Distribución Chile S.A., Santiago, Chile

Chapter outline

Low Carbon Energy Technologies in Sustainable Energy Systems. http://dx.doi.org/10.1016/B978-0-12-822897-5.00008-0

Nomenclature

i Customer

N Number of customers of the sustainable block

t Time (min)

A Year

Δt Time interval of 15 min

d_{it} Power demand of customer i in period t in [kW]

D_t Total power demand from the sustainable block in period t in [kW]

P_{PV_t} Power generated by the photovoltaic plant in period t in [kW]

η_t Number of batteries connected

Q_{Bat_t} Battery capacity in period t [kWh]

P_{Bat_t} Battery charge/discharge power at period t in DC side [kW]

P_{b_t} Battery charge/discharge power on the AC side in period t in [kW]

E_{Bat_t} Energy in the battery in period t in [kWh]

E_{max}, E_{min} Maximum and minimum battery energy in [kWh]

η_c, η_d Efficiency of charging and discharging the battery

P_{grid_t} Power supplied from the utility in period t in [kW]

$P_{grid_{it}}$ Power supplied from the utility to customer i at period t in [kW]

P_{inGrid_t} Power injected into the utility in period t in [kW]

$P_{inGrid_{it}}$ Power injected to the utility by customer i in period t in [kW]

P_{max_t} Average of the two highest registered demands in the last 12 months, including the current month [kW]

$P_{max_{Pt}}$ Average of the two highest demands in peak hour (18:00 to 23:00) registered in the last winter period (April to September)

P_{pool_t} Power transferred between customers in period t in [kW]

$P_{pool_{it}}$ Power supplied by customers i to another customer in period t in [kW]

C_m Monomic energy price [$/KWh]

C_{mp}, $C_{m_{op}}$ Monomic energy price at peak and off-peak hours, respectively [$/KWh]

C_p, C_{op} Peak and off peak demand cost [$/kW]

C_{im} Cost to supply energy to customer i in month m [$]

C_{iY} Annual cost to supply energy to customer i [$ CLP]

C_{SB_m} Cost to supply energy to the sustainable block in month m [$]

C_B Energy supply cost from the utility [$/kWh]

C_{ingrid} Price of the power injected into de utility [$/kW]

C_{pool} Exchange between customer energy price [$/kWh]

MD_{ph} Monthly maximum demand in peak hours [kW]

MD_{oph} Monthly maximum demand off-peak hour [kW]

$month_p$ Winter month (from April to September)

$month_{op}$ Summer month, (from October to March)

h_p Peak hour, from 18:00 h to 23:00h in winter month

h_{op} Off-peak hour, from 23:00 h to 18:00 h

C_{BS_A} Cost to supply energy to the sustainable block in year A in [$ CLP]

1 Introduction

1.1 Sustainable energy systems

Redefining the role of energy efficiency (EF) and optimality in sustainable energy systems (SES), from an operational, systemic, and economic viewpoint

In electrical distribution systems, distributed generation (DG) can be beneficial for consumers, as well as for electric utilities for a number of reasons. This is specially so in places where the electric supply from centralized power generation plants is impracticable/unfeasible due to technical and/or economic reasons or when, as in Chile's case for the most part, the electricity distribution networks infrastructure is frail, vulnerable and without the appropriate backup systems should natural disasters or environmental threats strike all of a sudden. The epitome of distributed generation systems (DGS) is the microgrid employing renewable and non-renewable energy sources. The main generation resources that comprise a microgrid are small gas turbines, microturbines, fuel cells, wind and solar energy, as well as biomass, and small hydroelectric power plants, in a variety of combinations and arrangements, depending on the local energy sources available and the scale of the project. All of these types of energy generation resources can be supplied at two levels: at the local level (specific location) and at the end point level (installed by the individual energy consumers themselves) [1,2].

Energy homeostaticity is that property of sustainable energy systems (SES) by which the energy system has both the capability and the capacity to respond to environmental challenges and perturbations very rapidly and effectively (in fractions of a second), so as to attain optimal equilibrium between the amount of power supplied by the energy system and the demand for energy from the loads [3,4]. This systems engineering concept is key to the whole nature of SES in regards to electric power systems in general and particularly so when discussing distributed generation systems tied to the grid. Furthermore, it is essential for the SES to be able to exercise this capability from a design and architectural engineering viewpoint in order to preserve systems stability and continuity of operations at any point in time in electric power systems' operation management. Reactive homeostasis—as the name suggests—is a feedback-enabled mechanism that alerts and prompts the energy system to act upon an imbalance between energy supply and demand, in order to attain homeostaticity. As an example, and in order to illustrate the above concepts more clearly, consider the case where loads are classified according to their type of consumption (fixed, variable), and according to the time of day they operate (transferable or non-transferable). Likewise, customer satisfaction is a measurable system performance parameter in electric power distribution services that can be further studied and modeled by electric utilities so as to explore the concept from a different angle. Thus customer satisfaction can also be modeled taking into account energy consumption trends and how these change based on particular habits and location conditions. For example, if there are incentives to use electric power for processes that can be scheduled for different times and days of the week, causing a surplus of green energy in the system, especially during peak hours, this reschedualing benefits can increase if the energy-consuming processes finish their work before the agreed time interval as set by the electric utility, rendering benefits for both parties, the consumers and the very same electric utility.

In addition, and pertaining to the example modeled here, the solar photovoltaic power plant is modeled so that its excess energy can be sold to the grid with the benefits stipulated in the Chilean law that regulates self-generation [3,5]. A variable electricity tariff time-of-use (TOU) is considered here and the model's objective function is to minimize the net cost (purchase of energy), taking into account the customer's satisfaction, for which the start operating time of appliances such as washing machines or electric dryer can be deferred [6,7]. As a result, economic savings for customers and for the electric utility are realized and a better use of the installed capacity of photovoltaic panels in the microgrid is reached [6].

1.2 The important, albeit not yet fully understood role of energy storage for electric utilities' power distribution systems

The key role of energy storage and why it is important for electric utilities to advance this concept in the distribution of electricity sector

To improve this situation, a hybrid tariff has been proposed which mixes an hourly rate (TOU) with a rate based on the deviation of the system frequency—something which may affect voltage levels—and whose value represents the imbalance between generation and consumption. This hybrid rate, which can be calculated every minute, can provide secondary frequency regulation and it becomes an incentive for better solar management and use of energy storage systems. Hence, the case of a hybrid tariff system designed for standard residential customers living in high rise apartment buildings is applied to this example. The hybrid tariff system to be analyzed in this example, is to be applied to a community of residential customers in a large residential building in Santiago de Chile. In this hypothetical case, the community of residents in the building are all customers of ENEL Distribucion, the largest electric utility in Chile and one of the largest in Latin America. The microgrid installed in the building by ENEL has photovoltaic generation and energy storage both of which are analyzed under various operation conditions. The model simulation shows that for residential customers, the hybrid rate is cheaper than both the hourly rate and the flat rate, something which encourages the behavior of prosumers to be aligned with the correct functioning of the electric power system (the grid-tie microgrid and the grid supply) while actively encouraging the use of energy storage. In this situation, frequency control can be offered as an ancillary service by the prosumer [8].

Other studies, aimed at achieving some cooperation among customers of a residential community being serviced by a local electric utility have proposed a distributed management system of electricity demand, based on game theory, for a group of residential customers [9]. The model proposes a dynamic pricing strategy (DPS) in which the electricity rate to be charged is a function of the global power demand of the pool of customers. Under this scheme, multiple customers choose time periods within the day as a window of opportunity where the electric rate is cheaper, thus obtaining certain economic benefits being offered by the electric utility, while meeting their daily needs [9]. The system arrives at a Nash equilibrium point without intervention of the central operator. Simulations results using real energy consumption data show that peak system reduction is near 20%, thus decreasing the CAPEX necessary to supply the growth of energy consumption [9].

The present energy homeostaticity model, which was presented to ENEL for the purpose of its assessment and future implementation in the electricity distribution sector in Chile and other parts of the world, is not only aimed at enhancing, complementing and supporting a variety of electricity distribution services through smart distributed generation systems tied to the grid for residential and commercial applications. It is also a means to creating a new market with new incentives for those customers who have very different consumption patterns and who value green energy and are willing to support the incorporation and sustainment of green energy in the distribution system. Thus, these types of customers emulate the trends that have successfully been taking place in California, USA; Norway, Denmark and other countries that are moving towards a green energy matrix. The idea behind the premise of this project is that, knowing how new trends are favorably adopted in Chile and given the strong sense of commitment to a greener environment, such customers will want to move towards a new energy matrix for the country altogether. Homeostaticity makes this possible by providing a choice on how to consume more wisely that is convenient for the utility and for other energy consumers who have different habits. Moreover, homeostaticity makes it possible for the utility grid to operate much more securely and to be assisted by a microgrid that is running parallel with a control and energy management system that allows the DG system to be both highly efficient and also watchful, ready to support with ancillary services and back up if needed. These modular DG plants are to be installed and managed by ENEL in the near future as part of the plan to diversify in the electricity distribution sector in line with the Smart Grid transformation [10,11].

In this chapter, the analysis of the most convenient tariff and cost sharing between the customers of an apartment block is performed. Section 2 present the case study and the implied parts. In Section 3, the case of clustered clients is analyzed following two possible criteria for sharing renewable production: equal sharing and merit-based sharing. Section 4 present the case of separate customers and its implications. Simulation results are presented in Section 5. Finally, Section 6 contains the conclusions of the study.

1.3 Case study

The study consists of grouping 60 residential households, which along with the PV energy plant comprise a sustainable block for the electric distribution company. This community of clients will be connected to the electrical grid installed and managed by ENEL Distribucion. The solar PV plant—the microgrid—is especially designed to meet the needs of the community in a percentage of its total electricity consumption needs with renewable energy. As a whole, the DG plant will seek to offer the optimal rate that is possible for the electric utility to provide to its customers, subject to the energy system's conditions and constraints. The arrangement should result in economic benefits (incentives) for the residential consumers of the sustainable block in exchange for maintaining a scheme of efficient, sustainable electricity consumption, aligned with the needs of the entire community (aggregate demand). The expressions (1) and (2) for the energy equilibrium of the grid-tie microgrid is given in terms of the total power supply and the homeostasis regulation mechanisms discussed in Ref. [3]:

$$E_{equilb} = P_{supply}(x)PH(u)RH(v)S(\alpha) = E_{consump}(u,v,\alpha) + \frac{d}{dt}E_{consump}(u,v,\alpha) \tag{8.1}$$

$$P_{supply} = Real\ Power + Reactive\ Power = (P+Q) - Losses \tag{8.2}$$

Where x represents the internal state of the energy systems at time t_0 and Energy equilibrium E_{equilb} is dependent upon several factors operating adequately in the SES. Both u and v represent

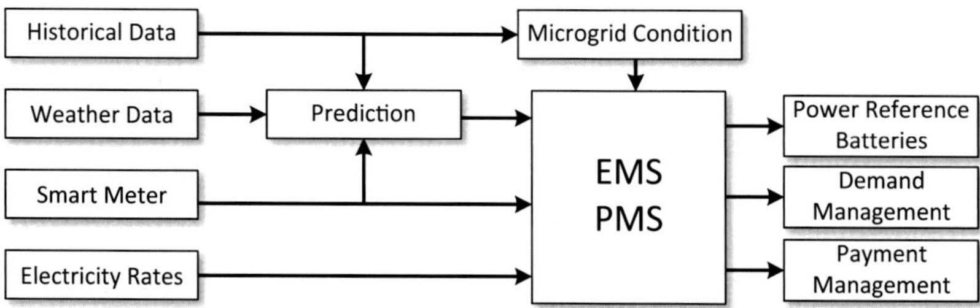

Figure 8.1. The figure shows supervisory control scheme for the sustainable block with energy management and power management systems.

the specific predictive and reactive homeostasis variables respectively, which are designed in the HC model. These are designed based on extensive data modeling to incorporate as much accuracy in the system's response as possible [3].

$S(\alpha)$ is the conditioning function of the SES and operates to alert and condition the HC system's adaptive mechanisms recursively in order to respond to a wide range of stimuli. Its actions are based on the situational awareness and degree of criticality being experienced by the system and the solution incorporates artificial intelligence and intelligent control. Thus $S(\alpha)$ is a function of adverse conditions and environmental challenges being sensed by the energy system and represented by the awareness and criticality variable α. All these three variables: u, v, α are incorporated as key constitutive elements of the intelligent algorithms built in the HC system and as such, are the equivalent of metabolic variables in living organisms like the physiological and endocrine systems' variables that affect the energy expenditure and storage of such systems [3].

Fig. 8.1 shows the general structure of the control system, where EMS (energy management system) and PMS (power management system) which will be designed following homeostatic control strategy. The EMS/PMS, receives as input the electric power generation predictions (based one predictive homeostasis data carried out by the HC system's assessment of internal and external variables) [3]. This is done taking into account the photovoltaic generation plant and the electricity consumption ranges in terms of demand side projection in order to decide on the magnitude and the energy flow. In addition, the storage status of the batteries must be monitored. Thus, pursuing the objective of minimizing operating costs, the homeostatic controller will have the following attributions:

A. Battery management

Defines when and how much energy to charge/discharge. The control system will charge the batteries when the demand is low and will draw energy from the batteries when the tariff of electricity is more expensive, depending on the electric tariff that is being implemented.

B. Active control of the energy demand

It is determined by how much energy is consumed by each client of the microgrid as recorded by the smart meters. Those customers who are not "solidary" or choose not to align their electricity consumption with the needs of the rest of the community, will be notified through an interface and/or alarm and their loads which exhibit constantly high electric power consumption (e.g., washing machine, charger or heating) will be disconnected by smart switches (Smart plug), leaving them with the grid-only option.

C. Payment management

This unit is responsible for prorating payments between users and the electric company. Customers who have low consumption of the microgrid supply (those that exhibit a thrifty consumption behavior) have the right to receive economic compensation (reward). Such reward is made possible by those who have a higher consumption of electrical energy, particularly those that use power consumption more often. This arrangement is being considered by ENEL Distribucion as a means to entice a sustainable and more manageable energy consumption in light of constraints imposed by DG plants generating mostly renewable energies. This, ENEL hopes, will in turn reinforce a frugal or thrifty electricity consumption behavior and an easier stabilization of the system if it were needed.

2 Deployment of distributed generation systems

Deployment of distributed generation systems for green energy integration in buildings and condos: a pending challenge for electric utilities that needs to be addressed

2.1 Electricity tariffs

Tariff calculation and the assignation of tariffs to clustered customers

The tariff calculation and the assignation of such tariffs to clustered customers in different scenarios evaluate a homeostatic control strategy that permits an efficient energy management in a residential building connected to the main grid, with a photovoltaic generation plant installed on it, plus energy storage and an energy management control system which has a homeostaticity model built in. For the local electric company ENEL Distribucion, the different customers of the building are considered as a single load, which is called SB. These customers ought to reach consensus in order to choose from tariffs that were previously exclusively reserved for the commercial and/or industrial sector. Based on this scenario, supervisory control strategies based on HC adapted to the specific requirements of the clients are applied, in order to obtain an efficient energy management.

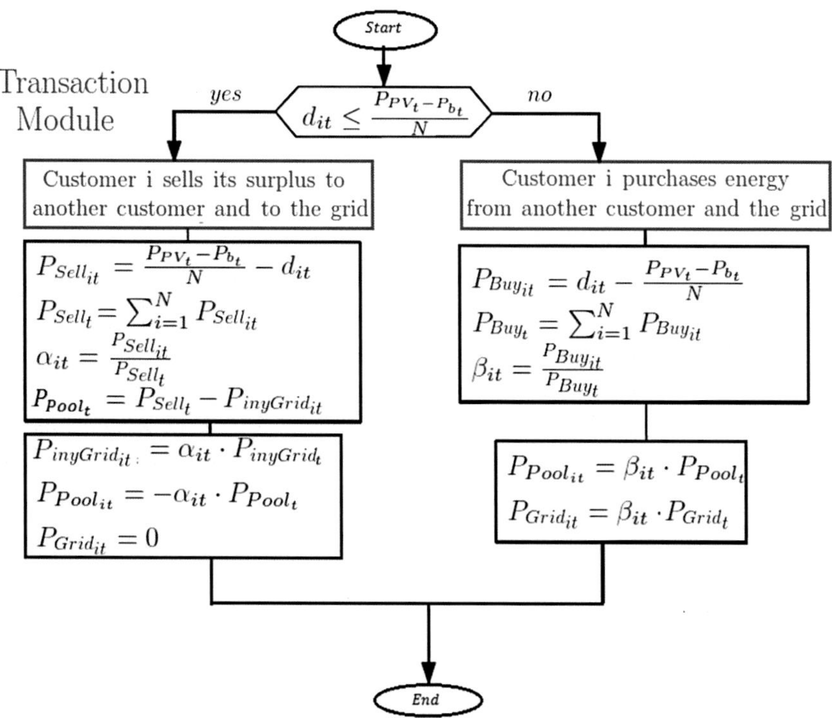

Figure 8.2. Flow diagram to supply renewable energy under equality criteria.

Such strategy should encourage a particular behavior of consumers in order to achieve and maintain a more flexible, adaptable and sustainable state of equilibrium. In this way, both supply and demand respond to each other in a cooperative way with mutual benefit.

The electricity tariffs that should be implemented for the SB will be BT4.3 or AT4.3, depending on the voltage level. These tariffs are common in the industrial sector, having the lowest price for the electric energy consumed, in addition to a charge for the maximum demand for power in peak hours. For this reason, the control strategy to be implemented should limit demand charges in order to maximize the benefits that can be granted to customers, and at the same time, the main network will operate at a higher efficiency point. The strategy incorporates both reactive and predictive homeostasis in such a way that achieves an efficient and sustainable balance. On the other hand, these systems can be seen as a complex sociotechnical system, in which energy users must play a crucial role as active loads within the sustainable block in order to make the most out of the microgrid's supply while always having the grid as a backup option.

The transaction module (Fig. 8.2) is in charge for assigning the energy quote for each client and calculates the energy flow among clients and the grid. In order to accomplish the above mention procedure, criteria A and B could be chosen.

2.2 Criteria A: Customers share the Nth part of generated renewable energy

The diagram in Fig. 8.2 highlights the strategy to be used, where each client owns one Nth part of the renewable energy produced and, for simplicity, all clients' charges and discharges the battery equally.

The module begins by discriminating between clients with energy excess or deficit, by using condition (8.8).

$$d_{it} =\leq \frac{P_{PV_t} - P_{b_t}}{N} \tag{8.3}$$

The client i that satisfies the condition (8.3) has excess of energy, expressed by (8.4), and can sell it to clients with energy deficit or to the network, as convenient.

$$P_{Sell_{it}} \leq \frac{P_{PV_t} - P_{b_t}}{N} - d_{it} \tag{8.4}$$

$$P_{Sell_t} \sum_{i=1}^{N} P_{Sell_{it}} \tag{8.5}$$

The sum of all the excess, given by (8.5), corresponds to the total energy available for selling, a fraction of this energy will feed the requirements of the customers with deficit P_{pool_t} and the rest will be injected into the network P_{inGrid_t}. The energy contribution of each client will be identified defining a factor according to (8.6).

$$\alpha_{it} = \frac{P_{Sell_{it}}}{P_{Sell_t}} \tag{8.6}$$

Then the energy supplied to customers with energy deficit and injected into the grid by the customers with excess i will be given by (8.7) and (8.8), respectively.

$$P_{pool_t} = P_{Sell_t} - P_{ingrid_t} \tag{8.7}$$

$$P_{pool_{it}} = \alpha_{it} \cdot P_{Pool_t} \tag{8.8}$$

$$P_{inGrid_{it}} = \alpha_{it} \cdot P_{inGrid_t} \tag{8.9}$$

If the condition (8.8) is not met, there is a client with energy deficit. This customer must use energy from the grid P_{grid_t} and/or from the excess of other customers with renewable energy P_{pool_t}. The procedure that describes the energy flow under this condition, is depicted in flow diagram (Fig. 8.2) to supply renewable energy under equality criteria included.

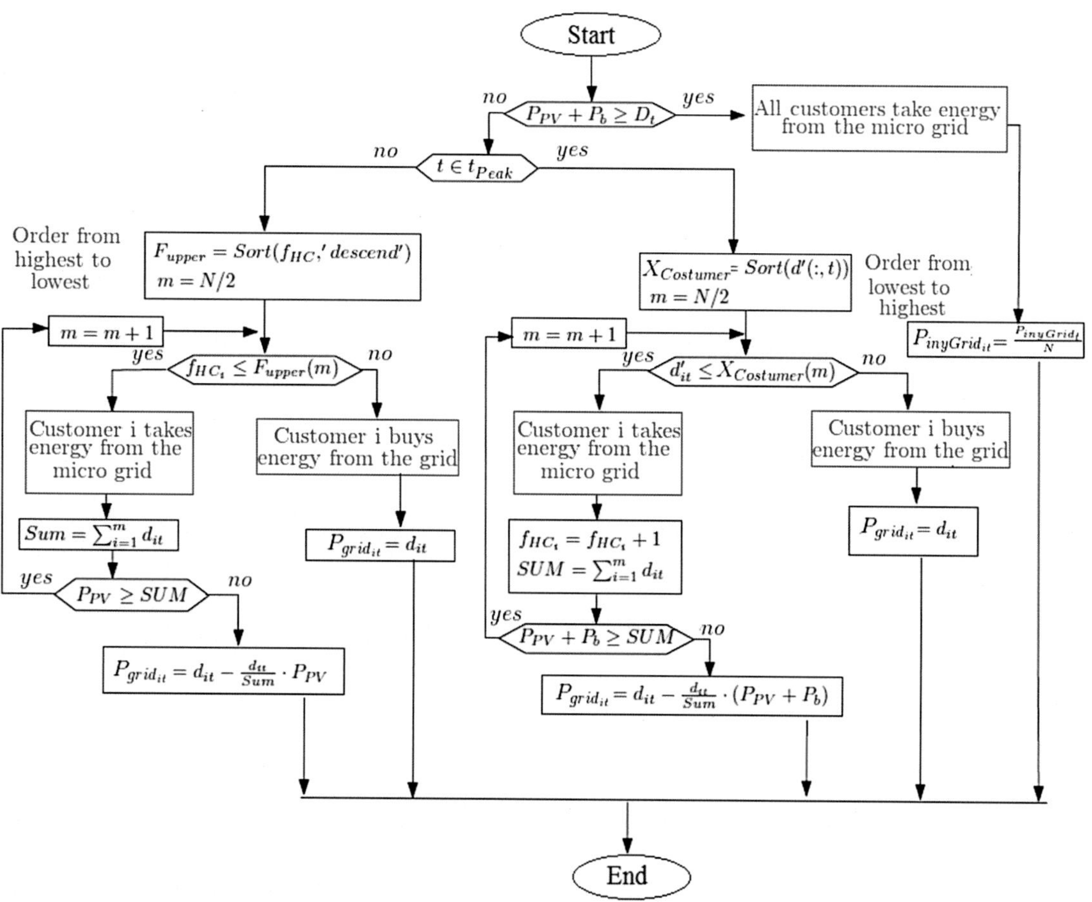

Figure 8.3. A particular flowchart depicting an energy homeostasis algorithm built to provide renewable energy to customers based on merit. Merit in this vase means adjusting consumption so as to make enough green energy available in the system for all. Those that exhibit energy consumption restraint and favor thriftiness will enjoy cheap electrical energy supply.

2.3 Criterion B: Substantial renewable energy supply according to customer merit

The flow diagram of Fig. 8.3 shows the control strategy of the power supply from the microgrid. Following this strategy, customers will get renewable energy as a reward by having an efficient and low consumption that enables to bounds the maximum demand.

The algorithm begins by checking if there is enough power from the microgrid to satisfy the demand of the SB by applying condition (8.10). If (8.10) is satisfy, all customers will take energy from the microgrid achieving 100% of their energy consumption. The excess of energy is injected into the main grid and clients will receive an equal income for that contribution.

$$P_{PV_t} + P_{b_t} \geq D_t \tag{8.10}$$

If condition (8.10) is not met, it is understood that the energy available in the microgrid is not enough to satisfy the demand. Therefore, this energy must be administered and delivered, as a reward for clients that have a low consumption during peak hours.

The module algorithm designed for peak hours is in charge of organizing the customers according to their energy consumption from lowest to highest. The first m customers will have the right to receive energy from the microgrid in proportion to their consumption, as indicated in (8.11).

$$P_{grid_{it}} = d_{it} - \frac{d_{it}}{Sum} \cdot \left(P_{PV_t} + P_{b_t}\right) \tag{8.11}$$

Where *Sum* corresponds to the sum of the consumptions of the first m customers. On the other hand, the remaining $N - m$ customers must satisfy 100% of their energy consumption from the utility. Customers who are allowed to receive energy from the microgrid at peak hours will increase an index called, "Homeostatic Index (f_{HC})" [4]. This index will be used to distribute the renewable energy in non-peak hours of the next day. In addition, customers will be ranked from higher to lower according to the mentioned index, so the first m clients with higher f_{HC} will obtain energy from the microgrid.

Due to the high cost of the peak hour demand, the control system must encourage customers to consume during off-peak hours. To meet this goal, the cost of the electricity supply is transferred to consumers through an internal tariff, which differentiates between low and high demands. Fig. 8.3 and Fig. 8.4 illustrate the applied internal tariff; this will be based on the monomic energy price, which consists of a single equivalent price per kWh that considers both, the energy and the power charge.

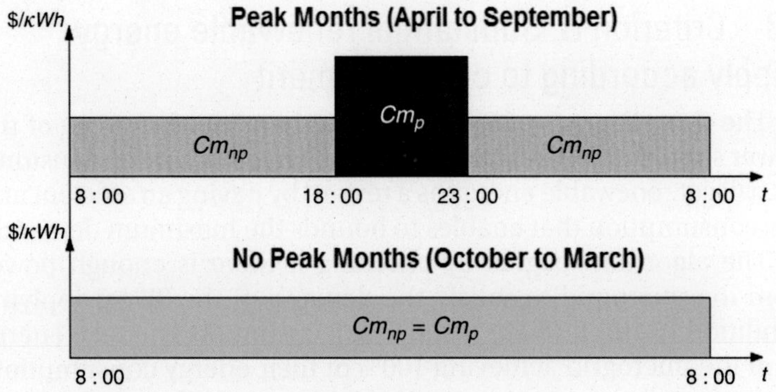

Figure 8.4. Internal tariff for sustainable block customers.

$$C_{m_p}(month_p)=\frac{\sum_{t\in h_p} P_{grid_t}\cdot\Delta t\cdot C_e+MD_{h_p}\cdot C_p}{\sum_{t\in h_p} P_{grid_t}\cdot\Delta t} \quad (8.12)$$

$$C_{m_{op}}(month_p)=\frac{\sum_{t\in h_{op}} P_{grid_t}\cdot\Delta t\cdot C_e+P_{max_t}\cdot C_{op}}{\sum_{t\in h_{op}} P_{grid_t}\cdot\Delta t} \quad (8.13)$$

$$C_{m_{op}}=C_{m_p}(month_{op})=\frac{\sum_{t\in month} P_{grid_t}\cdot\Delta t\cdot C_e+P_{max_t}\cdot C_{op}+P_{max_{pt}}\cdot C_p}{\sum_{t\in month} P_{grid_t}\cdot\Delta t} \quad (8.14)$$

Eqs. (8.12) and (8.13) correspond to the monomic price that should be used at peak hours and off peak hours, respectively. The monomic price for peak hours (C_{Mp}) is calculated on the basis of energy consumed and the maximum read demand, both in peak hours. On the other hand, the other monomic price is calculated in a similar way, but considering off peak hours. Monomic prices are calculated monthly together with the billing cycle. During the months that do not contain peak hours (October to March), the monomorphic peak and non-peak costs are the same and the calculation is according to Eq. (8.14).

After 1 year of evaluation ($t = 35040$, $\Delta t = 15$ min), the monthly and annual costs are calculated for each clients and the BS according to Eqs. (8.15) to (8.19) (prices are shown in Ref. [12]).

$$C_{SB_{month_{op}}} = \sum_{t\in Month} P_{grid_t}\cdot\Delta t\cdot C_e+ \underset{m\in month_p}{P_{max\,m}}\cdot C_p+P_{max_t}\cdot C_{op}+C_{fix} \quad (8.15)$$

$$C_{SB_{month_{op}}} = \sum_{t\in Month} P_{grid_t}\cdot\Delta t\cdot C_e+P_{max_{pt}}\cdot C_p+P_{max_t}\cdot C_{op}+C_{fix} \quad (8.16)$$

$$C_{SB_{year}} = \sum_{month=1}^{12} C_{BS_{Month}} \quad (8.17)$$

$$C_{i_{month}} = \left\{ \sum_{t\in Month} P_{grid_{it}}\cdot C_{m_t} + P_{pool_{it}}\cdot C_{pool} - P_{inGrid_{it}}\cdot C_{inGrid} \right\}\cdot\Delta t + C_{fix} \quad (8.18)$$

$$C_{i_{year}} = \sum_{month=1}^{12} C_{i_{Month}} \quad (8.19)$$

The algorithm is repeated up to 20 years (PV lifetime), each year a loss of efficiency in photovoltaic panels equal to 0.6% is added and a linear reduction of battery capacity is also considered, so that the final battery capacity is 80%. The depth of discharge of the battery is adjusted so that no intermediate replacements occur.

3 Analysis on Chilean potential case scenario

How to deploy distributed generation systems for green energy integration in buildings and condos: a Chilean potential case scenario

3.1 Separate customers' scenario

Upon analyzing the separate customers' scenario, one can notice the freedom of choice, one of the key aspects of grid flexibility policy that electric utilities are currently considering. Here customers are free to choose between different electric rates in the corresponding voltage level. Among the rates offered by the local electricity company, described in a previous section, only BT-1 tariffs and THR are competitive for levels and consumption characteristics of individual customers. In this scenario the option of incorporating a photovoltaic plant in the common roof of the building and an energy storage system is evaluated. Since the energy meter of each customer is operated by the electricity utility, the only option is to deliver the renewable energy to the common services of the building and / or to the main grid. As illustrated in Fig. 8.5, the meter of the customer will effectively record its electricity consumption but will not discriminate if it is supplied by the main network or the micro-grid, generating a conflict between the Electricity Company and customers of the building.

Therefore, in agreement with current regulations specified in [13], the convenient strategy is to inject renewable energy into the Common services of the building and then to the main network.

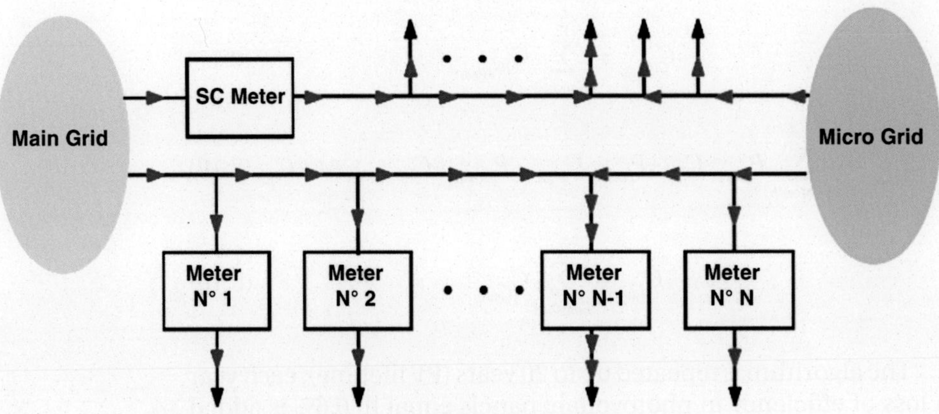

Figure 8.5. Energy flux and metering for separate customers.

The most common tariff used BT-3 and in addition, it is assumed that all the renewable energy is self-consumed. The project income should be calculated as the savings on the electricity common services bill paid by customers on a monthly basis.

3.2 Simulation results

In this section, results obtained by simulations are presented and analyzed. These results validate the homeostatic control strategy used to manage the energy of the customers, taking into consideration the benefits that they would receive under different alternatives.

A common practice to reduce the maximum demand during peak hours, is to charge the batteries from the main grid during the low demand hours based on weather forecasts and algorithms to predict the photovoltaic generation, the aim is to get into peak hours with the batteries fully charged. In this way, the benefits of the electric tariff to be used (BT4.3 or BT4.3) are maximized. As an example, the Fig. 8.6 shows the power flows between the different elements during one day considering the existence of a battery and applying hourly tariff.

Whether the microgrid has or not a storage device, the control system shall be provided with a set of controllable loads which

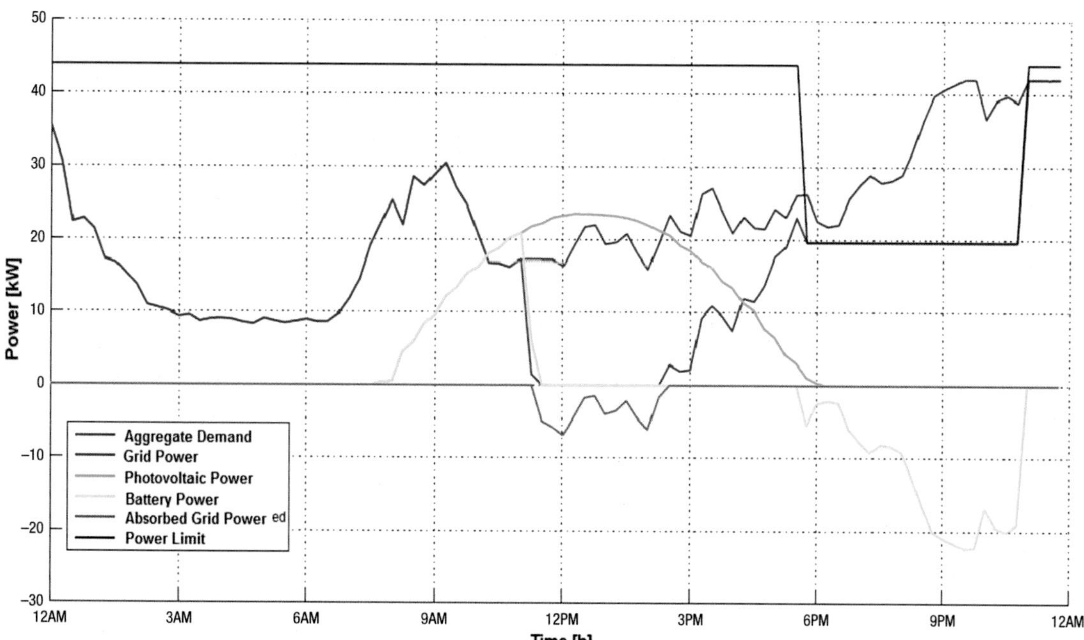

Figure 8.6. SB Power flow with battery and hourly tariff.

can be remotely disconnected, so the maximum demand can be maintained bellow a specified limit. In addition, customers will be notified automatically that their behavior is not being solidary with the needs of the community, and they will be penalized. Thus, the microgrid can be seen as a socio-technical complex system, in which energy users play a crucial role as active loads.

Internal electric tariff for microgrid customers is shown in Fig. 8.7. This tariff, based on the monomic energy cost is employed with the aim to achieve an efficient energy consumption and to transfer the energy directly the cost to customers. It can be observed that the energy cost at peak time is considerably higher than that of non-peak hours, so customers are expected to adapt and move part of their consumption to low demand hours, where the energy cost is lower. If, however, there is a shortage of supply from the microgrid for whatever reason, the grid supply will automatically take over and supply for the deficit.

Figs. 8.8 and 8.9 represent the two different criteria for allocating renewable energy. Fig. 8.8, defined as criterion A, responds to a logic in which customers own the N-th share of renewable energy available, indifferent to their consumption patterns, being able to sell its surplus to other clients and to the network. Fig. 8.9, defined as criterion B, corresponds to the allocation of renewable energy under a scheme based on merit. Customers who have a low consumption in the peak hours are entitled to obtain the renewable energy proportionally to their consumption. Choosing between one criterion and another, will depend on the degree of

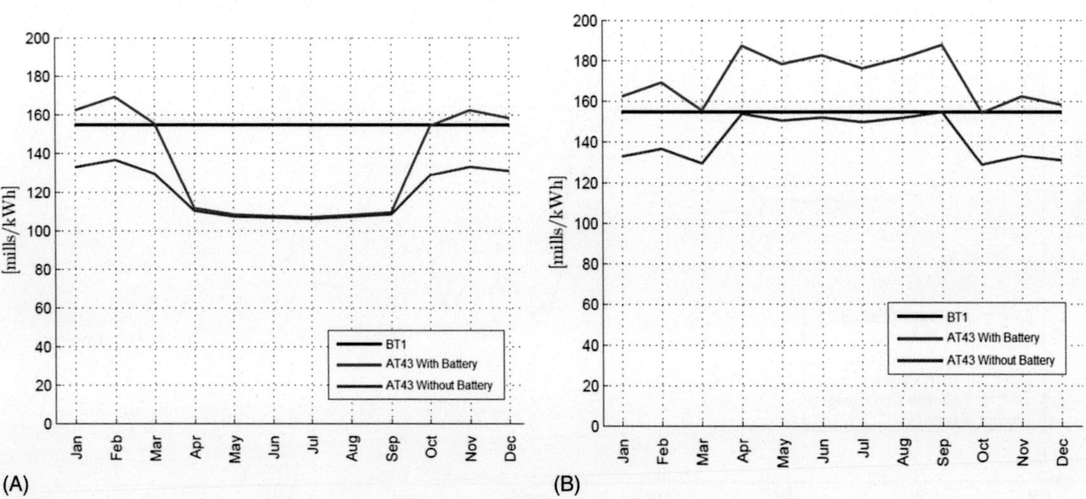

Figure 8.7. Energy monomic cost. (A) Off-peak hours. (B) Peak hours.

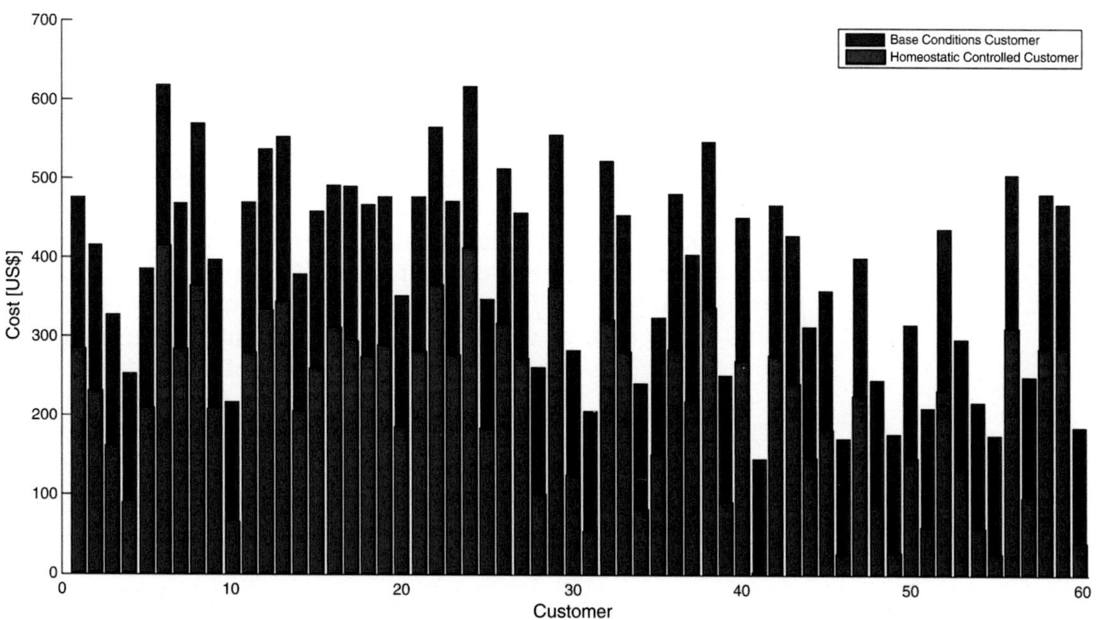

Figure 8.8. Annual energy cost per client under A criterion.

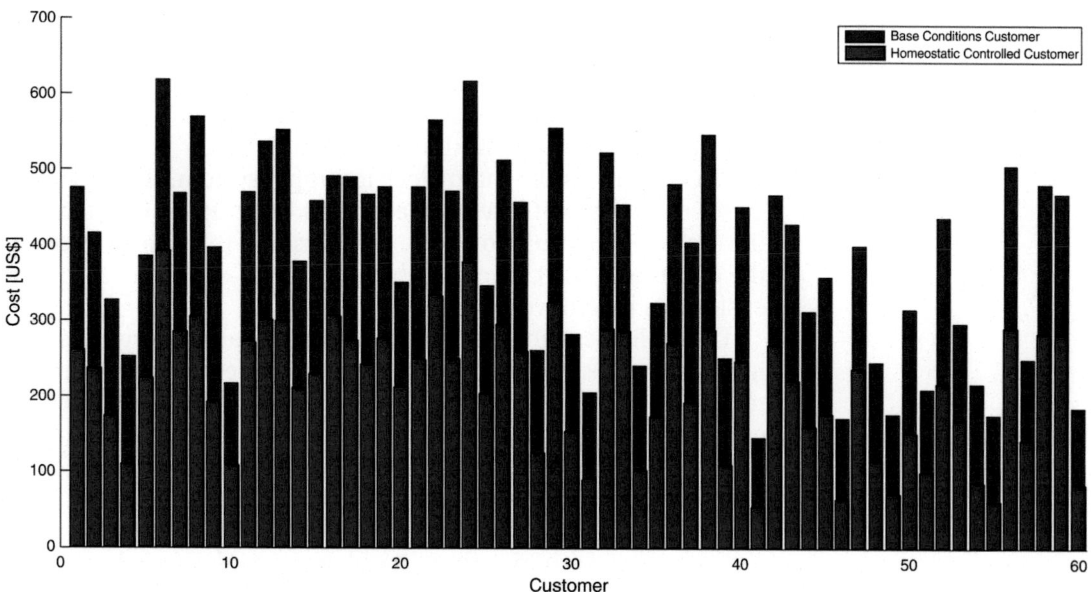

Figure 8.9. Annual energy cost per client under B criterion.

commitment that customers may have and their tendency to save energy in order to use more green energy rather than taking more from the grid.

Criterion A has more sense to be used in case there is a renewable generation and energy storage in accordance with the consumption of the sustainable block. On the other hand, criterion B might be more sensible to use in case you have a microgrid with a generation well below the needs of customers as a whole, so their allocation should be in basis of merit. Under criterion B, the homeostaticity factor f_{HCi} represents a measure of energy efficiency and energy savings, where the clients who accumulate a higher index are the ones who deserve the allocation of renewable energy. For purpose of this work, this factor was reset daily when simulation was carried out, in a real implementation could be reset weekly, monthly, and even annually, so that customers who have a sustained efficient consumption can pay off their consumption with renewable energy.

4 Conclusions

Upon concluding this chapter, it is useful to offer some insight and reflections from industry and academia in regards to what the future holds with climate change and severe drought scenarios. If anything one can hear an urgent call for action on the part of government and electric utilities in order to change how electric power systems infrastructure and electricity consumption have been managed for over a century. When looking at the case study and the different scenarios that it presents, each with distinct tariffs, it is clear that there is an aim designed to favor particular consumption profiles that better help maintain the DG supply as the main electric supply system, thus favoring green energy albeit having the grid as back up. Particularly, in the last set of simulations, the results show that whether using the criterion A or B, a very high reduction in annual cost for the customers is achieved.

Using the described techniques for renewable energy sharing, it is easier to establish and understand how every customer will receive their part of the generated energy, avoiding discordances and problems between them. Moreover, the case of separated (also called non-clustered) customers is considered, being possible to put them apart from the common billing and energy sharing system.

Using these criteria, some problems regarding renewable resources sharing can be solved, encouraging customers to install these systems on their blocks.

Finally, it is important to realize the fragility of today's electric power distribution infrastructure, particularly in Chile where seismic activity is recurrent. Thus it is crucial that government authorities, industry regulators and main industry players like ENEL Distribucion in Chile plan ahead and work on a Smart Grid Transformation Roadmap as Chile is doing, in order to advance and pave the way for utilities to embrace grid integrated distributed generators investments. There is much new technology and techniques in the market today to go that route safely [14].

Acknowledgments

Dr. Fernando Yanine is grateful to CONICYT of Chile for his scholarship endowment of his Postdoctoral FONDECYT N 3170399 "BUILDING SUSTAINABLE ENERGY SYSTEMS TIED TO GRID: A RESEARCH PROJECT PARTNERSHIP WITH CHILECTRA TO INSTALL HOMEOSTATICALLY CONTROLLED MICROGRIDS IN APARTMENT BUILDINGS" and to ENEL Distribucion S.A. for its auspice, sponsorship and support. Thanks are also given to Universidad Finis Terrae for its ongoing support in this research initiative and to the Department of Electrical Engineering of UTFSM.

Dr. Antonio Sanchez-Squella is supported by a research grant from FONDECYT #11150911 of Chile.

References

[1] ENEL. Microgrids: The Future of Energy, Available from: https://www.enel.com/media/news/d/2014/05/microgrids-the-future-of-energy.

[2] World Economic Forum. The Future of Electricity: New Technologies Transforming the Grid Edge, Available from: http://www3.weforum.org/docs/WEF_Future_of_Electricity_2017.pdf.

[3] F. Yanine, A. Sanchez-Squella, A. Barrueto, J. Tosso, F.M. Cordova, H.C. Rother, Reviewing homeostasis of sustainable energy systems: How reactive and predictive homeostasis can enable electric utilities to operate distributed generation as part of their power supply services, Renew. Sustain. Energy Rev. 81 (2017) 2879–2892.

[4] M. Manfren, P. Caputo, G. Costa, Paradigm shift in urban energy systems through distributed generation: Methods and models, Appl. Energy 88 (4) (2011) 1032–1048.

[5] F.F. Yanine, A. Sanchez-Squella, A. Barrueto, Cordova, S.K. Sahoo, Engineering sustainable energy systems: how reactive and predictive homeostatic control can prepare electric power systems for environmental challenges, Proc. Comput. Sci. 122 (2017) 439–446.

[6] S. Gottwalt, W. Ketter, C. Block, J. Collins, C. Weinhardt, Demand side management—A simulation of household behavior under variable prices, Energy Policy 39 (12) (2011) 8163–8174.

[7] P. Palensky, D. Dietrich, Demand side management: Demand response, intelligent energy systems, and smart loads, IEEE Trans. Indus. Inform. 7 (3) (2011) 381–388.

[8] S. Hambridge, N. Lu, A.Q. Huang, R. Yu, A frequency based real-time electricity rate for residential prosumers, in: Power and Energy Society General Meeting, 2017 IEEE, 2017, pp. 1–5.

[9] A. Barbato, A. Capone, L. Chen, F. Martignon, S. Paris, A power scheduling game for reducing the peak demand of residential users, in: Online Conference on Green Communications (GreenCom), 2013 IEEE, 2013, pp. 137–142.
[10] ENEL. World Energy Outlook, electricity is central, Available from: https://www.enel.com/stories/a/2018/12/world-energy-outlook-2018-presented-in-rome.
[11] World Economic Forum. The Future of Electricity New Technologies Transforming the Grid Edge, Available from: http://www3.weforum.org/docs/WEF_Future_of_Electricity_2017.pdf.
[12] S. Nasirov, C. Silva, C. Agostini, Investors' perspectives on barriers to the deployment of renewable energy sources in Chile, Energies 8 (5) (2015) 3794–3814.
[13] Net-Metering/Billing in Chile, Available from: https://energypedia.info/wiki/Net-Metering_/_Billing_in_Chile.
[14] P. Paliwal, N.P. Patidar, R.K. Nema, Planning of grid integrated distributed generators: A review of technology, objectives and techniques, Renew. Sustain. Energy Rev. 40 (2014) 557–570.

Further reading

[1] International Energy Agency. World Energy Outlook 2018, Available from: https://www.iea.org/weo2018/.
[2] International Energy Agency. World Energy Outlook 2017: A world in transformation, Available from: https://www.iea.org/weo2017/.
[3] ENEL. Smart City Santiago, Available from: http://www.smartcitysantiago.cl/smart-grid, 2019.
[4] S. Jain, S. Kalambe, G. Agnihotri, A. Mishra, Distributed generation deployment: State-of-the-art of distribution system planning in sustainable era, Renew. Sustain. Energy Rev. 77 (2017) 363–385.
[5] Y.S. Bhavsar, P.V. Joshi, S.M. Akolkar, Simulation of Microgrid with energy management system. 2015 International Conference on Energy Systems and Applications, 2015, pp. 592–596. doi:10.1109/ICESA.2015.7503418.
[6] Y. Cheng, S. Zhang, C. Huan, M.O. Oladokun, Z. Lin, Optimization on fresh outdoor air ratio of air conditioning system with stratum ventilation for both targeted indoor air quality and maximal energy saving, Build. Environ. 147 (2019) 1122, doi: 10.1016/j.buildenv2018.10.009.
[7] ENEL. ENEL Operates World's First "Plug And Play" Micro-Grid Powered By Solar PV and Hydrogen-Based Storage in Chile, Available from: https://www.enel.com/media/press/d/2017/05/enel-operates-worlds-first-plug-and-play-micro-grid-powered-by-solar-pv-and-hydrogen-based-storage-in-chile.
[8] ENEL. Enel Distribuzione: Italy's First Smart Grid in Isernia, Available from: https://www.enel.com/media/press/d/2011/11/enel-distribuzione-italys-first-smart-grid-in--isernia.
[9] W. Li, L. Yang, Y. Ji, P. Xu, Estimating demand response potential under coupled thermal inertia of building and air-conditioning system, Energy Build. 182 (2019) 19–29, doi: 10.1016/j.enbuild.2018.10.022.
[10] M. Brettel, N. Friederichsen, M. Keller, M. Rosenberg, How virtualization, decentralization and network building change the manufacturing landscape: An industry 4.0 perspective, Int. J. Mech. Ind. Sci. Eng. 8 (1) (2014) 37–44.
[11] M. Wolsink, The research agenda on social acceptance of distributed generation in smart grids: Renewable as common pool resources, Renew. Sustain. Energy Rev. 16 (1) (2012) 822–835.
[12] K. Moslehi, R. Kumar, A reliability perspective of the smart grid, IEEE Trans. Smart Grid 1 (1) (2010) 57–64.

[13] Future Market Insights. Microgrid Market: Global Industry Analysis and Opportunity Assessment 2014–2020, Available from: https://www.futuremarketinsights.com/reports/global-microgrid-market.
[14] P. Basak, S. Chowdhury, S.H. nee Dey, S.P. Chowdhury, A literature review on integration of distributed energy resources in the perspective of control, protection and stability of microgrid, Renew. Sustain. Energy Rev. 16 (8) (2012) 5545–5556.
[15] H. Jiayi, J. Chuanwen, X. Rong, A review on distributed energy resources and microgrid, Renew. Sustain. Energy Rev. 12 (9) (2008) 2472–2483.
[16] F.F. Yanine, F.I. Caballero, E.E. Sauma, F.M. Córdova, Homeostatic control, smart metering and efficient energy supply and consumption criteria: A means to building more sustainable hybrid micro-generation systems, Renew. Sustain. Energy Rev. 38 (2014) 235–258 doi:https://doi.org/10.1016/j.rser.2014.05.078.
[17] F.F. Yanine, F.I. Caballero, E.E. Sauma, F.M. Córdova, Building sustainable energy systems: Homeostatic control of grid-connected microgrids, as a means to reconcile power supply and energy demand response management, Renew. Sustain. Energy Rev. 40 (2014) 1168–1191 doi: https://doi.org/10.1016/j.rser.2014.08.017.
[18] F.F. Yanine, F.M. Córdova, L. Valenzuela, Sustainable hybrid energy systems: an energy and exergy management approach with homeostatic control of microgrids, Proc. Comput. Sci. 55 (2015) 642–649 doi:https://doi.org/10.1016/j.procs.2015.07.060.
[19] F.F. Yanine, E.E. Sauma, Review of grid-tie micro-generation systems without energy storage: Towards a new approach to sustainable hybrid energy systems linked to energy efficiency, Renew. Sustain. Energy Rev. 26 (2013) 60–95.
[20] F. Yanine et al. Smart energy systems: the need to incorporate homeostatically controlled microgrids to the electric power distribution industry: an electric utilities' perspective. Int. J. Eng. Technol. 7(2.28) (2018) 64–73. Available from: https://www.sciencepubco.com/index.php/ijet/article/view/12883/5131. http://dx.doi.org/10.14419/ijet.v7i2.28.12883.
[21] E.F. Caballero, Sauma, F. Yanine, Business optimal design of a grid-connected hybrid PV (photovoltaic)-wind energy system without energy storage for an Easter Island's block, Energy 61 (2013) 248–261.
[22] F.M. Cordova, F.F Yanine, Homeostatic control of sustainable energy grid applied to natural disasters. Int. J. Comput. Commun. Control 8(1) (2012) 50–60.
[23] F. Yanine, F.M. Córdova., Homeostatic control in grid-connected micro-generation power systems: A means to adapt to changing scenarios while preserving energy sustainability, in: Renewable and Sustainable Energy Conference (IRSEC), 2013 IEEE International, pp. 525–530.
[24] B. Singh, J. Sharma, A review on distributed generation planning, Renew. Sustain. Energy Rev. 76 (2017) 529–544.
[25] H. de Faria Jr., F.B. Trigoso, J.A. Cavalcanti, Review of distributed generation with photovoltaic grid connected systems in Brazil: Challenges and prospects, Renew. Sustain. Energy Rev. 75 (2017) 469–475.
[26] A. Thornton, C.R. Monroy, Distributed power generation in the United States, Renew. Sustain. Energy Rev. 15 (9) (2011) 4809–4817.
[27] E. Londono, Chile's Energy Transformation is Powered by Wind, Sun and Volcanoes; New York Times, Available from: https://www.nytimes.com/2017/08/12/world/americas/chile-green-energy-geothermal.html, 2017.
[28] P. Morgan, D. Martinez, A. Maxwell, From Good To Great: The Next Step in Chilean Energy Efficiency. Natural Resources Defence Council (NRDC). https://www.nrdc.org/sites/default/files/chile-energy-efficiency-report.pdf, 2014.

[29] N. Fitzgerald, A.M. Foley, E. McKeogh, Integrating wind power using intelligent electric water heating, Energy 48 (1) (2012) 135–143.
[30] C.A.G. Garcez, What do we know about the study of distributed generation policies and regulations in the Americas? A systematic review of literature, Renew. Sustain. Energy Rev. 75 (2017) 1404–1416.
[31] M. Muratori, G. Rizzoni, Residential demand response: Dynamic energy management and time-varying electricity pricing, IEEE Trans. Power Syst. 31 (2) (2016) 1108–1117.
[32] C. Silva, S. Nasirov, Chile: Paving the way for sustainable energy planning, Energy Sources Part B Econ. Plan. Policy 12 (1) (2017) 56–62.
[33] J. Ye, J. Xue, D. Wang, X. Zhou, W. Wang, H. Liu, Research on optimal scheduling strategy for household loads considering rooftop photovoltaic, in: Chinese Automation Congress (CAC), IEEE, 2017, pp. 1454–1459.
[34] W. Muneer, K. Bhattacharya, C.A. Canizares, Large-scale solar PV investment models, tools, and analysis: The Ontario case, IEEE Trans. Power Syst. 26 (4) (2011) 2547–2555.
[35] ENL Distribucion. "Tarifas", Available from: https://www.chilectra.cl/tarifas.
[36] Yanine F. Fernando, Felisa M. Córdova, Lionel Valenzuela, Sustainable hybrid energy systems: an energy and exergy management approach with homeostatic control of microgrids, Proc. Comput. Sci. 55 (2015) 642–649.
[37] F.F. Yanine, E.E. Sauma, F.M. Cordova, An exergy and homeostatic control approach to sustainable grid-connected microgrids without energy storage, Appl. Mech. Mater. 472 (2014) 1027–1031.
[38] J.C. Romero, P. Linares, Exergy as a global energy sustainability indicator. A review of the state of the art, Renew. Sustain. Energy Rev. 33 (2014) 427–442.
[39] A. Rearte-Jorquera, A. Sánchez-Squella, H. Pulgar-Painemal, A. Barrueto-Guzmán, Impact of residential photovoltaic generation in smart grid operation: real example, Proc. Comput. Sci. 55 (2015) 1390–1399 doi:https://doi.org/10.1016/j.procs.2015.07.129.
[40] H. Lund, A.N. Andersen, P.A. Østergaard, B.V. Mathiesen, D. Connolly, From Electricity smart grids to smart energy systems–a market operation based approach and understanding, Energy 42 (1) (2012) 96–102.
[41] H. Lund, P.A. Østergaard, D. Connolly, B.V. Mathiesen, Smart energy and smart energy systems, Energy 137 (2017) 556–565.

9

Integrating green energy into the grid: how to engineer energy homeostaticity, flexibility and resiliency in electric power distribution systems and why should electric utilities care

Fernando Yanine[a], Antonio Sanchez-Squella[b], Aldo Barrueto[b], Sarat Kumar Sahoo[c], Antonio Parejo[d], Dhruv Shah[e], Felisa Cordova[a]

[a]Faculty of Engineering, Universidad Finis Terrae, Santiago, Chile; [b]Department of Electrical Engineering, Universidad Técnica Federico Santa María (UTFSM), Santiago, Chile; [c]A Constituent College of Biju Patnaik Technological University, Parala Maharaja Engineering College, Department of Electrical Engineering, Govt. of Odisha, Berhampur, Odisha, India; [d]Department of Electronic Technology, Escuela Politécnica Superior, University of Seville, Seville, Spain; [e]Mukesh Patel School of Technology Management & Engineering (MPSTME), Mumbai Campus, India

Chapter outline

Low Carbon Energy Technologies in Sustainable Energy Systems. http://dx.doi.org/10.1016/B978-0-12-822897-5.00009-2

1 Introduction

In today's world, electric utility services are facing a number of threats, from harsh, and unexpected weather events like torrential rains, winds, or heavy snow in regions that did not normally have these happenings, as well as natural disasters like earthquakes and large fires, right along with unexpected malicious acts. The issue has hit the United States of America quite hard in later years and measures have been taken about these threats. In 2014 the US Senate passed a bill supporting grid-connected distributed energy systems like microgrids to confront the growing climate threats, so as to make electric power distribution more resilient and proactive towards the unexpected, particularly harsh weather like storms and floods. This is in line with Smart Grid transformation and modernization in the US, Canada and Europe. Such initiatives being led by big players like ENEL, have spurred several pieces of legislation not only in North America but also in South America, where changes in the electric law are being studied by industry stakeholders, legislators, and also by several local and federal authorities to promote a range of technologies and policies that can make the grid more reliable, resilient and cyber-secure [Web-1, Web-2]. All of these steps fall, in one way or another, on the path set forth by President Obama in 2013, when he introduced the Energy Independence Roadmap for the country [Web-2]. However, the pace towards more decisive and concrete steps to bring about badly needed industry changes have somehow sped up even more, following the catastrophic natural disasters that the United States has had to endure in recent years [1]. Among such initiatives are grants, microgrid technologies' demonstration projects, university initiatives, and various studies to determine the costs involved in the industry transformation and to define the precise scope of action in order to shape federal microgrid policy in the immediate future [2]. A significant step forward was the Energy Policy Modernization Act of 2015 that was introduced in the US Senate on September 2015, marking it a significant step forward towards Grid modernization in the US.

In Chile, like in North America, authorities are very concerned at the gloomy outlook of having public utility services shutdown and transport infrastructure break down and collapse as a result of earthquakes and violent weather phenomena. Both types of phenomena are nothing new in Chile and are expected to become more severe in years to come [3]. In this context, some power and water utilities are already taking action by adopting plans to counteract or fend-off such adverse conditions and circumstances, like those that prompt wild fires. One of such utilities is ENEL Distribucion in Chile, which currently supports ongoing research on Smart Grid transformation to explore the potential incorporation of grid-tied microgrids in the already huge number of buildings in the city of Santiago. Santiago is Chile's capital and has a very large metropolitan area of over 5 million people, and a very large portion of the Metropolitan Region is serviced by ENEL Distribucion. Thus ENEL is looking at microgrids of various sizes to act as a complementary alternative to grid-only power distribution. Hence, the present chapter discusses such research initiative by first presenting a prescriptive energy and power management control system model (EPMCSM) which was presented to and assessed by ENEL and which employs energy homeostaticity (EH) by means of homeostatic control (HC) strategies for microgrids [4,5]. The present model is being considered by electric utilities like ENEL in Chile to operate grid-tied microgrids of different sizes that have the potential of being an means to incorporate green energy in the distribution sector. Such distributed energy systems are to be installed and operated by the electric utilities themselves and applied to the large stock of residential and commercial buildings in high density urban areas. ENEL Distribucion is part of ENEL Chile (https://www.enelchile.cl/ es.html.html), the largest electric utility in Chile and in the region.

The work presented here is part of an ongoing research program that was partially funded by FONDECYT 3170399 of ANID (Agencia Nacional de Investigación y Desarrollo) of Chile under the auspice and support of ENEL Distribucion in Chile. Section 2 discusses the energy homeostaticity concept and explains how homeostaticity in energy systems can be incorporated to electric power systems connected to the grid. The chapter also discusses its advantages and feasibility. Section 3 briefly shows and discusses the control engineering design, simulations and results. Finally, Section 4 presents the conclusions.

2 How to incorporate energy homeostaticity in electric power systems?

2.1 Thriftiness and resiliency in electric power systems

The concepts of thriftiness and resiliency in electric power systems in general and in distributed generation systems tied to the grid in particular: how to engineer them in green energy systems integration in buildings and condos

Ever since Cannon first formulated the concept of homeostasis over 80 years ago [6,7], attention has largely been focused on corrective responses initiated after the steady state of the living organism is perturbed. However, the concept of homeostasis not only encompasses reactive homeostasis (RH) but should also be extended to include predictive homeostasis (PH) mechanisms, operating recursively and in coordination with one another in the face of environmental or external challenges [4,5]. In a different context altogether, the concept of homeostaticity has been defined as a property of databases wherein the database has the ability "to restore its integrity constraints after ruinous disturbances of its medium" [6,8]. Yet, in the context of energy systems in general and particularly in the context of energy sustainability, homeostaticity is defined as a property of sustainable energy systems (SES) to be able to react and respond very rapidly and effectively (in fractions of a second) to challenges and perturbations brought upon the energy system, so as to restore stability and continuity of operation while, at the same time, striving to attain efficient equilibrium between the amount of electricity supplied by the energy system and the demand for electricity from the loads. The essential point here is not only the need to preserve systems stability and continuity of operations in electric power systems but also to reach and maintain an optimal equilibrium in the energy system. Reactive homeostasis—as the name suggests—is a feedback-enabled mechanism that alerts and prompts the system to act when there is an imbalance between supply and demand or some other type of perturbation, in order to attain homeostaticity [4,5]. The proposed application of energy homeostaticity for the microgrid is shown in Fig. 9.1.

3 Control engineering design

3.1 Why is energy efficiency (EF) not enough?

Lessons learned from technology initiatives, international applications, and future work projections so far

Sometimes we pause and think: why is energy efficiency not enough? Although EF was hailed some time back as the 5th energy source, and one that perhaps could really solve the energy shortage to a large extent, evidence has shown that it is simply not enough, and that without structural changes in the energy matrix and new technology innovations, we will never get far enough. Lessons learned from technology initiatives, international applications and future work projections so far have all pointed quite clearly to the need to do more in terms of transforming the grid and consider the role of green energy systems tied to the grid in order to complement and potentiate whatever can be achieved with EF. Indeed, EF all by itself is clearly not enough and this is especially true in large metropolitan areas where EF measures have fallen short of expectations, not delivering what was once thought to be enough.

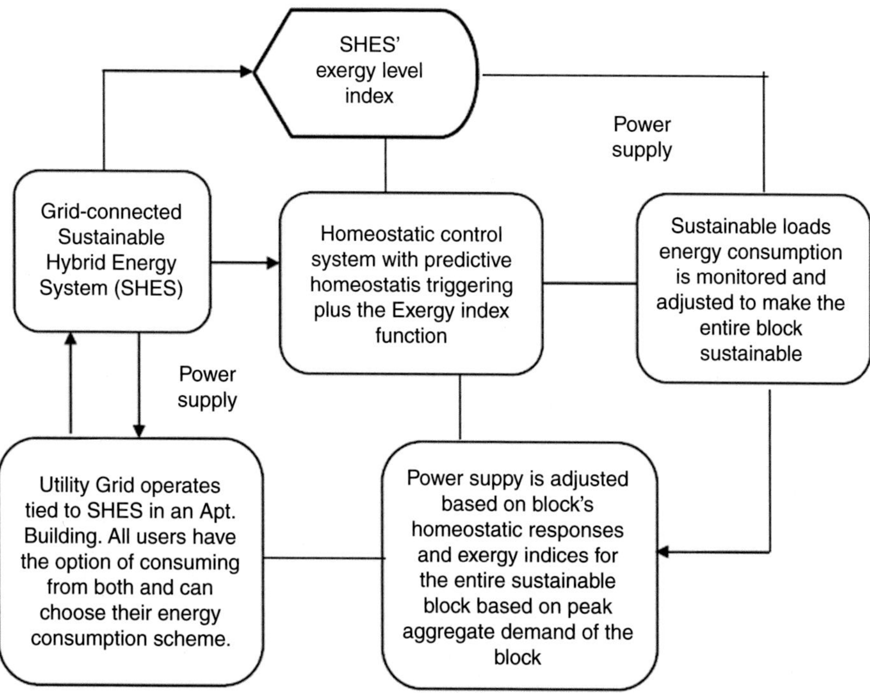

Figure 9.1. Diagram showing predictive and reactive homeostasis mechanisms operating as part of the HC of the microgrid, this time with energy storage [2,9]. Source: own elaboration.

This section details the control strategy designed to efficiently manage the available energy of both the microgrid and the network (utility grid) as well as their dynamic interaction, in a way that reduces consumers' electricity costs as well as helping the electric utility to manage a more orderly and predictable hourly load. This is done to incentivize and enhance EF practices among customers in residential settings and to enable the electric utility to better assign its energy supplying resources under any given operational circumstances, with the incorporation of localized sustainable hybrid energy systems (SHES). Such SHES will usually take the form of grid-tied microgrids of their own devise and operation, particularly in urban and semi urban areas where the electric distribution grid is always present and, at the same time, providing incentives to change toward a more efficient and sustainable energy consumption. It is important to realize that the utility's customers can benefit from special pricing mechanisms given the time-of-use (TOU) of electricity so that pricing serves as an efficient and cost-effective way of adequately managing electricity demand response during different hourly tariffs under normal conditions. This is something which aims at relieving peak electricity demand from the electric utility especially during steep peak hourly consumption [9]. In this regard the customers' participation is crucial for the success of such pricing mechanisms [10]. In order to fulfill the potential of such programs, customers must be able to access electricity tariffs and understand their terms [10]. Thus, strategies for integrating renewable generation employing SHES are important not only because of peak load shifting, but also because of the need to tap into localized onsite solar photovoltaic generation along with solar thermal energy for heating and AC, providing the utility grid the support and backup it needs.

Under this operation scheme, there is no such thing as energy independence from the grid, unless of course, there is no grid available as it occurs in remote and isolated communities and tiny villages located outside of the utility's electricity distribution range. Rather, what is expected from a strategic perspective is that both SHES and the utility grid complement and support each other, and both should rightly be operated by the electric utility itself, given the technical expertise, operational support, as well as considering the electric power distribution's industry infrastructure and ownership. Some authors have advanced the idea of electric power distribution decentralization by means of electric utilities installing localized electrical hubs as an effective way to integrate renewable energy resources and to provide additional services with little or no impact to the utility grid's operation [11,12].

3.2 Energy prosumers

The concept and role of "prosumers" and the concept of the sustainable block: applications in residential systems

The prosumer is a customer that is willing to "produce" energy by expending less energy in his/her daily consumption and in doing so, he/she makes possible that more energy is available for others that consume more and thus, pay more. The benefit for the prosumer is clearly economic and it is reflected in his/her electric bill. The residential communities in which this sustainable practice is implemented, along with the operation of SHES tied to the grid as a main supply of electricity to the consumers and where there is a concerted effort by the consumers to sustain the SHES green supply instead of relying solely on the grid, is termed a sustainable block. The homeostatic control strategy based on energy homeostaticity, unlike other control systems, considers the role that these prosumers—the energy users (residential and commercial electricity consumers)—can play as "active loads," conscientiously responding to the network needs and to the possibilities of the microgrid's supply, both of which seek to motivate and entice consumers by economic incentives as well as social and technical ones. Moreover, energy storage plays a key role in the model as it enhances and strengthens thriftiness in energy consumption.

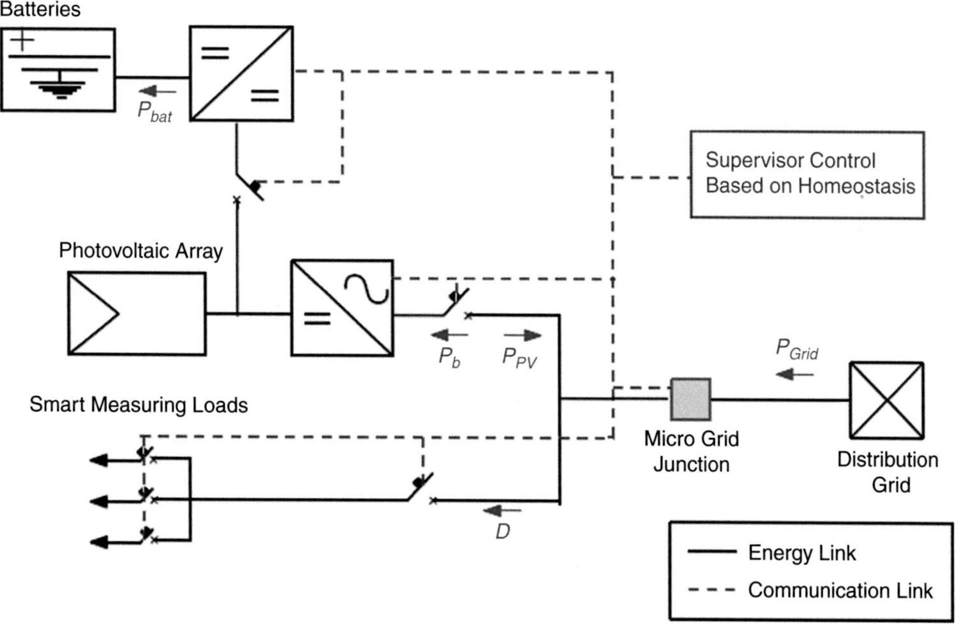

Figure 9.2. Global SHES diagram scheme to be operated by ENEL Distribucion for the new electricity pricing program strategy to incorporate renewables to the grid's supply taking into account the efficiency of the battery charging/ discharging process. Source: own elaboration.

This type of energy system is able to self-regulate and adapt itself when integrating and interacting with other higher level complex systems (meta system) like the utility grid, where the customers' load profile and current demand is included. The supervisory control strategy considered here is inspired in homeostasis of living organisms, which consists on the ability of living beings to keep certain variables in a steady, self-regulated, dynamic equilibrium state or within certain predefined limits to guarantee such equilibrium [9]. For this, the SHES must be adaptive and be capable of changing some parameters of their internal structure whenever is necessary to maintain energy homeostaticity. The structure of the simulated network can be seen in Fig. 9.2. Its characteristics are detailed in Table 9.1.

The simulation of 50, 100, and 150 kWh batteries on the microgrid has been performed. The results obtained regarding the change of energy cost in every case are presented next.

3.3 50 kWh battery

It is very important here to note the paramount role that energy storage plays in the operation of the energy homeostaticity model. The first case to consider is the battery with capacity of 50 kWh. The electricity tariff (or electric rate) being charged for each of the

Table 9.1 Components comprising the grid-tied microgrid.

Component	Size
PV plant	40 kWp
Inverter DC/AC	40 kW
Battery bank	0-50-100-150 kWh

apartments of the 12 stories residential building can vary according to the protocol they choose (Fig. 9.3). In the graph of Fig. 9.4, you can see what happens when you take the BT1 protocol *(in black)*, AT43 without battery *(in blue)* and AT43 with 50 kWh battery *(in red)*. It should be noted that the left axis is in US$ per kWh. Each electric rate varies with time of day and gets more expensive as the day moves to peak hours, which are between 6 p.m. and 11 p.m.

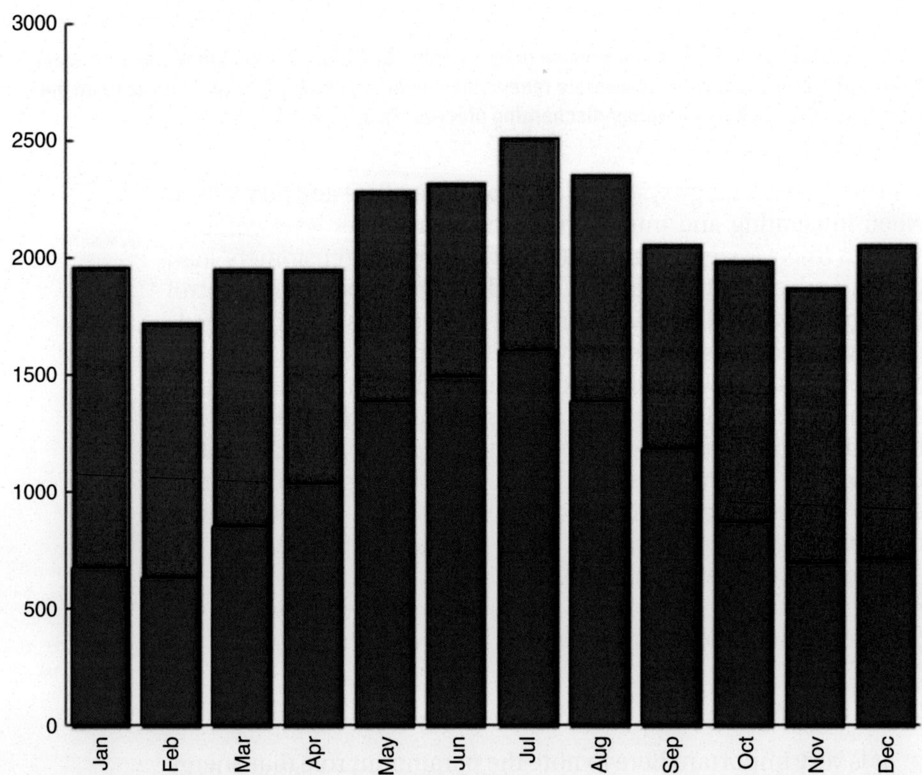

Figure 9.3. Electricity cost in US Dollars with battery of 50 kWh.

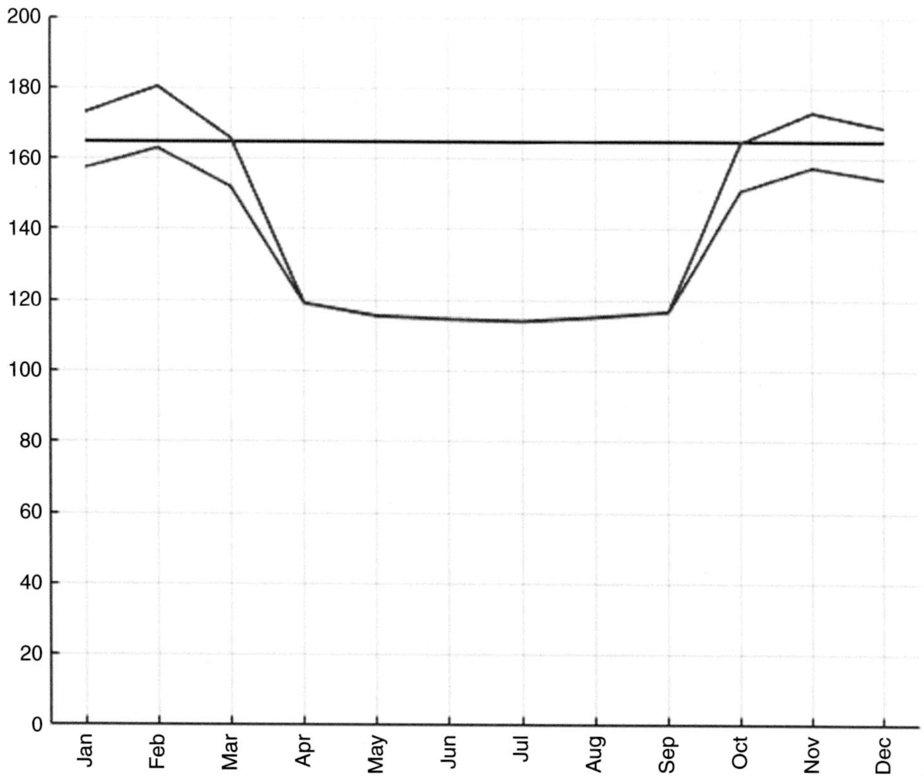

Figure 9.4. Electric rate in US Dollars with battery of 50 kWh.

Fig. 9.5 shows the power flow with a 50 kWh battery in operation: in red the aggregate demand, in blue the power from the electrical distribution network, in green the photovoltaic power, in yellow the power of the battery, in pink the power absorbed by the network and in black the power limit.

For the simulation of the system proposed in the example shown as an illustration of the system's model, three types of batteries were chosen and tested by the homeostatic control system's simulation for the grid-tied microgrid installed in a residential building in Las Condes, Santiago de Chile. Each one of the three types alters the graphs as you can see further. In the following graph, you can see the difference in annual costs of clients who do not control their electricity consumption by means of the energy homeostaticity model in use (in red) and those that do use it with a 50 kWh battery as energy storage. The electricity rate applied by the electric utility is the AT- 43.

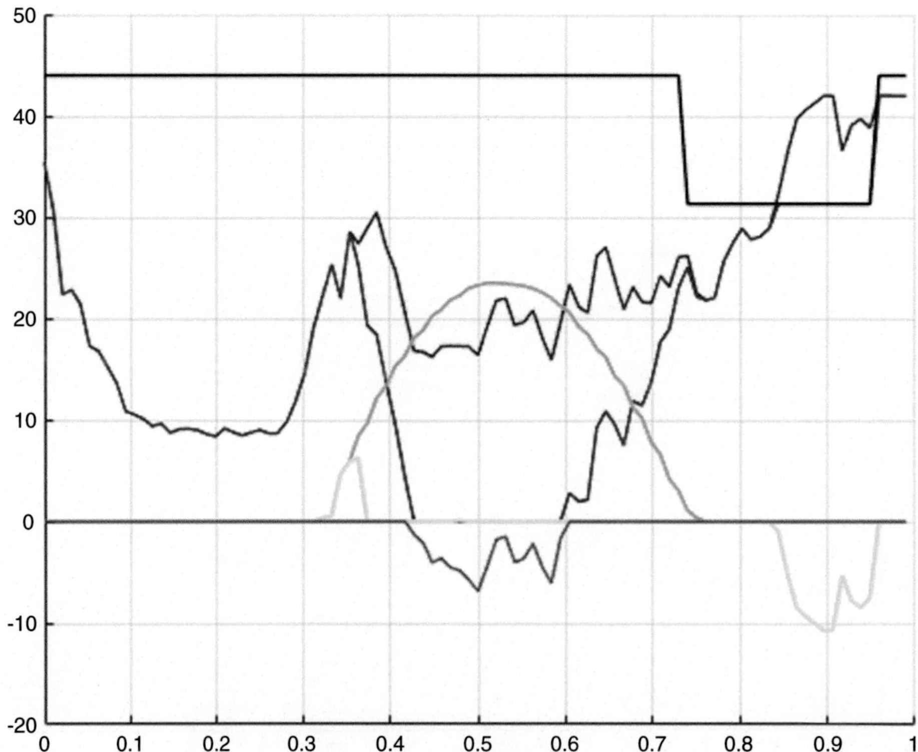

Figure 9.5. Power flow with a 50 kWh battery in operation.

3.4 The case with 100 kWh battery

As in the case of the 50 kWh battery, the difference in annual costs (Fig. 9.6) can be seen for customers who opt for not having their electricity consumption being controlled by the homeostatic control system of the microgrid with their energy use *(in red)* and those who do use the system this time with a 100 kWh. The electric rate applied is AT-43.

Fig. 9.7 shows the power flow with a 100 kWh battery in operation: in red the aggregate demand, in blue the power from the distribution network, in green the photovoltaic power from the grid-tied microgrid, in yellow the power of the battery, and in pink the power absorbed by the network and black the power limit set. Here, the simulation shows how increasing the size of the battery by doubling the energy storage capacity has an effect that typifies perfectly well what the model seeks to instill in the energy consumers: the more energy the system can save the higher the benefits for all of the parties involved and the more sustainable and resilient the energy system becomes, especially when considering the role of the utility's electricity distribution network which is acting as the meta system of the sustainable block [4,5].

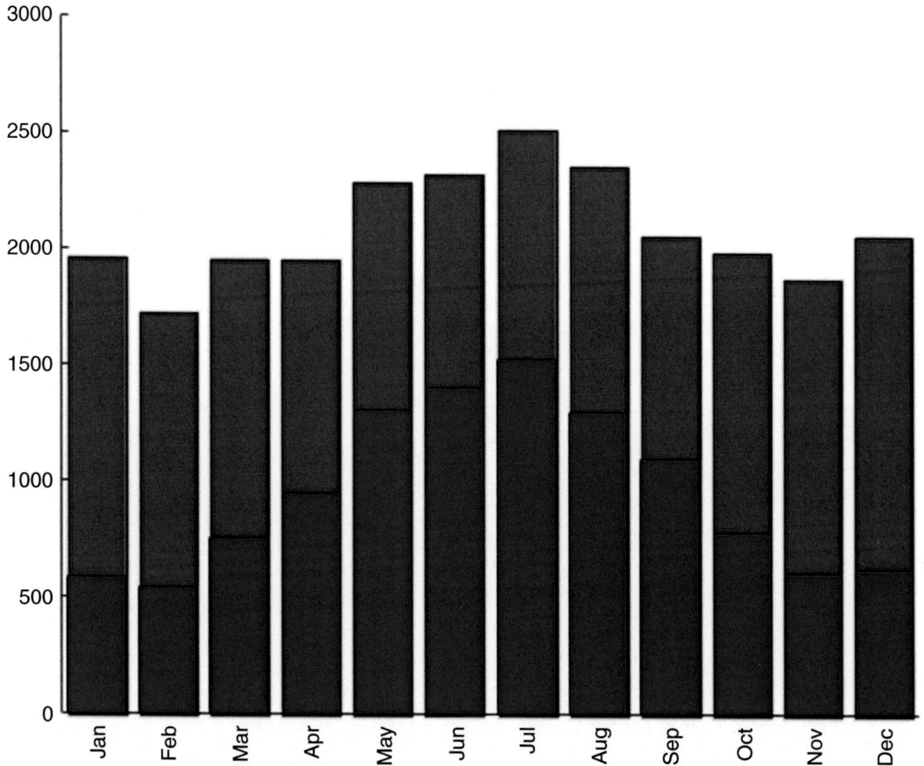

Figure 9.6. Cost in US Dollars battery of 100 kWh.

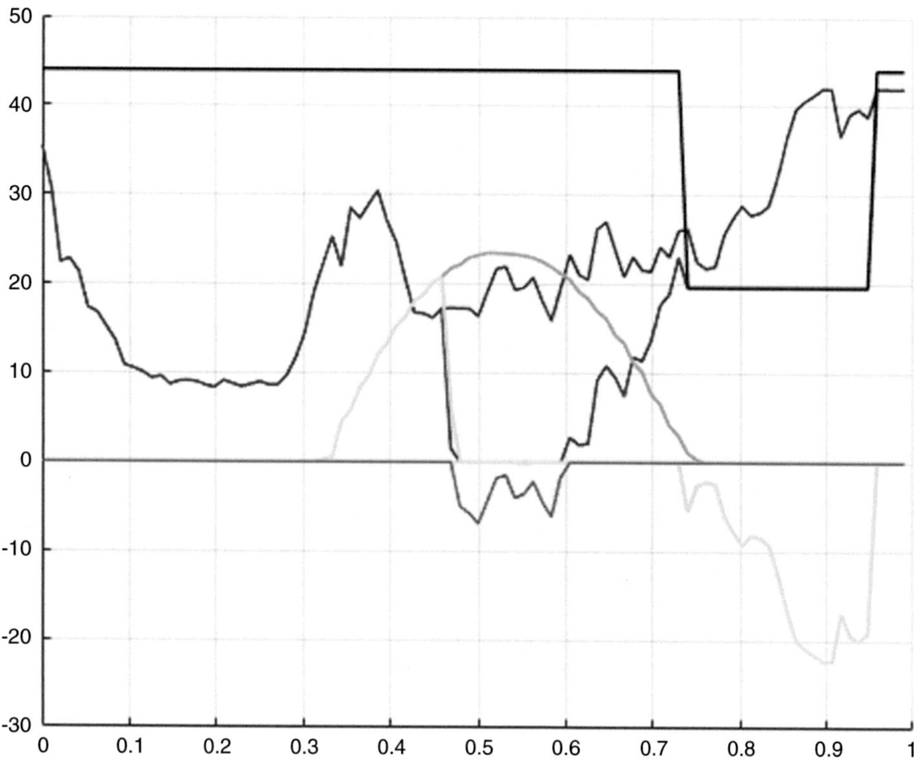

Figure 9.7. Power flow with a 100 kWh battery.

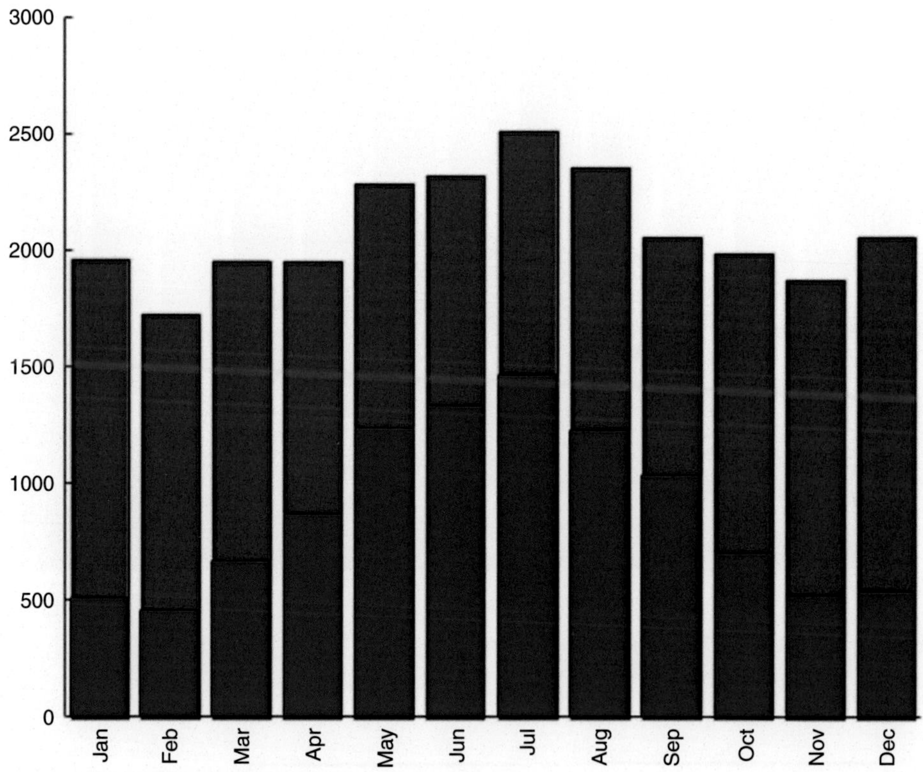

Figure 9.8. Cost in US Dollars with battery of 150 kWh.

3.5 The case with 150 kWh battery

The rate used is AT-43 for the chart showing the cost in US dollars of residential customers that use the homeostatic control system, this time with a 150 kWh battery in place, against those who do not control their energy use and choose to draw their electric supply directly and entirely from the distribution network, without the input of the microgrid's supply (Fig. 9.8).

4 Conclusion

No doubt, much needed changes in the law are needed to advance distributed generation (DG) in the Electricity Distribution Industry. It represents both a challenge and a pending task, which remains still unresolved, which begs the question of what are the industry actors and pundits to do in order to bring about such change. Well, a glimpse of the solution has been presented in this chapter as the results of the simulations show that the addition of energy storage capacity and its judicious management through the energy homeostaticity model proposed here for a residential building in Santiago, Chile may result in a high impact on the energy cost, hence the advantages are evident. But not only that, energy storage acts also as an enabler and enhancer of energy homeostaticity in grid-tied microgrids, thus impacting energy efficiency in a way that is not altogether obvious, at least for the untrained observer. Given the very nature of energy systems in general, their complexity invariably determines that an inherent synergistic divergence is to be expected, particularly in very diverse residential as well as commercial communities of energy users, such as those found in any large urban area. This ongoing systemic evolution subsequently provides an additional adaptive capacity of the metasystem, that is the distribution network, and its relationship with the grid-tied microgrid and with the consumers, all of which constitutes a whole, a super system which in turn, comprises what has been termed a sustainable block.

Acknowledgments

Dr. Fernando Yanine is grateful to CONICYT of Chile for his scholarship endowment of his Postdoctoral FONDECYT N° 3170399 "BUILDING SUSTAINABLE ENERGY SYSTEMS TIED TO GRID: A RESEARCH PROJECT PARTNERSHIP WITH CHILECTRA TO INSTALL HOMEOSTATICALLY CONTROLLED MICROGRIDS IN APARTMENT BUILDINGS" and to ENEL Distribucion for its auspice and support. Special thanks are also given to Universidad Finis Terrae for its ongoing support for this research initiative and to the Department of Electrical Engineering of UTFSM.

Dr. Antonio Sanchez-Squella is supported by a research grant from FONDECYT #11150911 of Chile.

D. Antonio Parejo is supported by the research grant "Formacion de Profesorado Universitario" by the Ministry of Education, Culture and Sport (Government of Spain).

Websites

Web-1: https://microgridknowledge.com/hybrid-microgrids, consulted December 31, 2019.
Web-2: http://www.iea.org/Textbase/npsum/US2014sum.pdf, consulted December 31, 2019.

References

[1] E.S. Blake, T.B. Kimberlain, R.J. Berg, J.P. Cangialosi, J.L. Beven II, National Hurricane Center (February 12, 2013). Hurricane Sandy: October, 22, p. 29. (PDF) (Tropical Cyclone Report). United States National Oceanic and Atmospheric Administration's National Weather Service, 2013. Archived from the original on February 17.

[2] U.S.N. Oceanic, A.A.N.W. Service, Hurricane/Post-Tropical Cyclone Sandy, October 22-29, 2012 (PDF) (Service Assessment), May 2013, p. 10. Archived from the original on June 2, 2013. (Retrieved December 2, 2017).
[3] A. Becena, M. Diaz, J.C. Zagal, Feasibility study of using a small satellite constellation to forecast, monitor and mitigate natural and man-made disasters in Chile and similar developing countries, 2012.
[4] F.F. Yanine, A. Sanchez-Squella, A. Barrueto, Cordova, S.K. Sahoo, Engineering sustainable energy systems: how reactive and predictive homeostatic control can prepare electric power systems for environmental challenges, Proc. Comput. Sci. 122 (2017) 439–446.
[5] F.M.F. Yanine, A. Sanchez-Squella, A. Barrueto, J. Tosso, Cordova, H.C. Rother, Reviewing homeostasis of sustainable energy systems: How reactive and predictive homeostasis can enable electric utilities to operate distributed generation as part of their power supply services, Renew. Sustain. Energy Rev. 81 (2017) 2879–2892.
[6] W.B. Cannon, Organization for physiological homeostasis, Physiol. Rev. 9 (3) (1929) 399–431.
[7] W.B. Cannon, Stress and strains of homeostasis, Am. J. Med. Sci. 189 (1935) 1–14.
[8] M.I. Dekhtyar, A.J. Dikovsky, On homeostatic behavior of dynamic deductive data bases, in: International Andrei Ershov Memorial Conference Perspectives System Informatics, Springer, Heidelberg, Berlin, 1996, pp. 420–432.
[9] F.M. Cordova, F.F. Yanine, Homeostatic control of sustainable energy grid applied to natural disasters, Int. J. Comput. Commun. Control 8 (1) (2012) 50–60.
[10] C. Uckun, Dynamic Electricity Pricing for Smart Homes, The University of Chicago, (2012).
[11] H.A. Rearte-Jorquera, A. Sánchez-Squella, Pulgar-Painemal, A. Barrueto-Guzmán, Impact of residential photovoltaic generation in smart grid operation: real example, Proc. Comput. Sci. 55 (2015) 1390–1399.
[12] S.L. Wearing, M. Wearing, M. McDonald, Understanding local power and interactional processes in sustainable tourism: Exploring village–tour operator relations on the Kokoda Track, Papua New Guinea, J. Sustain. Tourism 18 (2010) 61–76.

10

Multi energy systems of the future

Vasileios C. Kapsalis
National Technical University of Athens, School of Mechanical Engineering, Section of Industrial Management and Operational Research, Athens, Greece

Chapter outline

1 Introduction

Environmental degradation and resource depletion aspects have been led in designing multi energy systems which exploit the several energy carriers and interaction between them in a sustainable manner. The enormous increase in energy demand is urgently looking for a corresponding potential to be matched within the framework of an environmental balance. The intensified problems that they have been derived from the environmentally impacts of conventional energy sources utilization turned into renewable resources and the accompanied policies of social, economic, and financial instruments.

Low Carbon Energy Technologies in Sustainable Energy Systems. http://dx.doi.org/10.1016/B978-0-12-822897-5.00010-9

2 Multi energy supply chain

2.1 Supply chain

The world energy demand model forms the backbone of the future scenario outlook of the future energy systems and simultaneously, we are seeing an unprecedented energy transformation with renewables uptake. In our world, different infrastructures provide the daily multi energy demands in residential, commercial and industrial consumers using different forms to serve the network flows. The main drivers correlated with this evolution are the Gross Domestic Product per income, the population, the commodities price and the technological change. The generation, transmission and distribution of energy carriers are challenging from new types of energy and technologies to be optimized in future systems (Fig. 10.1).

2.2 Transactive multi energy systems

The primary purpose of a transactive energy system is to balance the demand and supply of the grid via the price signals. This could be managed by a smart grid which is a typical grid with a digital layer added, which contributes to five key elements, which has to do with the control of equipment and devices, the sense

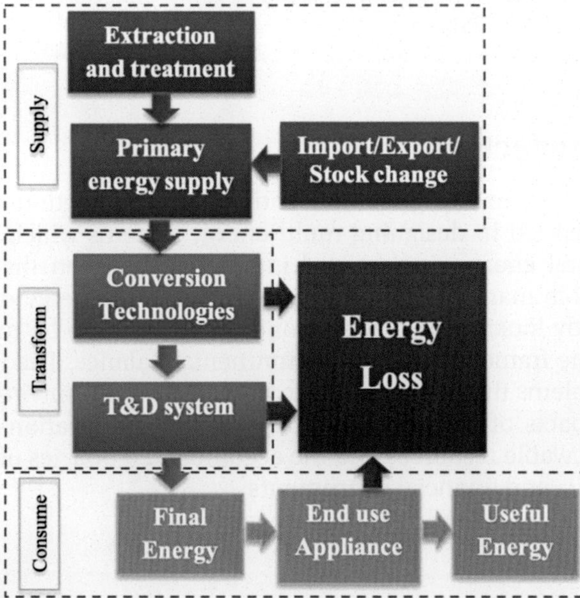

Figure 10.1. Energy supply chain [1].

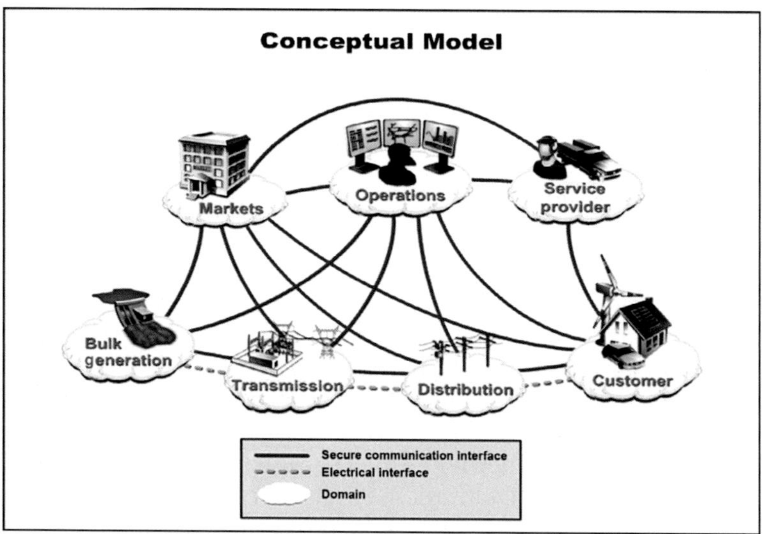

Figure 10.2. The concept of multi energy systems and the smart grid interconnections [2].

capabilities, the communication, the input data and the software integration to better manage the demand responses and the interconnected thermostat to billing strategies (Fig. 10.2).

The decline of the prices due to the scale effect of the installed capacity and the increased growth rate affected to the energy generation problems, due to the randomness of the renewable sources [3]. On the other hand, the smart metering may provide the data analytics methodology that the distributors, the system operators and the customer side of view to adjust the production and consumption [2]. A transactive system seems to be expandable and unlimited, so it is possible to incorporate more resources and outputs of the multi energy systems in the future. According to the discrete events signals which constitute a transactive energy, they occur when a predetermined limit in wholesale prices or another form of initiatives warning, for example, a peak load curtailment. Since the renewables are expected to increase their proportion in the energy resources inputs, the stochastic nature of the forecasting day ahead supply would face more risks in matching the demand and phenomena like the ducky curve formation need to be handled [4].

3 Multi forms of energy storage systems

3.1 General

Energy storage technologies are used to manage all types of energy, such as mechanical, biological, magnetic, chemical and thermal. And are found in buildings and industrial processes but also there is an increasing interest in micro and nano-scale applications due to its dominant role in the efficiency enhancement.

Although today's forecasting methods and simulation tools have been reaching at high accuracy levels, the thermal energy storage is the most proper technology, we use to achieve the goals of the efficiency enhancement in energy systems. These systems stock thermal energy by heating or cooling a storage medium so that the stored energy can be used at a later time for heating and cooling applications and power generation or for electricity storage in combination with conventional and hybrid configurations. The system's economic performance substantially depends on its specific application and operational need, including the number and frequency of storage cycles. The thermal energy storage may be separated in sensible, latent and thermochemical mechanisms regarding the driving force causes of activation. Similar cyclic processes are also found in biological photosynthetic systems. A comprehensive analysis of the thermal storage characteristics and behavior may be essential to the understanding of the process and the optimization of its performance under different conditions and requirements.

Typically, there are three driving mechanisms for the corresponding thermal energy storage types which are analyzed here. They are based on the same procedure which is related to the charging, storing and discharging process, aiming to match the accumulated energy in a later time to better serve the demand. This seasonal variation of energy may apply to different scale capacities, depending on the specific applications. Since the thermal energy storage cycle always refers to the circular process, it can be defined by two instances: the decrement factor (DF) and the phase shift of the period or the time lag (τ). The former refers to the percentage reduction of the amplitude of the sinus curve and the later to the delay of the angle of the oscillations before and after the implementation of the thermal energy system. The significant role of thermal energy storage technologies in our everyday life can be realized by the implementation in a wide range of human activities as well as in natural and biological systems. Thermal energy systems are used to provide efficient energy supply which leads to sustainable and cost-effective resources. The balance between supply and demand, in turn, provides a better performance of the demand and supply curves due to the effective use of the thermal equipment. The specific application of the thermal storage technology depends on many factors, such as temperature and environmental conditions, space requirements and availability, heat losses and economic evaluation, the construction and operation costs and the energy efficiency, to name some [5].

Energy harvesting technologies is another way of exploitation the energy losses within a system and it occurs in terms of heat forms which usually it is dissipated to the environment. Therefore, the capture and utilization of the waste energy is essential to the optimized process and the increased efficiency of the multi energy systems. The energy harvest, which efficiently store energy for a later usage, have been reported as a breakthrough technology in industrial recovery systems [6] and in renewable processes [7].

3.2 Forms and key properties

3.2.1 *Sensible multi energy storage systems*

When a change in a closed, adiabatic system's energy state conditions is experienced, a thermal equilibrium of heat transfer rates is achieved. Nevertheless, the steady state consideration of the systems is useful for a rather bulk analysis. In the static and time independent state where no spontaneous change occurs the macroscopic quantities remain unchangeable, therefore the sum of entropy changes is equal to zero. According to the second law of thermodynamic the system gives the inequality such that the entropy is greater than zero. Therefore, the behavior analysis during the charging and discharging cycles is a challenging task and reveals the influence between the key parameters during the process [8]. Thus, the rate of heat released and extracted of a thermal storage system is another critical property to consider about and is expressed with the thermal conductivity of the material. Thermal conductivity is depended of the electron transport of energy (thermal diffusion) and the phonon interactions. The thermal conductivity is a macroscopic or aggregate property of the materials, which reveals what occurs in a molecular level. Moreover, the electron conduction theory and the quantum mechanics result in an analogy consideration between the thermal and the electrical conductivity, through the Wiedemann-Franz Law. The electrons (or phonons in insulators), which contribute to the thermal conductivity, are those with velocities close to the Fermi velocity. For example, and especially in metals, electrons do not obey Maxwellian statistics. They rather follow the Fermi-Dirac statistics and present a fraction of the absorbed energy.

Thermophysical properties of the materials, such as the heat transfer coefficients, density, diffusivity, viscosity, and the specific heat capacity are involved with the potential of thermal energy storage [9–12]. The conductivity and the specific heat capacity are defined by the free electrons' mobility and the thermal vibrations of the crystalline lattice, the energy quanta which called phonons. The conduction electrons play a dominant role in metals and their movement is impeded by scattering, which is the result of the interaction with phonons or impurities or other imperfections. In semiconductors the thermal conduction is caused by the electron where they are stimulated and go over the conduction band. In insulators, the dominant role attributes to the phonons where they provide the mechanism for the thermal flux transportation through the material. Time-dependent conduction in materials obeys the first law of thermodynamics.

The transient process is induced by surface convection conditions, surface radiation conditions, a surface temperature or heat flux, and internal energy generation. Heat transfer mechanisms act simultaneously in an infinitesimal control volume element and are time dependent. Thickness-based Ra critical value on thermal boundary layer triggers convection which draws the energy from the surface. The temperature distribution depends on velocity distribution, the fluid type and the flow regimes. The radiation can proceed even in the absence of continuous medium.

Obviously, the sensible thermal energy storage of a given material depends on the value of specific heat capacity $\left(c_p\right)$ or the volumetric energy density $\left(\rho c_p\right)$. The former characterize, for example, the high capability of water and the later the excellent properties of the iron which is also found to present reversible latent heat transition and controlled conductivity [13], as we will see later. Beside the heat transfer rates and the high specific capacity, it is desirable, the long-term stability, the compatibility of the containment, and the cost effectives of the cycling process.

The efficiency in a stratified storage system is affected by many factors. Due to the significance of these technologies regarding the efficiency improvement, there are plenty of investigations using dimensionless numbers, which are designed to include the basic parameters of the heat transfer and the thermo physical properties of the process and the materials used. Richardson number is the best measure to define the stratification in a water tank and the MIX number characterize some problems and bad behavior [14]. Since the analysis and the investigation of the stratification is very important technology, many developments have been performing advanced analytical procedures to provide a better understanding of the performance parameters which influence the charging and the discharging process. Separation of stratified temperature gradients due to a smart exploitation of buoyancy forces led to the thermocline thermal energy storage technologies. The investigation of this method is in an infant stage and many researches have to be done in a lab and industrial scale in order to be better understood. The key point in this technology is the combination of water storage and solid quartzite rocks at the same tank to cut costs.

The storage efficiency of this technology is depended of the mass flow, the inlet velocity and the size of the particles. In general, the smallest the size of the particle the better the storage efficiency due to the better heat transfer between the fluid and the solid [15]. Recent advances in this technology include the use of techniques to improve the stratification in the tank by delaying the mixing process. For example, an equalizer in the dynamic inlet of the water tank may improve the stratification and the fill efficiency while at the same time reduce the mixing process. Other dimensionless numbers have been investigating in the boundaries of the buildings with air, as the heat transfer fluid and the dependence on the temperature gradients are the Rayleigh (Ra) and the Prandtl (Pr) number showing the dependence of the stratification process to the convection and the viscosity of the interactive media [16].

The sensible based technologies may be classified on the basis of the heat storage media. Therefore, they are liquid (e.g., water, oil-based fluids, molten salts), solid (e.g., bricks, rocks, metals, and others) and gas technologies. The thermal storage with liquid media is achieved by heating accumulation of the bulk material (pressurized water, molten salt, etc.) without state changing during the accumulation and later energy recovery which is used as heat source to drive the demand.

High temperature thermal storage for solar power plants or industrial process heat technologies may use liquid media or two-phase heat transfer fluids. Development of high temperature heat storage and ceramic heat exchangers for gaseous heat transfer fluids increase the energy efficiency of power plant and process technologies. For example, the later technology in power plants increase the flexibility of the of combined cycle gas turbine plants through deploying high-temperature heat storage tanks.

The low temperature fluids are used for thermal energy storage, instead of water, around the temperature of 4°C or below and refer to aqueous solutions containing chemical additives or no aqueous chemicals. They support low temperature air conditioning and some food process applications while they have performed good behavior against corrosion and microbiological control properties.

Aquifers when they are combined with several forms of multi energy storage provide space heating and cooling and potentially electricity when they are coupled with heat pumps and or other devices. The climatic conditions, the availability of the aquifer, and the feasibility of the specific application are the most critical factors for the implementation of such a technology. Aquifers are used as a sink heat pump or sources to store energy ambient air, waste heat or renewable sources. These systems use natural water in a saturated and permeable underground layer as the storage medium. The extraction of the water from a well and the reinjection of it in an appropriate temperature in a nearby well is the main principle of this technology. They may be divided to open and closed or borehole systems. The former is cheaper and provides a greater transfer capacity than the latter and is preferred for a longer period. The volume of storage is dependent of the thickness and the porosity of the aquifer. The cost effectiveness of the technology is based on the avoided equipment and the lower operation costs while the specific application provides flexibility to the designer incorporating augmentation facilities and combination with dehumidification or desiccant systems. In deep sedimentary basins, the temperature gradient from the earth crust or the confined hot water or vapor provides thermal energy storage opportunities and exploitation with several technologies. The heat flow of the hot fluids or the magma structures which are surrounded by low thermal conductivity sediments, results in hydrothermal conduction and convection systems of temperature and pressure gradient, looking for passage toward the surface through the rock permeability and pores. The thermal energy storage potential in geothermal resources is primarily exists in rocks and secondarily in fluids that fill the pores and the structure. The three regions of temperatures (low at below 90°C, medium between 90°C and 150°C, and high above 150°C) classify the storage technologies to low, medium, and high temperature. The couple of these storage technologies with heat exchangers and heat pump systems is a well-known renewable technology and may be configured in vertical, horizontal, or hybrid scheme. A review in concepts and applications of similar systems may be found in the literature [17].

Solar ponds are shallow bodies of water in which an artificially maintained salt concentration gradient prevents convention. The combination of heat collection, through the radiation adsorption passing the water layers, with long-term storage can provide sufficient heat for the entire year. It is interesting to observe that the insulating layer is the water itself. The vertical salinity which is created in the pond makes the deeper layers to contain more salt and become correspondingly denser. In this way, it is possible to impede convection and to achieve high bottom temperatures. Instead of polymer covers some others use gels, sufficiently viscous to impede convection, too. This technology is used as a heat sink in large areas or may be coupled with solar energy applications in roof ponds, in agriculture, as well as in thermally driven separation processes with sustainable desalination membranes [18].

The variations of temperature distribution, called stratification, in the direction of the implemented gradients. For example, in a quiescent fluid reservoir which exhibit a temperature gradient in any direction, such as a core fluid in an enclosure heated from any side or the bulk air in a sealed room. Stratification of multi storage systems has a substantial effect on the heat transfer efficiency [19,20].

3.2.2 Latent multi energy storage systems

In most cases variations of enthalpy with temperature depends on the direction considered and is different for melting and solidification the changing phase. Recently, solidification process is further examined as a heat transfer process for internal convection and external conduction under a statistical thermodynamic kinetic theory in the interfacial boundaries. The distinct regimes of convection, nucleation, transient, and film solidification are explored [21]. In the melting point with no heat gain or loss no growth or dissolution occurs. Added crystals above the melting point, where the solution is saturated, tend to dissolve. Convective heat transfer dominates the process here and superheating phenomena are associated with. In encapsulated phase change materials contact melting occurs when the solid is free moving within the capsule due to the density difference between the solid and the liquid phase. The shrinking solid affects the geometrical shape and size and, therefore, the melting process. This phenomenon affects the time of the process [22]. Recently, Ho [23] reported the complexity of the transient transport, which includes density gradients at early stages and free moving boundaries. Lane [24] presents some empirical evidence, which connects the tendency to super cool with the viscosity of the melt in the melting point. Materials with high viscosity in the liquid state have low diffusion coefficients for their constituent atoms (or ions) and these are unable to rearrange themselves to form a solid and instead the liquid super cools. The advantage of the material to store energy is reduced because the melt does not solidify at the thermodynamic melting point. Proper nucleating agents return the solution in equilibrium, immiscible fluid, or metallic surfaces minimize the effect of supercooling [25,26] proposes the employment of metallic surfaces to promote heterogeneous nucleation and this reduces super cooling as well, while stability corresponds to a certain critical size of crystal nuclei [27] or the vapor pressure of the material surface [28–30]. Conductivity enhancement in latent multi energy storage technologies in order to reduce the charging and discharging response time of the process and consequently the response time of the system is a significant factor. Another common method that plays a key role in the electricity production as well as the industrial heat management is the specific design of finned heat exchangers tubes which improve the heat transfer to the used phase change material. In the technology, is of great significant the exact investigation of the role of the conduction and the convection dominated mechanisms to define in a better way the impact of the fins design. Moreover, the uses of bimetallic materials in the design of the tubes and the fins have performed advantages regarding to the operation conditions in medium temperature range and the stresses.

3.2.3 Thermochemical multi energy storage systems

Chemical reactions accompanied with energy changes. They absorb or release energy as heat with sorption or without sorption process. These processes may be considered and analyzed by thermochemistry which is the study of the energy transformations and transfers accompanying chemical and physical changes [26]. Absorption is the penetration of the adsorbate through the surface layer of an absorbent with a change of composition. The adsorption desorption cycle-based technologies are well established in several processes. The hysteresis between the reversible cycles provides thermal heat storage potential and is widely used in the industry.

3.2.4 Hybrid multi energy storage systems

The utilization of excess energy in a storable manner that could efficiently and effectively balance the demand and the supply curve has been extensively using in engineering processes, such as mechanical [31], biological [32,33], magnetic [34,35], and chemical [36]. The technology developments of combined thermal systems boost the efficiency and result in increased renewable fraction coefficient within the system operation while at the same time reduced running life cycle costs. The prerequisite for this configuration is to integrate the waste energy from one resource within the heat pump operation. Therefore, we can recognize plenty of hybrid configurations in the literature. In fact, the innovative utilization of different multi forms of energy technologies is limited only from the creativity of the designer and the cost effectiveness of the specific applications. For example, the building integration of renewables may provide multifunctional prefabricated elements in façade, windows, wallboards, and roofing combined with heat pumps, PV/T, and thermoelectricity.

4 Assessment, economic issues, and perspectives

4.1 Technological and economic issues

4.1.1 Productivity and costs of multi energy systems

From an economic perspective and the enterprises point of view in multi energy systems, like all other production systems,

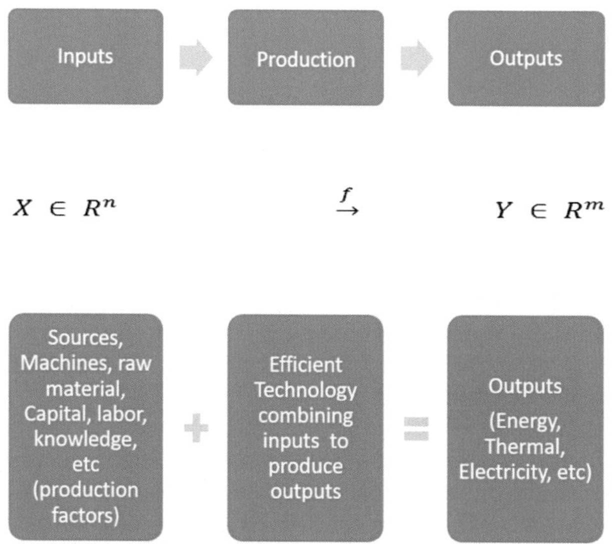

Figure 10.3. Production factors efficient transformations to provide different multi energy forms.

they are interested in the appropriate combination of inputs and technology in order to effectively produce the outputs. The way the new technologies are developed and diffused in the market, it is well described by the different and complex parameters (Fig. 10.3).

Typically, the adoption of the appropriate energy-modeling concept is determined by an explicit approach, which is dominated by exogenous parameters; by an implicit derived definition, which utilize endogenous drivers; and by a combined methodology.

4.1.2 The exogenous method

The energy technology evolution in a decision tree, incorporating uncertainties, simulated scenarios and probabilities distribution under constraints, using statistic and operation research programming framework. The assumptions made, are using projections and experts' knowledge, regarding the decision parameters, the future costs, the corresponding technological values, and the demand meeting [37,38], while the results are sensitive to the selection of parameters [39]. It is often used under the light of high uncertainty and hidden deterministic factors [38].

4.1.3 The endogenous method

This method is based on the experience curve, with an important parameter, the progress ratio (PR), which represents the percent reduction in cost with every doubling of the cumulative production quantity [40].

$$C_{Cum} = C_O Cum^b$$

$$PR = 2^b$$

$$LR = 1 - PR$$

where, C_O the cost of first unit of production, Cum the cumulative unit production at present, C_{Cum} the cost per unit at present, LR *and* PR the learning and progress ratio respectively, and b the experience index and $b < 0$.

Experience curve effect is a common concept on technology evolution and widely used to predict costs of energy technologies by establishing the relationship with the cumulative production quantity [41,42]. The construction of experience curves incorporates all production cost, the resources input values, such as capital, labor, energy, direct economic costs as well as indirect economic expenses which they are defined as opportunity costs. The methodology of the progress ratio analysis is a top down methodology and usually would be better performed by bottom up approaches from field engineering measurements. It can be used from past data driven processing methodologies but also is extrapolated in future production and costs with corresponding regression analysis. The influence of exogenous factors and prices fluctuations are excluded in order to have a clear sense of the impact and the pure sensitivity from the endogenous factors. The standard error from the experience index b, σ_b, is used to measure the error in $PR, \sigma_{PR} = \ln 2 PR \, \sigma_b$, which is derived from the chi-square fitting curve.

4.1.4 Combined methodology

A more advanced methodology includes the combined consideration of endogenous curves, namely the learning curve which is coming from the production, with externally defined parameters, such as the aggregated market demand satisfaction, in order to model the energy systems technology evolvement. The technological advancements which they are considered through the learning curve or the cumulative and continuous production method, are also based on the law of diminishing average and the marginal productivity and costs, which means that the unit cost is reduced as the cumulative production curve increased, it is also known as learning curve effect, while the diffusion into other contexts/sectors lead to innovations [43].

4.2 Optimization and decision making

4.2.1 Budget constraints, opportunity cost, and prioritization for resource allocation

There are different perspectives on prioritization and resource allocation followed by choosing the best (or most attractive) combination of options as part of a portfolio construction task [44]. On the other hand, the cost benefit analysis is based on the principles of social welfare economics and the axiom of rational behavior [45] using willingness-to pay to translate non-monetary to monetary values with risk often incorporated through discount rates (HM [46]).

4.2.2 Multi criteria decision analysis (MCDA) assessment

Since the decision making is also a multi-criteria process, the corresponding assessment with the family of multi criteria decision analysis (MCDA) is proven useful to investigate certain parameters to design the values, the risks, and the weighting ratios which affect the system and the complexity of the appropriate preferences [47]. The plurality of these methods belongs in different patterns such as the value measurement, the outranking and the sacrificing/aspiration methods, mainly depending on different preference taxonomy [47–50]. The modeling of costs and benefits threshold is constraint by the available budget and incorporates the probabilities, indices and criteria as well as alternative options [51–57] or multi stream frameworks [58–60].

4.2.3 The mean variance models framework

Based on the original theory [61] which aims at the efficient portfolio mixtures selection that provide the minimum risk for any given return, the expected returns are derived from the historical time series and relied on the means and covariance values. This model provides a valuable tool, especially for technologically changing environments such as the multi energy systems and capture the critical factors of the evolutions process under risk [62–64].

Especially for multi energy systems, the willingness to pay in the retail market operator broadcasts the clearing price, as well as the estimated mean price, and standard deviation price over a future time window of duration. Active thermostats provide their willingness-to-pay per unit of energy price by computing the bid price which is derived from the Gaussian distribution and the state of temperature (dimensionless) regarding the upper and lower state values. In more advanced considerations the stochastic expectations of higher moments are used (skeweness and kurtosis). A variety of optimization problems with quadratic objective functions and linear constraints are solved using Lagrange multipliers and the Kuhn–Tucker basic theorem algorithm.

4.2.4 Linear programming

The generalized linear problems and the solving techniques are inspired from the pioneering works on multistage commodity network problems and the decomposition principle [65–67]. Later, Bender's algorithm approach generalized to nonlinear problems under the convex duality theory [68] and dynamic programming, taking into account the decomposition techniques to reduce the computational complexity [69].

The main objective functions are formulated in subproblems with constraints. The generated solutions after each cycle add new basic decision variables to the column of the main problem until finite iterations and typical optimality test. Artificial and slack variables are used to transform the mathematical model in a standard type. Under this procedure, the equivalent matrix results in reduced feasible solutions in extreme points, and among them to the optimized one, for example with the simplex algorithm or the duality problem. Post optimal analysis is used to study of the shadow prices and the sensitivity of the objective functions.

A similar approach in LP is used in the formulation of the transportation problem, which define the optimized way to supply homogeneous commodity from m sources to n destinations, meeting the requirements of the demand and supply balance. The problem here may be balanced or unbalanced with inequality of demand and supply. As in simplex case, artificial and slack variables are used towards the formulation of the standard matrix. The algorithms they have developed here are based on the fact that an optimal solution remains optimal if a constant is subtracted from all unit costs associated with a particular source or destination. The stepping-stone algorithm or the modified distribution methods are the most popular and work efficiently because they follow the simplex logic without tableau and inverse of the basis.

4.2.5 Multi objective approaches

Many times, the formulating design variables of the objective functions and the underlined constraints are depicting conflicting interests and the typical optimization procedure is replaced with the Pareto optimality [70] and multi objective optimization problems. According to this concept, a point is Pareto optimal between the feasible solutions of a problem if there is no other point that improves at least one objective without detriment of at least one other objective function. The formulation of the problems consists of a vector of objective functions, which is called objective space, and a vector space which is composed of points defined by design-variables values and is called design space. There are many methods to generate the Pareto optimal set points. In general, they use three step techniques, namely the construction of normalized and ideal point of the optimal Pareto sets [71], the finding of the weights [72] and finally a single scalar objective function to minimize and determine the Pareto optimal solution [73].

4.2.6 Nonlinear programming and learning algorithms

In general, multi energy systems there are further a number of nonlinear algorithms which have been developing and include small total number of nonlinear variables or large number of linear variables with some nonlinear added. The approximation programming method, the separable programming and the quadratic programming are the main representatives of the latter. The latter has a very interesting implementation in unconstraint optimization in combined with Levenberg method. A method of finding the global optimum to some very special convex problem is that of dynamic programming. It has effectively used in many practical problems under the sequential approach of stage divided and a recurrence relation. According to Bellman's optimality, a multi stage decision making of n-years can be considered as a n−1 years process plus the nth year. The corresponding problem is transformed to a linear one, after implementing the decomposition principle.

In most aspects, there is a need in the long run to the need for further economic reforms and the nonlinear regression analysis, following the model structure, utilizes traditional time-series methods such as ARIMA (Auto-Regressive Integrated Moving Average) and Generalized AutoRegressive Conditional Heteroskedasticity (GARCH)-type models and their variations or by machine learning methods such as Artificial Neural Networks (ANNs) to determine the functions. Latest advances regarding the more accurate simulation of chaotic behaviors include two fundamental learning strategies, namely the deep learning and the ensemble learning. The former filters out the noise of the input data by exploiting specific features based on neural networks. The latter reduces the variance of error by exploiting multiple learners to generate strong prediction models [74–80].

5 Conclusions

This chapter presents a generic formulation of the future multi energy systems and the storage forms capabilities taking into account some key parameters from the technical point of view. The production perspectives and the assessment methodologies are analyzed and provide a basis for a comprehensive research orientation under specific features of energy system.

References

[1] N.H. Motlagh, M. Mohammadrezaei, J. Hunt, B. Zakeri, Internet of Things (IoT) and the energy sector, Energies 13 (2) (2020) 494.

[2] NIST, Draft NIST Framework and Roadmap for Smart Grid Interoperability Standards, Release 2.0. National Institute of Standards and Technology. Special Publication for public comment, October 2011 (accessed November 2011), 2011. http://www.nist.gov/smartgrid/.

[3] D. Pudjianto, C. Ramsay, G. Strbac, Virtual power plant and system integration of distributed energy resources, IET Renew. Power Gener. 1 (1) (2007) 10–16.

[4] CAISO, in: Long Term Resource Adequacy Summit. Presentation on February 26, 2013 by Mark Rothleder of the California Independent System Operator, 2013, (accessed November 11, 2013), http://www.caiso.com/Documents/Presentation-Mark_Rothleder_CaliforniaISO.pdf.

[5] B. Rezaie, et al., Energy analysis of thermal energy storages with grid configurations, Appl. Energy 117 (2014) 54–61.

[6] A. Gutierrez, et al., Advances in the valorization of waste and by-product materials as thermal energy storage (TES) materials, Renew. Sustain. Energy Rev. 59 (2016) 763–783.

[7] Z. Yu, et al., Numerical study based on one-year monitoring data of groundwater-source heat pumps primarily for heating: a case in Tangshan, China, Environ. Earth Sci. 75 (14.) (2016).

[8] B. Rezaie, et al., Thermodynamic analysis and the design of sensible thermal energy storages, Int. J. Energy Res. 41 (1) (2017) 39–48.

[9] A.I. Fernandez, et al., Selection of materials with potential in sensible thermal energy storage, Solar Energy Mater. Solar Cells 94 (10) (2010) 1723–1729.

[10] M.E. Navarro, et al., Selection and characterization of recycled materials for sensible thermal energy storage, Solar Energy Mater. Solar Cells 107 (2012) 131–135.

[11] S. Khare, et al., Selection of materials for high temperature sensible energy storage, Solar Energy Mater. Solar Cells 115 (2013) 114–122.

[12] G. Li, Sensible heat thermal storage energy and exergy performance evaluations, Renew. Sustain. Energy Rev. 53 (2016) 897–923.

[13] Y. Grosu, et al., Natural magnetite for thermal energy storage: excellent thermophysical properties, reversible latent heat transition and controlled thermal conductivity, Solar Energy Mater. Solar Cells 161 (2017) 170–176.

[14] A. Castell, et al., Dimensionless numbers used to characterize stratification in water tanks for discharging at low flow rates, Renew. Energy 35 (10) (2010) 2192–2199.

[15] J.F. Hoffmann, et al., Experimental and numerical investigation of a thermocline thermal energy storage tank, Appl. Thermal Eng. 114 (2017) 896–904.

[16] V. Kapsalis, D. Karamanis, On the effect of roof added photovoltaics on building's energy demand, Energy Build. 108 (2015) 195–204.

[17] K.S. Lee, A review on concepts, applications, and models of aquifer thermal energy storage systems, Energies 3 (6) (2010) 1320.

[18] K. Rahaoui, et al., Sustainable membrane distillation coupled with solar pond, Energy Procedia 110 (2017) 414–419.

[19] M.A. Abdoly, D. Rapp, Theoretical and experimental studies of stratified thermocline storage of hot water, Energy Convers. Manage. 22 (3) (1982) 275–285.

[20] B. Rezaie, et al., Configurations for multiple thermal energy storages in thermal networks. IEEE International Conference on Smart Energy Grid Engineering, SEGE 2013, 2013.
[21] H.-S. Roh, Heat transfer mechanisms in solidification, Int. J. Heat Mass Transf. 68 (0) (2014) 391–400.
[22] I. Dincer, M.A. Rosen, Thermal Energy Storage: Systems and Applications, J. Willey & Sons Ltd, England, (2002).
[23] C.J. Ho, et al., Melting processes of phase change materials in an enclosure with a free-moving ceiling: An experimental and numerical study, Int. J. Heat Mass Transf. 86 (0) (2015) 780–786.
[24] G.A. Lane, Phase change materials for energy storage nucleation to prevent supercooling, Solar Energy Mater. Solar Cells 27 (1992) 135–160.
[25] M. Sokolov, Y. Keizman, Performance indicators for solar pipes with phase change storage, Solar Energy 47 (5) (1991) 339–346.
[26] Y. Tian, C.Y. Zhao, A review of solar collectors and thermal energy storage in solar thermal applications, Appl. Energy 104 (2013) 538–553.
[27] J.W. Gibbs, New Heaven, Yale University Press, 1948.
[28] S. Behzadi, M.M. Farid, Long term thermal stability of organic PCMs, Appl. Energy 122 (0) (2014) 11–16.
[29] E. Efimova, P. Pinnau, M. Mischke, C. Breitkopf, M. Ruck, P. Schmidt, Development of salt hydrate eutectics as latent heat storage for air conditioning and cooling, Thermochimica Acta 575 (2014) 276–278.
[30] S.A. Memon, Phase change materials integrated in building walls: A state of the art review, Renew. Sustain. Energy Rev. 31 (0) (2014) 870–906.
[31] M.F. Lumentut, I.M. Howard, Electromechanical analysis of an adaptive piezoelectric energy harvester controlled by two segmented electrodes with shunt circuit networks, Acta Mechanica 228 (2017) 1321–1341.
[32] T. Ebenhard, et al., Environmental effects of brushwood harvesting for bioenergy, Forest Ecol. Manage. 383 (2017) 85–98.
[33] S.H. Yan, et al., Advances in management and utilization of invasive water hyacinth (*Eichhornia crassipes*) in aquatic ecosystems–a review, Critic. Rev. Biotechnol. 37 (2) (2017) 218–228.
[34] C. Ung, et al., Electromagnetic energy harvester using coupled oscillating system with 2-degree of freedom. Proceedings of SPIE—The International Society for Optical Engineering, 2015.
[35] D.C. Chen, et al., Study of piezoelectric materials combined with electromagnetic design for bicycle harvesting system, Adv. Mech. Eng. 8 (4) (2016) 1–11.
[36] F. Khademi, et al., Advances in algae harvesting and extracting technologies for biodiesel production, in: Progress in Clean Energy, Vol. 2, Novel Systems and Applications, pp. 65–82, 2015.
[37] C. Azar, K. Lindgren, B.A. Andersson, Global energy scenarios meeting stringent CO_2 constraints—cost-effective fuel choices in the transportation sector, Energy Policy 31 (10) (2003) 961–976.
[38] Y. Chen, Y. Fan, Coping with technology uncertainty in transportation fuel portfolio design, Transp. Res. Transp. Environ. 32 (2014) 354–361.
[39] S. Yeh, A. Farrell, R. Plevin, A. Sanstad, J. Weyant, Optimizing U.S. mitigation strategies for the light-duty transportation sector what we learn from a bottom-up model, Environ. Sci. Technol. 42 (22) (2008) 8202–8210.
[40] W. Hettinga, H. Junginger, S. Dekker, M. Hoogwijk, A. McAloon, K.B. Hicks, Understanding the reductions in us corn ethanol production costs: an experience curve approach, Energy Policy 37 (1) (2009) 190–203.
[41] M. Weiss, M. Junginger, M.K. Patel, K. Blok, A review of experience curve analyses for energy demand technologies, Technol. Forecast. Soc. Chang. 77 (3) (2010) 411–428.

[42] R. Wand, F. Leuthold, Feed-in tariffs for photovoltaics: learning by doing in Germany?, Appl. Energy 88 (12) (2011) 4387–4399.
[43] N. Rosenberg, Keynote address: challenges for the social sciences in the new millennium, OECD Social Sciences and Innovation, OECD, Paris, (2001).
[44] L.D. Phillips, C.A. Bana e Costa, Transparent prioritisation, budgeting and resource allocation with multi-criteria decision analysis and decision conferencing, Ann. Oper. Res. 154 (2007) 51e68.
[45] R.S.J. Tol, Is the uncertainty about climate change too large?, Climatic Change 56 (2003) 265–289.
[46] H.M. Treasury, The Green Book: Appraisal and Evaluation in Central Government: Treasury Guidance, TSO, London, (2003).
[47] V. Belton, T. Stewart, Multiple Criteria Decision Analysis: An Integrated Approach, Kluwer Academic Publishers, Dordrecht, (2002).
[48] J. Hammond, R. Keeney, H. Raiffa, Smart Choices: a Practical Guide to Making Better Decisions, Harvard University Press, Cambridge, (1999).
[49] V. Diaby, R. Goeree, How to use multi-criteria decision analysis methods for reimbursement decision-making in healthcare: a step-by-step guide, Expert Rev. Pharmacoecon. Outcome. Res. 14 (2014) 81e99.
[50] J.G. Dolan, Multi-criteria clinical decision support: a primer on the use of multiple criteria decision making methods to promote evidence-based, patient centered healthcare, The Patient 3 (2010) 229e248.
[51] S.J. Peacock, J.R.J. Richardson, R. Carter, D. Edwards, Priority setting in health care using multi- attribute utility theory and programme budgeting and marginal analysis (PBMA), Soc. Sci. Med. 64 (2007) 897e910.
[52] R.T. Clemen, Making Hard Decisions; an Introduction to Decision Analysis, Belmont, Duxbury, (1996).
[53] R.L. Keeney, H. Raiffa, Decisions with Multiple Objectives: Performances and Value Trade-offs, Wiley, New York, (1976).
[54] M. Weinstein, R. Zeckhauser, Critical ratios and efficient allocation, J. Public Econ. 2 (1973) 147e157.
[55] D. Postmus, T. Tervonen, G. van Valkenhoef, H.L. Hillege, E. Buskens, A multi-criteria decision analysis perspective on the health economic evaluation of medical interventions, Eur. J. Health Econ. 15 (2014) 709e716.
[56] M.J. Al, T.L. Feenstra, B.A.V. Hout, Optimal allocation of resources over health care programmes: dealing with decreasing marginal utility and uncertainty, Health Econ. 14 (2005) 655e667.
[57] T. Tervonen, G. van Valkenhoef, E. Buskens, H.L. Hillege, D. Postmus, A stochastic multicriteria model for evidence-based decision making in drug benefit-risk analysis, Stat. Med. 30 (2011) 1419–1428.
[58] A. Knaggard, The multiple streams framework and the problem broker, Eur. J. Polit. Res. 54 (2015) 450–465.
[59] R. Zohlnhöfer, F.W. Rüb, Decision-making under ambiguity and time constraints, Assessing the Multiple-Streams Framework, Colchester, ECPR Press, (2016).
[60] S. Blum, The multiple-streams framework and knowledge utilization: argumentative couplings of problem, policy, and politics issues, Eur. Policy Analysis 4 (2018) 94–117.
[61] H.M. Markowitz, Portfolio selection, J. Finance 7 (1) (1952) 77–91.
[62] K.B. Leggio, D.L. Bodde, M.L. Taylor, Managing Enterprise Risk—What the Electric Industry Experience Implies for Contemporary Business. Elsevier, 2006.
[63] J.J. Smit, Trends in emerging technologies in power systems, IEEE, Amsterdam, The Netherlands, 2005.

[64] K. Claxton, J. Posnett, An economic approach to clinical trial design and research priority- setting, Health Econ. 5 (1996) 513–524.
[65] L.R. Ford, D.R. Fulkerson, Suggested computations for maximal multicommodity network flows, The Rand Corporation, Paper P-1114, Manage. Sci. 5 (1) (1958) 97–101.
[66] W.S. Jewell, Optimal Flow Through Networks, Interim Technical Report No. 8, on Fundamental Investigations in Methods of Operation Research, Massachusetts Institute of Technology, Cambridge, Massachusetts, (1958).
[67] G.B. Dantzig, Lineal Programming and Extensions, Princeton University Press, Princeton, N.J., (1965).
[68] A.M. Geoffrion, Generalized Benders decomposition, J. Optim. Theory Appl. 10 (1972) 237–260 doi: 10.1007/BF00934810.
[69] R.M. Van Slyke, R. Wets, L-shaped linear programs with applications to optimal control and stochastic programming SIAM, J. Appl. Math 17 (4) (1969) 638–663.
[70] V. Pareto, Manuale di Economica Politica, Societa Editrice Libaria, Milan (A.S. Schwier, Trans.), in: A.S. Schwier, A.N. Page, A.M. Kelley (Eds.). Manual of Political Economy, New York, 1971, 1906.
[71] M. Ehrgott, D.M. Ryan, Construction robust crew schedules with bi criteria optimization, J. Multi-Crit. Decis. Anal. 11 (2002) 139–150.
[72] Shannon CE, A mathematical theory of communications, Bell. Syst. Tech. J. 27 (1948) 379–423.
[73] R.T. Marler, J.S. Arora, Transformation methods for multi objective optimization methods for engineering, Struct. Multidiscip. Optim. 26 (6) (2005) 369–395.
[74] I.E. Livieris, E. Pintelas, S. Stavroyiannis, P. Pintelas, Ensemble deep learning models for forecasting cryptocurrency time-series algorithms 2020 13 (2020) 121 doi: 10.3390/a13050121.
[75] G. De Luca, N. Loperfido, A skew-in-mean GARCH model for financial returns, in: M. Corazza, C. Pizzi (Eds.), Skew-Elliptical Distributions and Their Applications: A Journey Beyond Normality, CRC/Chapman & Hall, Boca Raton, FL, USA, 2004, pp. 205–202.
[76] G. De Luca, N. Loperfido, Modelling multivariate skewness in financial returns: A SGARCH approach, Eur. J. Financ. 21 (2015) 1113–1131.
[77] A.S. Weigend, Time Series Prediction: Forecasting the Future and Understanding the Past, Routledge, Abingdon, UK, (2018).
[78] E.M. Azoff, Neural Network Time Series Forecasting of Financial Markets, John Wiley & Sons, Inc, Hoboken, NJ, USA, (1994).
[79] B. Oancea, S.C. Ciucu, Time series forecasting using neural networks. arXiv 2014, arXiv:1401.1333. Algorithms (13) (2020) 121.
[80] W. Van Ackooij, I. Danti Lopez, A. Frangioni, et al. Large-scale unit commitment under uncertainty: an updated literature survey, Ann. Oper. Res. 271 (2018) 11–85 https://doi.org/10.1007/s10479-018-3003-z.

11

Bibliometric analysis of scientific production on energy, sustainability, and climate change

Theodore Kalyvas, Efthimios Zervas
Laboratory of Technology and Policy of Energy and Environment, School of Science and Technology, Hellenic Open University, Parodos Aristotelous, Patra, Greece

Chapter outline

1 Introduction

The global community is experiencing several immanent environmental threats such as climate change, biodiversity loss, pollution and the overexploitation of natural resources [1]. Because climate change is an issue of global importance, it is of major concern to scientists and the relevant scientific publications increase year after year. The evolution of the scientific research regarding climate change can be monitored through the analysis of the scientific research productivity, and thus investigate the dominant trends of this scientific research. As climate change is related to

Low Carbon Energy Technologies in Sustainable Energy Systems. http://dx.doi.org/10.1016/B978-0-12-822897-5.00011-0

energy and sustainability cannot be assumed without the mitigation of climate change, these three notions are closely linked.

One of the most widely used and accepted tools to measure the scientific research productivity in any particular filed of research is bibliometric analysis. Bibliometrics, firstly introduced by Pritchard [2], is considered as a well-established research method for conducting systematic analyses [3]. Bibliometrics uses quantitative analysis and statistics to analyze the bibliometric characteristics of a given field, evaluate the performance of authors, academic institutions or countries, discover the hot topics, reveal the research tendency in future and help researchers to recognize novel schemes among research [4–9].

This study investigates and analyzes the scientific production related to climate change, sustainability, and energy, through a bibliometric analysis. To accomplish this, an analysis of publications, journals, institutions, subject areas, source types, document types, and countries in relation to publication year is performed for each one of these three keywords.

2 Data and methodology

2.1 Data

Data are derived from Elsevier's Scopus database, which covers a significant part of the world scientific production. Scopus was selected for its vast abstract and citation collection of over 22,000 journals from 5,000 international publishers.

2.2 Methodology

Three different searches, using Scopus for topics containing: (1) the keywords climat* AND change*, (2) sustainability, and (3) energy, were conducted.

These searches were restricted to material published until 2019, as 2020 is ongoing and the number of works changes every day. Also, only the works published in English are selected (more than 95%, 95%, and 93% respectively for the three searches, of the total documents). Moreover, the source types defined as "trade publications" and "undefined," and document type defined as "erratum," were excluded. The research was performed from 1 to 20 of April 2020.

The initial search returned 354,722, 193,200 and 4,104,703 documents, for the three searches, respectively. The following data were extracted and analyzed: year of publication, document type, source type, source title, subject area, country, and institution. The impact factors (IFs) of the journals were obtained from the Journal Citation Reports (JCR) Science Edition 2018. The total publications and citations per country were obtained by SCImago Journal & Country Rank. It is a portal including journals and country scientific indicators developed from the information contained in the Scopus database.

3 Results

3.1 Analysis of publications per year

Climate change: In order to have a comprehensive overview of the research production on climate change, the publication of each year is shown (Fig. 11.1). According to the current Scopus documents coverage the first publication on the topic was published in 1837 and only 199 documents were published until 1945. The number of publications began to increase and show an overall exponential trend: from 398 in 1980, they reached 1,275 in 1990, 4,012 in 2000, 15,869 in 2010, and 34,086 in 2019.

Sustainability: The first publication on the topic was published just in 1970, only very few documents were published until 1980, showing that "sustainability" is a recent scientific term. Since then the number of publications began to increase and show an overall

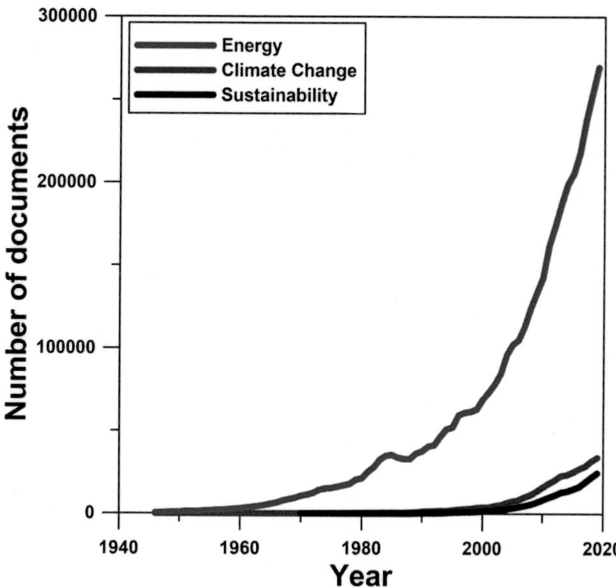

Figure 11.1. Number of documents per year for the three searches performed.

Table 11.1 Number and percentage of papers published totally, after 1980, after 2000, and after 2010.

	Climate Change		Sustainability		Energy	
	Papers	**%**	**Papers**	**%**	**Papers**	**%**
Total	354,722		193,200		4,104,703	
After 1980	352,007	99.23	193,196	100	3,861,389	94.07
After 2000	323,468	91.18	186,842	96.71	3,031,958	73.87
After 2010	248,008	69.91	152,470	78.92	2,051,092	49.967

exponential trend after 1990: they reached 155 papers in 1990, 1,418 in 2000, 9,287 in 2010, and 24,545 in 2019.

Energy: Energy is a very old scientific term; the first publication on the topic was published in 1841. After 1945 there is a steady upward trend, but since 1970 the number of publications began to increase exponentially: already 10,754 papers in 1970, 21,373 in 1980, 37,766 in 1990, 69,036 in 2000, 142,614 in 2010, and 269,733 in 2019.

Table 11.1 shows that the majority of these works are published only the recent years: the 94%–100% after 1980, the 74%–97% after 2000 and the 50%–79% after 2010.

3.2 Subject area

Climate change: 27 subject areas appear in Scopus results. The first 15 subject areas with 579,926 related documents (94,9% of total related documents), are grouped in decreasing order from the most "productive" to the least "productive" subject area (Table 11.2). It must be mentioned that a paper may belong to more than one subject areas and for this reason the total number of documents obtained by adding the documents appearing in each subject area is higher than the total number of papers found (which is 354,722).

The "Environmental Science," with 140,371 publications, it is the first subject area in the list, covering the 22.9% of the total publications; while the second area is "Earth and Planetary Sciences" with 123,603 publications (20.2%), followed by "Agricultural and Biological Sciences" (101,386 documents, 16.6%).

Sustainability: 27 subject areas appear in Scopus results. The first 15 subject areas with 337,499 related documents (92,9% of

Table 11.2 Climate change. Ranking of the 15 first subject areas in terms of publications.

Subject area	Number of documents
Environmental Science	140,371
Earth and Planetary Sciences	123,603
Agricultural and Biological Sciences	101,386
Social Sciences	50,247
Engineering	34,209
Biochemistry Genetics and Molecular Biology	20,398
Energy	19,806
Medicine	16,554
Multidisciplinary	15,974
Arts and Humanities	12,827
Computer Science	11,868
Economics Econometrics and Finance	9,136
Business Management and Accounting	8,959
Physics and Astronomy	7,880
Mathematics	6,708

total related documents), are grouped in decreasing order from the most "productive" to the least "productive" subject area (Table 11.3).

The "Environmental Science" is the first subject area in the list, covering the 18.3% of the total publications (66,495 articles); while the second area is "Social Sciences" with 55,790 publications (15.4%), followed by "Engineering" (41,558 documents, 11.4%).

Energy: 27 subject areas appear in Scopus results. The first 15 subject areas, with 7,209,424 related documents (95,8% of total related documents), are grouped in decreasing order from the most "productive" to the least "productive" subject area (Table 11.4).

The "Physics and Astronomy" subject area is the first in the list, covering 1,533,390 publications (20.4% of the total publications); while the second area is "Engineering" with 1,237,891 publications (16.4%), followed by "Materials Science" (863,894 documents, 11.5%).

Table 11.3 Sustainability. Ranking of 15 subject areas in terms of publications.

Subject area	Number of documents
Environmental Science	66,495
Social Sciences	55,790
Engineering	41,558
Agricultural and Biological Sciences	29,434
Business Management and Accounting	28,208
Energy	26,817
Economics Econometrics and Finance	15,963
Computer Science	15,568
Medicine	13,803
Earth and Planetary Sciences	11,989
Materials Science	7,235
Chemical Engineering	6,401
Arts and Humanities	6,364
Biochemistry Genetics and Molecular Biology	6,064
Decision Sciences	5,810

Table 11.4 Energy. Ranking of 15 subject areas in terms of publications.

Subject area	Number of documents
Physics and Astronomy	1,533,390
Engineering	1,237,891
Materials Science	863,894
Chemistry	724,395
Energy	463,600
Computer Science	364,444
Biochemistry Genetics and Molecular Biology	355,118
Medicine	302,453
Chemical Engineering	290,689
Environmental Science	274,895
Mathematics	255,332
Earth and Planetary Sciences	224,805
Agricultural and Biological Sciences	174,962
Social Sciences	82,490
Multidisciplinary	61,066

3.3 Document and source type

Climate change: Of the 354,722 publications recorded to Scopus in our search, 15 document types are identified. The peer-reviewed journal articles is the most common type (Table 11.5) (264,075 papers or 74.4% of all publications), followed by conference papers (34,392 papers, 9.7%), reviews (19,747 papers, 5.6%), book chapters (19,389 chapters; 5.5%), notes (4,247, 1.2%), and books (4,148, 1.2%). The other document types are minor.

As a consequence of the previous results, the majority of documents are published in journals (Table 11.6) (295,832, 83.4%), followed by conference proceedings (26,917, 7.6%), books (22,412, 6.3%), book series (9,546, 2.7%), while the other media are minor.

Sustainability: Of the 193,200 publications recorded to Scopus from 1970 to 2019, 15 document types are identified. The peer-reviewed journal articles are the most common type (Table 11.7) (122,820 papers or 63.6% of all 193,200 publications), followed by conference papers (34,014 papers, 17.6%), book chapters (13,839 chapters, 7.2%), reviews (13,399 papers, 6.9%), books (3,163 books, 1.6%), and editorials (1,958 editorials, 1%). The other document types are minor.

Table 11.5 Climate change. Total documents of the most common document types.

Document type	Total number of document type (1837–2019)	Percentage of document type (1837–2019) (%)
Article	264,075	74.4
Conference Paper	34,392	9.7
Review	19,747	5.6
Book Chapter	19,389	5.5

Table 11.6 Climate change. Most common source types.

Source type (1837–2019)	Total number of source type (1837–2019)	Percentage of source type (1837–2019) (%)
Journals	295,832	83.4
Conference Proceedings	26,917	7.6
Books	22,412	6.3
Book Series	9,546	2.7

Table 11.7 Total documents of the most common document types.

Document type	Total number of document type (1970–2019)	Percentage of document type (1970–2019) (%)
Article	122,820	63.6
Conference Paper	34,014	17.6
Book Chapter	13,839	7.2
Review	13,399	6.9

As a consequence of the previous results, the majority of documents are published in journals (Table 11.8) (141,885, 73.4%), followed by conference proceedings (28,272, 14.6%), books (15,122, 7.8%), book series (7,911, 4.1%), while the other media are minor.

Energy: Of the 4,097,906 publications recorded to Scopus from 1841 to 2019, 15 document types are identified. The peer-reviewed journal articles are the most common type (Table 11.9) (2,960,799 papers or 72.3% of all 4,097,906 publications), followed

Table 11.8 Most common source types.

Source type (1970–2019)	Total number of source type (1970–2019)	Percentage of source type (1970–2019) (%)
Journals	141,885	73.4
Conference Proceedings	28,272	14.6
Books	15,122	7.8
Book Series	7,911	4.1

Table 11.9 Total documents of the most common document types.

Document type	Total number of document type (1841–2019)	Percentage of document type (1841–2019) (%)
Article	2,960,799	72.3
Conference Paper	886,971	21.7
Review	115,846	2.8
Book Chapter	56,674	1.4

Table 11.10 Most common source types.

Source type (1841–2019)	Total number of source type (1841–2019)	Percentage of source type (1841–2019) (%)
Journals	3,193,331	77.9
Conference Proceedings	749,796	18.3
Book Series	93,205	2.3
Books	60,455	1.5

by conference papers (886,971 papers, 21.7%), reviews (115,846 reviews, 2.8%), and book chapters (56,674 chapters, 1.4%). The other document types are minor.

As a consequence of the previous results, the majority of documents are published in journals (Table 11.10) (3,193,331, 77.9%), followed by conference proceedings (749,796, 18.3%), book series (93,205, 2.3%), books (60,455, 1.5%), while the other media are minor.

For the three searches, the order of magnitude is the same for all document types: the articles correspond to 64%–74% of the total documents, the conference papers from 10% to 22%, the reviews from 3% to 7% and the book chapters from 1.5% to 7%. The same is also valid in the case of source types: 74%–83% for the journals, 8%–18% for the conferences and 1.5%–8% for the books.

3.4 Analysis of the major sources of publication and citation

There is an export limit when searching to Scopus; consecutively, all journals, which published the articles of our research, cannot be extracted. The following analysis covers, for the three searches (climate change, sustainability, and energy) the first 156, 153, and 157 journals, respectively, contained in Scopus' results file, based on the number of publications:

- Climate change: The 40.6% (143,946 papers) of the total number of papers have been published in the first 156 journals.
- Sustainability: The 28.4% (55,070 papers) of the total number of papers have been published in the first 153 journals.
- Energy: The 33.5% (1,373,472 papers) of the total number of papers have been published in the first 157 journals.

Fig. 11.2 shows the cumulative percentage of articles covered by these journals as a function of the number of journals that publish them, in decreasing order of journals, according to the

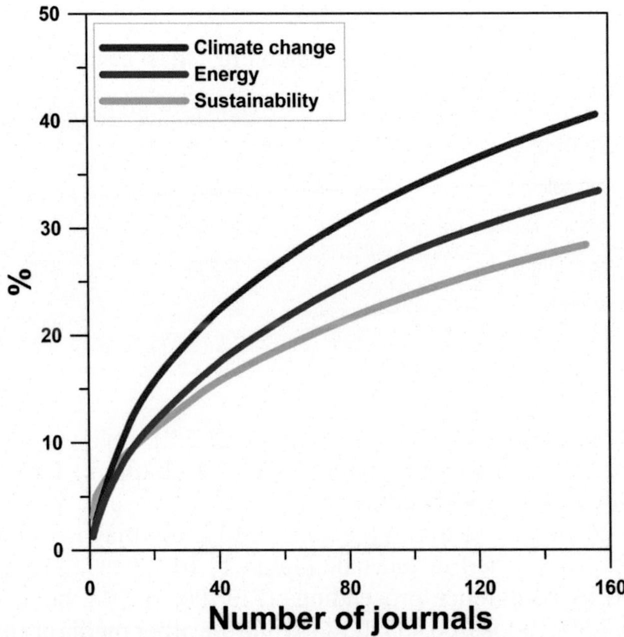

Figure 11.2. Cumulative percentage of papers published as a function of the number of journals that have published them.

number of articles they have published. It can be seen that, even if the journals with the highest number of publications occupy a significant part of the total articles published, a high predominance of a few journals is not observed. For example, the 20% of the total publications is covered, respectively for the three keywords, from 69, 33, and 52 journals.

The 20 first sources of publication are analyzed in the following tables.

Climate change: The 20 main sources of publication are grouped in decreasing order from the most productive to the least productive journal (Table 11.11). The total number of documents published, the percentage of the total documents, and the IF (of 2018) of these journals, are shown. In the case of this keyword, the 20 first sources of publication account for 15.6% of the total number of publications.

Sustainability: Table 11.12 shows the same results as in Table 11.11, in the case of the keyword sustainability. The 20 first sources of publication account for 11.3% of the total number of publications.

Energy: The 20 main sources of publication are grouped in decreasing order from the most productive to the least productive

Table 11.11 Climate change. Ranking of 20 journals in terms of publications.

Journal name	Number of documents	%	Impact factor 2018
Geophysical Research Letters	5,018	1.41	4.578
Journal Of Climate	5,009	1.41	4.805
Climatic Change	3,963	1.12	4.168
Nature	3,575	1.01	43.07
Plos One	2,436	0.69	2.776
Global Change Biology	3,393	0.96	8.88
Science	2,837	0.80	41.063
Science Of The Total Environment	2,797	0.79	5.589
Palaeogeography Palaeoclimatology Palaeoecology	2,684	0.76	2.616
Climate Dynamics	2,597	0.73	4.048
International Journal Of Climatology	2,545	0.72	3.601
Journal Of Geophysical Research-Atmospheres	2,517	0.71	3.633
Quaternary Science Reviews	2,501	0.71	4.641
Proceedings Of The National Academy Of Sciences Of The United States Of America	2,138	0.60	9.58
Quaternary International	1,953	0.55	1.952
Journal Of Hydrology	1,896	0.53	4.405
Environmental Research Letters	1,742	0.49	6.192
Scientific Reports	1,626	0.46	4.011
Proceedings Of SPIE The International Society For Optical Engineering	1,582	0.45	1.209
Forest Ecology And Management	1,512	0.43	3.126

journal (Table 11.13). The 20 first sources of publication account for 11.9% of the total number of publications.

The previous tables show that, in all three keywords, there is no correlation between the number of IF of each journal and the number of papers published.

3.5 Analysis of countries

Climate change: Table 11.14 shows the publication contribution of the top 10 productive countries. Undoubtedly, the most productive country is the United States in terms of the number

Table 11.12 Sustainability. Ranking of 20 journals in terms of publications.

Journal name	Number of documents	%	Impact factor 2018
Sustainability Switzerland	6,041	3.13	2.592
Journal of Cleaner Production	3,716	1.92	6.395
Ecological Economics	1,036	0.54	4.281
Wit Transactions On Ecology And The Environment	972	0.50	0.21
Iop Conference Series Earth And Environmental Science	862	0.45	0.45
Science Of The Total Environment	844	0.44	5.589
Renewable And Sustainable Energy Reviews	819	0.42	10.556
Journal Of Environmental Management	764	0.40	4.865
Energy Policy	683	0.35	4.88
Acta Horticulturae	661	0.34	0.23
Ecological Indicators	614	0.32	4.49
ASEE Annual Conference And Exposition Conference Proceedings	600	0.31	0.4
Lecture Notes In Computer Science Including Subseries Lecture Notes In Artificial Intelligence And Lecture Notes In Bioinformatics	583	0.30	0.402
International Journal Of Sustainability In Higher Education	580	0.30	1.437
Resources Conservation And Recycling	576	0.30	7.044
Plos One	535	0.28	2.776
Energy	513	0.27	5.537
Marine Policy	510	0.26	2.865
Iop Conference Series Materials Science And Engineering	507	0.26	0.53
Applied Energy	505	0.26	8.426

of total publications from single-country articles and international collaborations (111,579 papers, more than the double of the second country), followed by United Kingdom (45,658), China (34,788), Germany (29,426), Australia (27,457), Canada (24,418), France (18,855), Italy (13,883), Spain (13,614), and Netherlands (12,698).

Table 11.13 Energy. Ranking of 20 journals in terms of publications.

Journal name	Number of documents	%	Impact factor 2018
Journal Of Chemical Physics	52253	1.27	2.997
Proceedings Of SPIE The International Society For Optical Engineering	46465	1.13	1.209
Physical Review B Condensed Matter And Materials Physics	37029	0.90	3.736
Journal Of Applied Physics	32596	0.79	2.328
Physical Review Letters	31716	0.77	9.227
Physical Review B	28232	0.69	3.736
International Journal Of Hydrogen Energy	24864	0.61	4.084
Aip Conference Proceedings	23659	0.58	0.4
Applied Physics Letters	23251	0.57	3.521
Journal Of Physics Conference Series	22722	0.55	0.54
Journal Of The American Chemical Society	22546	0.55	14.695
Chemical Physics Letters	18631	0.45	1.901
Journal Of Physical Chemistry A	17178	0.42	2.641
Astrophysical Journal	16901	0.41	5.58
Surface Science	16390	0.40	1.849
Physical Review	16145	0.39	2.907
Physical Review D Particles Fields Gravitation And Cosmology	14608	0.36	4.368
Journal Of Physical Chemistry C	13975	0.34	4.309
Advanced Materials Research	13778	0.34	0.121
Physical Review A Atomic Molecular And Optical Physics	13597	0.33	2.907

Fig. 11.3 shows the evolution of the percentage of publications of the top seven most productive countries for the period 1990–2019. United States has a predominant role in publications, but with a sharp decrease after 2005. All other western countries had an increased percentage until the beginning of 2000s, but this percentage either decreased (like United Kingdom), or remained quite constant (like France or Germany). China has a very high penetration, as the corresponding percentage increases very quickly, especially after 2010. This is a result of the rapid economic and industrial development of this country.

Table 11.14 Contribution of the top 10 productive countries.

Country	Documents	Document ranking	Total documents (SCImago 2018, All subject areas)	Document ranking (SCImago 2018, All subject areas)	H-Index	Docs/All docs
United States	111,579	1	12,070,144	1	2222	0.0092
UK	45,658	2	3,449,243	3	1373	0.0132
China	34,788	3	5,901,404	2	794	0.0059
Germany	29,426	4	3,019,959	4	1203	0.0097
Australia	27,457	5	1,362,848	11	914	0.0201
Canada	24,418	6	1,744,508	7	1102	0.014
France	18,855	7	2,120,161	6	1094	0.0089
Italy	13,883	8	1,744,314	8	953	0.008
Spain	13,614	9	1,376,358	10	830	0.0099
Netherlands	12,698	10	966,986	14	957	0.0131

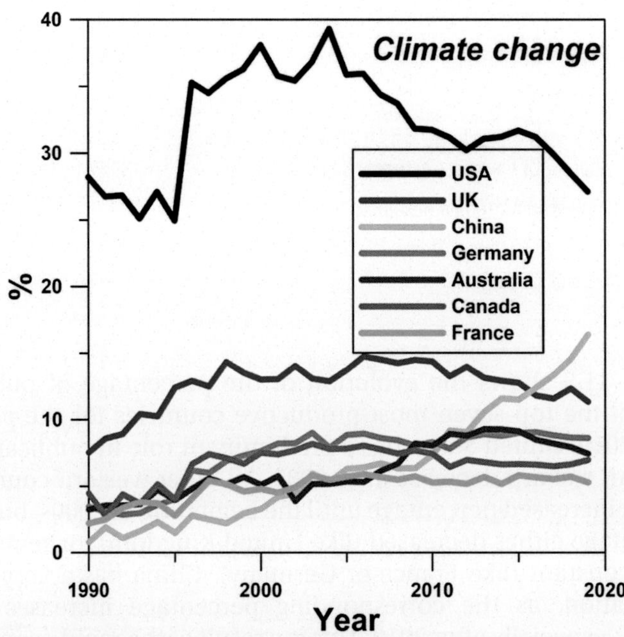

Figure 11.3. Climate change. Evolution of the percentage of publications of the top seven most productive countries (1990–2019).

Table 11.15 Sustainability. Contribution of the top 10 productive countries.

Country	Documents	Document ranking	Total documents (SCImago 2018, All subject areas)	Document ranking (SCImago 2018, All subject areas)	H-Index	Docs/All docs
United States	46,002	1	12,070,144	1	2222	0.0038
United Kingdom	22,116	2	3,449,243	3	1373	0.0064
Australia	13,954	3	1,362,848	11	914	0.0102
China	11,074	4	5,901,404	2	794	0.0019
Italy	10,768	5	1,744,314	8	953	0.0062
Canada	10,299	6	1,744,508	7	1102	0.0059
Germany	10,002	7	3,019,959	4	1203	0.0033
India	8,618	8	1,670,099	9	570	0.0052
Spain	7,535	9	1,376,358	10	830	0.0055
Netherlands	7,223	10	966,986	14	957	0.0075

Sustainability: Table 11.15 shows the publication contribution of the top 10 productive countries. The most productive country is United States in terms of the number of total publications from single-country articles and international collaborations (46,002 papers, again more than the double of the second country), followed by United Kingdom (22,116), Australia (13,954), China (11,074), Italy (10,768), Canada (10,299), Germany (10,002), India (8,618), Spain (7,535), and the Netherlands (7,223).

Fig. 11.4 shows the evolution of the percentage of the publications of the top seven most productive countries for the period 1990–2019. United States has a predominant role in publications but shows a sharp decrease after 2010. United Kingdom shows a similar trend in a slower rate. China's publications increase constantly, and the upward trend is bigger after the year 2015. Among the other countries only Germany and Italy show a tiny increase while the percentage of the other countries decreases.

Energy: Table 11.16 shows the publication contribution of the top 10 productive countries. Undoubtedly, the most productive country is the United States in terms of the number of total publications from single-country articles and international

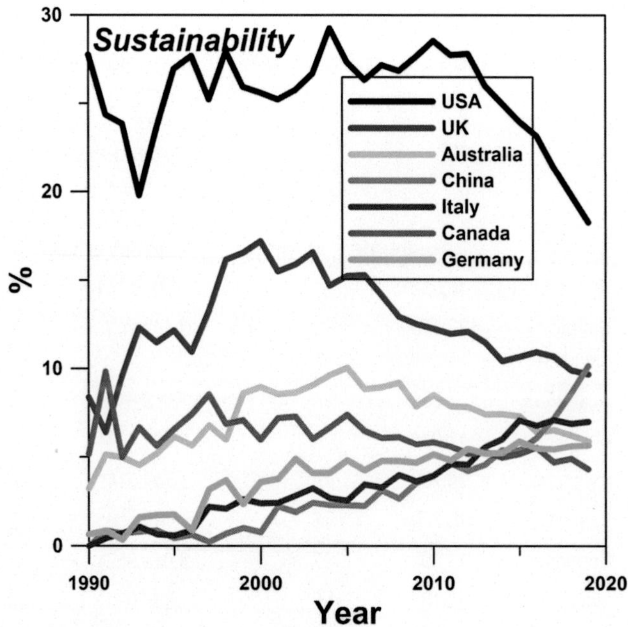

Figure 11.4. Sustainability. Evolution of the percentage of publications of the top seven most productive countries (1990–2019).

Table 11.16 Energy. Contribution of the top 10 productive countries.

Country	Documents	Document ranking	Total documents (SCImago 2018, All subject areas)	Document ranking (SCImago 2018, All subject areas)	H-Index	Docs/All docs
United States	1,106,550	1	12,070,144	1	2222	0.0917
China	514,620	2	5,901,404	2	794	0.0872
Germany	314,316	3	3,019,959	4	1203	0.1041
Japan	282,342	4	2,750,108	5	967	0.1027
United Kingdom	268,884	5	3,449,243	3	1373	0.078
India	212,240	6	1,670,099	9	570	0.1271
France	205,754	7	2,120,161	6	1094	0.097
Italy	175,718	8	1,744,314	8	953	0.1007
Russian Federation	153,719	9	1,076,966	13	540	0.1427
Canada	147,834	10	1,744,508	7	1102	0.0847

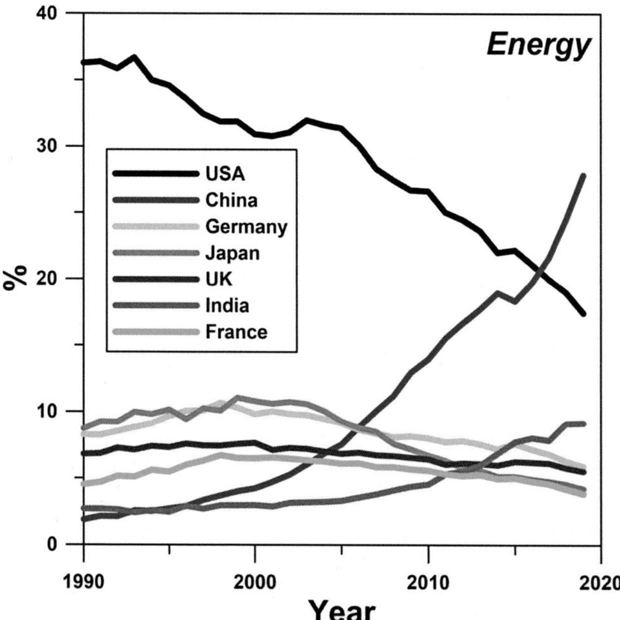

Figure 11.5. Energy. Evolution of the percentage of publications of the top seven most productive countries (1990–2019).

collaborations (1,106,550 papers, again more than the double of the second country), followed by China (514,620), Germany (314,316), Japan (282,342), United Kingdom (268,884), India (212,240), France (205,754), Italy (175,718), Russian Federation (153,719), and Canada (147,834).

Fig. 11.5 shows the percentage of the publications of the top seven most productive countries for the period 1990–2019. United States has a predominant role in publications but shows a constant decrease after 1990. China's publications increase rapidly, and have exceeded United States since 2017. The other countries' publications show constant decrease after 2000, except for India, which also shows a remarkable small upward trend.

Tables 11.14–11.16 also show the contribution of the top 10 countries in all scientific fields, as pointed out in the database of SCImago Journal and Country Rank [10] based on the total publications and the H-index. For the tree keywords used here, the position of each country in the list of the top 10 productive countries is quite close to the position in all scientific fields, extracted by SCImago 2018. Moreover, the number of publications of the three keywords versus the total number of publications shows a quite linear trend for all three keywords.

3.6 Analysis of institutions

Climate change: Among the first 160 most productive institutions extracted by Scopus, the Chinese Academy of Sciences is the most productive institution with 14,311 papers, followed by the French CNRS (Centre National de la Recherche Scientifique) with 6,805 works, the University of Chinese Academy of Sciences with 4,231, the National Oceanic and Atmospheric Administration with 4,199, the University of Colorado at Boulder with 3,730, the United States Geological Survey with 3,726, the University of Washington, Seattle with 3,438, the University of Oxford with 3,224, the ETH Zurich with 3,187 and the National Center for Atmospheric Research with 3,090 works. Most of these institutions are from United States of China (Table 11.17).

Sustainability: Among the first 160 most productive institutions extracted by Scopus, the Chinese Academy of Sciences is the most productive institution with 1,822 papers, followed by the Wageningen University and Research Centre with 1,801, the University of British Columbia with 1,156, the University of Queensland with 1,039, the Arizona State University with 1,025, the Delft University of Technology with 988, the Universidade de Sao Paulo—USP with 959, the University of Melbourne with 855, the University of Toronto with 829 and the CNRS Centre National de la Recherche Scientifique with 821. Contrary to the previous keyword, there is quite wide distribution of these institutions in several countries (Table 11.18).

Energy: Among the first 160 most productive institutions extracted by Scopus, the Chinese Academy of Sciences is the

Table 11.17 Climate change. Contribution of the top 10 productive institutions.

Institution	Number of documents
Chinese Academy of Sciences	14,311
CNRS Centre National de la Recherche Scientifique	6,805
University of Chinese Academy of Sciences	4,231
National Oceanic and Atmospheric Administration	4,199
University of Colorado Boulder	3,730
United States Geological Survey	3,726
University of Washington Seattle	3,438
University of Oxford	3,224
ETH Zürich	3,187
National Center for Atmospheric Research	3,090

Table 11.18 Sustainability. Contribution of the top 10 productive institutions.

Institution	Number of documents
Chinese Academy of Sciences	1,822
Wageningen University and Research Centre	1,801
The University of British Columbia	1,156
University of Queensland	1,039
Arizona State University	1,025
Delft University of Technology	988
Universidade de Sao Paulo - USP	959
University of Melbourne	855
University of Toronto	829
CNRS Centre National de la Recherche Scientifique	821

most productive institution with 74,150 papers, followed by the CNRS Centre National de la Recherche Scientifique with 53,996, the Russian Academy of Sciences with 45,323, the Ministry of Education China with 42,541, the University of Tokyo with 34,932, the University of California Berkeley with 29,703, the Massachusetts Institute of Technology with 27,965, the Tsinghua University with 23,784, the Lawrence Berkeley National Laboratory with 23,730 and the Kyoto University with 22,404. Again here, the majority of the institutions come from China or the United States (Table 11.19).

Table 11.19 Energy. Contribution of the top 10 productive institutions.

Institution	Number of documents
Chinese Academy of Sciences	74,150
CNRS Centre National de la Recherche Scientifique	53,996
Russian Academy of Sciences	45,323
Ministry of Education China	42,541
University of Tokyo	34,932
University of California Berkeley	29,703
Massachusetts Institute of Technology	27,965
Tsinghua University	23,784
Lawrence Berkeley National Laboratory	23,730
Kyoto University	22,404

4 Conclusions

A bibliometric research using the keywords "climate change," "sustainability," and "energy" is performed. This search returned 354,722, 193,200 and 4,104,703 documents, respectively for the three keywords. There is a remarkable increase of annual number of publications for all three keywords, mainly after 1990 and even more after 2010. Almost the 100% of the publications is produced after 1980, the 74%–97% after 2000 and the 50%–79% after 2010.

The works of climate change is mainly in the field of "Environmental Science," followed by "Earth and Planetary Sciences," and "Agricultural and Biological Sciences." Sustainability is mainly if the field of "Environmental Science," followed by "Social Sciences" and "Engineering." Energy, is mainly in "Physics and Astronomy", followed by "Engineering" and "Materials Science."

For the three keywords, the articles correspond to 64%–74% of the total documents, the conference papers from 10% to 22%, the reviews from 3% to 7%, and the book chapters from 1.5% to 7%. The first source type is journals (74%–83%), followed by conferences (8%–18%), and books (1.5%–8%). For all keywords, there is no clear dominance of some journals and the papers are quite largely distributed to many journals.

Among the countries, United States, United Kingdom, and China were the top three productive countries for the "climate change" research, United States, United Kingdom, and Australia, for the "sustainability" research and United States, China, and Germany for the "energy" research. For all three keywords, there is a constant decrease of the relative contribution of United States and of the other western countries, while the percentage of China constantly increases.

The most productive institution for all three issues was the Chinese Academy of Sciences. The only other institution that manages to be included among the top 10 productive institutions for all 3 issues was the CNRS Centre National de la Recherche Scientifique. Generally, the institutions with the highest productivity come from United States or China.

References

[1] J. Rockstrom, W. Steffen, K. Noone, A. Persson, F.S. Chapin, E.F. Lambin, et al., A safe operating space for humanity, Nature 461 (7263) (2009) 472–475.

[2] A Pritchard, Statistical bibliography or bibliometrics, J. Doc. 25 (1969) 348–349.

[3] A.F. van Raan, For your citations only? Hot topics in bibliometric analysis, Measurement 3 (2005) 50–62.

[4] J.A. Wallin, Bibliometric methods: pitfalls and possibilities, Basic Clin. Pharmacol. Toxicol. 97 (2005) 261–275.
[5] F. De Battisti, S. Salini, Robust analysis of bibliometric data, Stat. Methods Appl. 22 (2013) 269–283.
[6] H.Z. Fu, M.H. Wang, Ho Y.S., Mapping of drinking water research: A bibliometric analysis of research output during 1992-2011, Sci. Total Environ. 443 (2013) 757–765.
[7] R. Aleixandre-Benavent, J.L. Aleixandre-Tudo, L. Castello-Cogollos, J.L. Aleixandre, Trends in scientific research on climate change in agriculture and forestry subject areas (2005-2014), J. Clean. Prod. 147 (2017) 406–418.
[8] R. Haunschild, L. Bornmann, Werner M., Climate change research in view of bibliometrics., PLOS One 29 (2016). doi: 10.1371/journal.pone.0160393 2016.
[9] J. Li, M.H. Wang, Ho Y.S., Trends in research on global climate change: A Science Citation Index Expanded-based analysis, Global Planetary Change 77 (2011) 13–20.
[10] SCImago Journal and Country Rank. SCImago Research Group. Available from: http://www.scimagojr.com

12

Public acceptance of renewable energy sources

Zoe Gareiou, Efi Drimili, Efthimios Zervas
Laboratory of Technology and Policy of Energy and Environment, School of Science and Technology, Hellenic Open University, Parodos Aristotelous, Patra, Greece

Chapter outline

1 Introduction

The need for energy demands is constantly increasing globally, while stocks from conventional energy sources are finite. Locating and exploiting new sources of conventional energy is becoming increasingly difficult due to the unknown quantities of coal, oil, and gas reserves and their expensive and dangerous way of extraction [1].

In addition to the energy crisis, global warming, resulting in the potential threat of global climate change, is mainly due to Green House Gases (GHGs) mainly from CO_2 emitted from fossil fuel consumption emissions, but also from other gases contributing to greenhouse effect. Climate change causes several negative effects on nature and humans [1,2].

Low Carbon Energy Technologies in Sustainable Energy Systems. http://dx.doi.org/10.1016/B978-0-12-822897-5.00012-2

Renewable Energy Sources (RES) produce energy from natural processes (e.g. sun, wind), which are replenished at a higher rate than consumed [3]. According to [4], renewable energy is the energy that is drawn from the repetitive energy flows, which constantly occur in the natural environment. Renewable energy technologies have significant potential for development, as these resources are globally distributed globally, contrary to conventional sources (gas, coal, oil), which are geographically concentrated. All countries have at least one abundant renewable resource, while the majority have a rich resource portfolio. The role of RES is expected to increase significantly over time in all scenarios of the International Energy Agency (IEA), with the largest contribution to electricity generation, heating and cooling and transport [3].

The most important benefits from the penetration of RES are: (1) Reduction of environmental impacts, including greenhouse gas emissions and other pollutants, (2) Energy security, (3) Economic growth strategy, and (4) Storage of energy distributed inside and outside the network.

The main forms of RES are the following [5]:

1. Solar energy is the energy produced by technologies using the sun radiation. Solar energy technologies are divided into: (1) photovoltaic solar systems, which directly convert the solar energy to electricity, (2) active solar systems, which convert the solar radiation in heat, and (3) bioclimatic design and passive solar systems, which include architectural solutions and the use of appropriate building materials to maximize the direct utilization of solar energy for heating, air conditioning, or lighting.
2. Wind energy is the kinetic energy, which is produced by the force of the wind and is converted into mechanical or electrical energy.
3. Geothermal energy is the thermal energy, which comes from the interior of the earth and is contained in natural vapors, in surface and underground hot waters, as well as in hot dry rocks.
4. Hydroelectric energy is the energy, which is produced from waterfalls for the purpose of generating electricity.

The European Commission has set up a fair transition mechanism to finance the green reform to a climate-neutral economy [6]. This mechanism consists of three main sources of funding: (1) A Just Transition Fund, which will be financed with new Community funds of €7.5 billion, in addition to the Commission's proposal for the next long-term budget of the European Union, (2) a special regime of fair transition within the framework of InvestEU for the mobilization of investments and finding new sources of growth, and (3) a public sector lending mechanism in cooperation with the European Investment Bank.

It is obvious that according to the EU directions there is a need for energy transformation from fossil fuels to RES. Consequently, relevant policies have to be designed and energy RES projects need to be implemented. All these actions require the corporation of various actors on all policy levels.

One of the most important factors for the penetration of RES in the energy mix is their social acceptance. Public participation has been identified as a crucial factor to energy transformation from fossil fuels to RES [7]. In relation to the various forms of RES, the social acceptance of wind energy and the installation of wind farms are the two forms of RES that have been most investigated [8–11]. In addition, individual factors that influence the formation of social acceptance have been further examined, such as the level of knowledge of different forms of RES [8,12], the association of RES projects with NIMBY (Not In My Back Yard) syndrome [12], the impact of the information provided and the knowledge on the acceptance of RES [13], the attitude of citizens toward RES [8,12], as well as the willingness for additional payment for investments in RES technologies [8,13]. The role of economic, environmental and social impacts from the RES development on their social acceptance or rejection as well as the public perception of the benefits arising from the installation of RES projects has also been investigated [11]. An additional aspect that has also been considered is the ownership status of the bodies that implement and operate the RES projects, as well as the degree of the private sector involvement in the investment model of RES development [8].

Regarding Greece, some studies have focused on the issue of exploring the impact of demographic and socio-economic factors on the knowledge of different forms of RES [14], as well as the willingness for additional payment in order to promote the development of RES [15,16]. The social acceptance of specific forms of RES such as biomass has also been explored [17]. The most well-known forms of RES in Greece are solar and wind energy; biomass and biofuels are the least known forms and those that gains the least public support. The level of acceptance of different forms of RES differs between different regions depending on the past experience from RES projects installation. In areas where RES projects have been installed and are operating, in particular wind farms, a significant percentage of the residents are opposed to the installation; objections focus primarily on the visual disturbance and secondarily on the noise emitted by the operation of the wind turbines [18].

The present work investigates the social attitude toward RES by exploring their beliefs and perceptions, while highlighting the main parameters that affect their social acceptance. This works uses a structure questionnaire, as this type of research is effectively used to reveal the opinion of the citizens about environmental issues, such as climate change [19], energy use [20], water use [21], waste management [22], etc.

2 Materials and methods

2.1 Sample size and collection

The survey was conducted in Athens, the capital of Greece, between January and April 2018. In total, 536 valid questionnaires were collected using face-to-face interviews. The questionnaire was addressed to the adult residents of Athens (aged >18 years old). The collection of the questionnaires took place in 10 different places of the Athens Region at different times of the day and on different days of the week.

The error of using the sample of 536 valid questionnaires is estimated as follows. Statistical theory suggests that the connection between the sample size and the desired margin of error, in case of finite population, is given by the formula [23]:

$$n = \frac{z^2 * p * (1-p) * N}{e^2 * (N-1) + z^2 * P * (1-p)} \tag{12.1}$$

Since the sample size n is small compared to the size of the population N (adult population of Athens), the above equation can be used in a simplified form, omitting the correction/factor of the population. Thus, we applied the equation for the size of the sample for infinite population when we are interesting to estimate the proportion in the population [23–25]:

$$n = \frac{z^2 * p * (1-p)}{e^2} \tag{12.2}$$

- *n is the sample size (n = 536 in our case)*
- *z is the two-tailed value of the standardized normal deviate associated with desired level of confidence (for 95% confidence interval the value of z is equal to 1.96)*
- *e is the desired margin of error (for the desired reliability, the acceptable maximum error is 0.05, with an associated 95% confidence interval)*
- *p is the preliminary estimate of the proportion in the population (as the value of p was not known the maximum value of 0.50 was assumed)*

From the above equation, e = 0.04, or 4%, which is less than 5%.

2.2 Survey questionnaire

The survey was conducted using a properly designed questionnaire which was divided in four sections. The first section of the questionnaire was focused on the environmental sensitivity of the participants. Those questions were related to the assessment of the impact of environmental problems in everyday life, of the effectiveness of actions taken by the state for the environmental protection, as well as the assessment of the importance of specific environmental issues (air pollution, water pollution, solid waste management, climate change, depletion of natural resources-fossil fuels, ecosystem disruption).

The next section includes a series of questions aimed at exploring various aspects and components of RES. The knowledge of the concept of RES, the level and the source of information for each form of RES (solar, wind, geothermal, biomass, hydroelectric) were investigated. The participants' views on the environmental friendliness of five basic forms of RES was also requested, as well as the evaluation of the potential benefits derived from the construction of a RES project. The participants were also asked for their opinion about the cost of energy produced from RES, in comparison to the energy produced from conventional sources. Then, their degree of confidence in the government for the promotion and development of RES projects, as well as the level of their agreement with the promotion of RES and the prioritization on the form of RES were also investigated. Lastly, their views on the degree of involvement of various bodies (scientists, government, local government, residents, and investors) during the installation of RES projects were asked.

In the next section, a hypothetical scenario was given to the participants. They were asked about the degree of their annoyance in case that a RES project (wind turbine/wind farm, photovoltaic, geothermal, biomass, hydroelectric station/dam) was planned to be installed at a short distance from their residence. The desired distance for the installation of a RES project, the ownership status of RES management bodies at all stages of their development (design, construction, operation and also pricing) and the willingness for financial contribution for the promotion of RES (e.g., through the electricity bill) was asked; in case of agreement for additional payment, the determination of the desired price was also asked.

The last section of the questionnaire includes questions related to the sociodemographic characteristics of the respondents (gender, age, marital status, number of children, education, professional status, income, and general political opinion).

2.3 Data analysis

The data collected from the questionnaire were evaluated statistically using SPSS. The participants' responses were analyzed using descriptive statistics (frequency analysis, percentages, mean and standard deviation). Chi-square test for independence was used to determine whether the variables concerning the participants' views related to RES were statistically related to personal socio-demographic characteristics. A chi-square test is considered unreliable if more than 20% of the expected values are less than five. In the cases where tests were not reliable, variables were grouped in order to overcome this shortcoming. A 2-sided p-value less than 0.05 is considered to be significant.

3 Results and discussion

3.1 Socio-demographic characteristics of the sample

Table 12.1 shows the frequency distribution of the socio-demographic profiles of the respondents. The sample reflects the gender and age distribution of the corresponding official data of the Hellenic Statistical Authority as recorded in the last census of 2011 (Table 12.1).

Table 12.1 shows that the sample consists of 47.9% males and 52.1% females, the average age of both genders are 44.0 years. Also, almost half of the respondents are married (46.5%), while the 56.0% have children. The majority (47.6%) of the respondents have university education level, while the 28.4% have high school

Table 12.1 Socio-demographic profile of the respondents.

Variables	Value	% (Sample)	% (Census 2011)
Gender	Male	47.9	48.2
	Female	52.1	51.8
Age group	18–29	21.8	18.4
	30–39	21.5	19.5
	40–49	20.6	18.2
	50–59	17.4	15.9
	60–69	14.8	12.2
	> 70	3.9	15.8

Table 12.1 Socio-demographic profile of the respondents. (*Cont.*)

Variables	Value	% (Sample)	% (Census 2011)
Marital status	Married	46.5	50.2
	Single	37.9	39.1
	Divorced	8.5	3.1
	Widowed	7.1	7.6
Number of children	0	44.0	34.2
	1	14.2	14.4
	2	32.7	35.5
	3	7.8	11.2
	>4	1.3	4.7
Education	Secondary school	19.7	19.0
	High school	28.4	32.7
	Higher education	47.6	45.1
	Master Diploma/PhD	4.3	3.2
Occupation	State employee	12.1	7.0
	Private employee	30.0	22.8
	Self-employed	10.1	7.9
	Retired	19.6	22.7
	Student	8.6	16.4
	Farmer	0.2	3.5
	Home worker	9.9	11.5
	Unemployed	9.5	8.1
Employment status	Permanent employee	28.4	
	Undetermined duration contract employee	45.6	
	Limited duration contract employee	7.0	
	Employee with a service block	11.6	
	Employee with a project contract	3.2	
	Part-time employee	4.2	
Family income, Euros	0–5.000	11.2	
	5.001–10.000	20.2	
	10.001–20.000	40.1	
	20.001–30.000	18.4	
	30.001–40.000	5.6	
	40.001–50.000	1.9	
	>50.000	2.7	

education. Regarding occupation, the 30% of the respondents are private employees, the 19.6% are retired and 12.1% are state employees. The majority of the respondents are employees with undetermined duration contract (45.6%), while the 28.4% are permanent employees. Finally, the annual total family income corresponds to 10.001–20.000€ to 40.1% of respondents.

4 Descriptive analysis and the effect of socio-demographic characteristics

Participants' responses were analyzed using descriptive statistics (mean and standard deviation) (Table 12.2). Furthermore, a Chi-square test was performed to determine the existence of relationships between the participants' responses and the socio- demographic characteristics (Table 12.3). The results of descriptive statistics and Chi-square test are shown separately for each part: the environmental sensitivity, the opinions and the knowledge about the RES, and the hypothetical RES installation scenario.

Table 12.2 Description of variables.

Question/Variable	Variable's code	Mean	S.D.
Environmental Sensitivity			
Environmental problems affect daily life[a]	Env_Probl	3.90	± 0,909
The actions taken in Greece for the protection of the environment are effective[a]	Env_Act	1.86	± 0.784
Opinions and knowledge about the RES			
Knowledge about RES[a]	Know_RES	3.10	± 1.097
Knowledge about the wind energy (e.g., wind turbines)[a]	Know_Wind_Energy	3.04	± 1.036
Knowledge about the solar energy (e.g., photovoltaic)[a]	Know_Solar_Energy	3.30	± 0.988
Knowledge about the geothermal energy[a]	Know_Geo_Energy	1.95	± 1.136
Knowledge about the biomass[a]	Know_Bio_Energy	1.89	± 1.026
Knowledge about the hydroelectric energy (e.g., hydroelectric factory)[a]	Know_Hydro_Energy	2.89	± 1.111
Source of information *(Education system, family—social environment, TV—radio, Internet, newspapers—magazines)*	Source_Inform	2.47	± 1.064
Environmental friendly the wind energy (e.g., wind turbines)[a]	Env_Fr_Wind_Energy	3.85	± 0.954
Environmental friendly the solar energy (e.g., photovoltaic)[a]	Env_Fr_Solar_Energy	3.89	± 0.934
Environmental friendly the geothermal energy[a]	Env_Fr_Geo_Energy	3.57	± 0.970
Environmental friendly the biomass[a]	Env_Fr_Bio_Energy	2.76	± 1.026

Table 12.2 Description of variables. (*Cont.*)

Question/Variable	Variable's code	Mean	S.D.
Environmental friendly the hydroelectric energy (e.g., hydroelectric factory)[a]	Env_Fr_Hydro_Energy	3.55	± 1.039
RES installation creates new job positions[a]	RES_New_Job	3.36	± 0.984
RES installation degrades an area[a]	RES_Degr_Area	2.06	± 1.014
RES installation reduces dependence on fossil fuels[a]	RES_Red_FF	3.52	± 0.984
RES installation requires high initial investment cost[a]	RES_Cost	3.71	± 0.891
RES installation contributes to the economic development of an area[a]	RES_Econ_Area	3.47	± 1.024
RES installation binds large area of land[a]	RES_Large_Area	3.35	± 0.936
Electricity from RES is more expensive/cheaper than conventional resources[b]	RES_Electr	2.39	± 1.284
Confidence in the government for promotion and development of RES[a]	Confid_Govern	1.62	± 0.788
In the decision to install a RES (in an area) must participate specialist scientists[a]	Install_Scient	4.43	± 0.777
In the decision to install a RES (in an area) must participate the government[a]	Install_Govern	3.10	± 1.286
In the decision to install a RES (in an area) must participate the local government[a]	Install_LocGovern	3.67	± 1.065
In the decision to install a RES (in an area) must participate the residents of the area[a]	Install_Resid	3.67	± 1.121
In the decision to install a RES (in an area) must participate investors (public or private)[a]	Install_Investors	3.85	± 1.042
In favor of the development of RES for electricity production[a]	Electr_Prod	4.12	± 0.854
Should be given priority in a RES *(Wind, solar, geothermal, biomass, hydroelectric)*	RES_Priority	2.09	± 0.929
Hypothetical RES installation scenario			
Annoyance from installing a wind turbine/wind farm a short distance from the residential area[a]	Annoy_Wind	3,56	± 1.291
Annoyance from installing a photovoltaic system/park a short distance from the residential area[a]	Annoy_Photov	2.25	± 1.162
Annoyance from installing a geothermal project a short distance from the residential area[a]	Annoy_Geo	2.02	± 1.200
Annoyance from installing a biomass project a short distance from the residential area[a]	Annoy_Bio	2.12	± 1.274
Annoyance from installing a hydroelectric power station/dam a short distance from the residential area[a]	Annoy_Hydro	2.96	± 1.366
Public/Private the management of RES regarding the design[c]	Manag_Design	3.35	± 1.248
Public/Private the management of RES regarding the construction[c]	Manag_Constr	3.48	± 1.244
Public/Private the management of RES regarding the exploitation[c]	Manag_Expoit	2.90	± 1.251
Public/Private the management of RES regarding the setting prices[c]	Manag_Prices	2.74	± 1.221
Willing to pay for RES[a]	Willing_Pay	2.02	± 0.985

[a] Ranking scale from 1 to 5 (Not at all to Very Much).
[b] Ranking scale from 1 to 5 (Much Cheaper to Much Expensive).
[c] Ranking scale from 1 to 5 (Fully Public to Fully Private).

Table 12.3 The impact of the socio-demographic characteristics on public's perceptions.

Socio-demographic characteristics	Variable's name		No (%)	Yes (%)	Pearson Chi-square	*p* value
Gender	Env_Probl	Male	33.46	66.54	10.283	0.001
		Female	21.15	78.85		
	Know_RES	Male	56.42	43.58	13.590	0.000
		Female	71.68	28.32		
	Know_Geo_Energy	Male	43.70	56.30	3.967	0.046
		Female	52.35	47.65		
	Know_Hydro_Energy	Male	28.80	71.20	20.108	0.000
		Female	47.67	52.33		
	Env_Fr_Geo_Energy	Male	35.90	64.10	7.975	0.005
		Female	52.08	47.92		
	Env_Fr_Bio_Energy	Male	42.15	57.85	3.842	0.050
		Female	51.03	48.97		
	RES_Electr	Male	59.38	40.63	5.459	0.019
		Female	69.06	30.94		
	Annoy_Hydro	Male	40.83	59.17	5.053	0.025
		Female	51.03	48.97		
	Willing_Pay	Male	43.58	56.42	11.662	0.001
		Female	29.39	70.61		
Age	Know_RES	Younger	56.47	43.53	11.131	0.001
		Older	70.39	29.61		
	Know_Wind_Energy	Younger	28.45	71.55	5.177	0.023
		Older	37.83	62.17		
	Know_Solar_Energy	Younger	47.84	52.16	11.478	0.001
		Older	62.50	37.50		
	Know_Geo_Energy	Younger	41,30	58,70	7.752	0.005
		Older	53,49	46,51		
	Know_Bio_Energy	Younger	36,52	63,48	12.862	0.000
		Older	52,16	47,84		
	Install_Govern	Younger	64,66	35,34	6.768	0.009
		Older	53,47	46,53		
	Install_LocGovern	Younger	40,09	59,91	4.061	0.044

Table 12.3 The impact of the socio-demographic characteristics on public's perceptions. (*Cont.*)

Socio-demographic characteristics	Variable's name		No (%)	Yes (%)	Pearson Chi-square	*p* value
		Older	31,68	68,32		
	Install_Resid	Younger	43,97	56,03	6.164	0.013
		Older	33,44	66,56		
	Electr_Prod	Younger	57,76	42,24	6.071	0.014
		Older	68,09	31,91		
	Annoy_Wind	Younger	50,43	49,57	19.231	0.000
		Older	31,51	68,49		
	Annoy_Photov	Younger	58,08	41,92	23.921	0.000
		Older	36,61	63,39		
	Annoy_Geo	Younger	46,94	53,06	3.839	0.050
		Older	35,76	64,24		
	Annoy_Bio	Younger	47,44	52,56	14.144	0.000
		Older	26,53	73,47		
	Annoy_Hydro	Younger	61,06	38,94	33.515	0.000
		Older	34,55	65,45		
	Manag_Design	Younger	47,84	52,16	8.551	0.003
		Older	60,53	39,47		
	Willing_Pay	Younger	30,60	69,40	5.536	0.019
		Older	40,46	59,54		
Marital Status	Install_Govern	Married	53,82	46,18	3.885	0.049
		Unmarried	62,24	37,76		
	Install_LocGovern	Married	28,63	71,37	9.078	0.003
		Unmarried	41,11	58,89		
	Install_Resid	Married	33,47	66,53	4.063	0.044
		Unmarried	41,96	58,04		
	Manag_Design	Married	60,24	39,76	5.088	0.024
		Unmarried	50,52	49,48		
Children	Annoy_Hydro	No children	51,27	48,73	5.236	0.022
		With children	40,89	59,11		
	Manag_Design	No children	50,19	49,81	4.903	0.027

(*Continued*)

Table 12.3 The impact of the socio-demographic characteristics on public's perceptions. (*Cont.*)

Socio-demographic characteristics	Variable's name		No (%)	Yes (%)	Pearson Chi-square	*p* value
		With children	59,71	40,29		
	Willing_Pay	No children	27,00	73,00	18.916	0.000
		With children	45,05	54,95		
Education	Env_Probl	Lower	30,69	69,31	3.836	0.050
		Higher	23,17	76,83		
	Know_RES	Lower	76,17	23,83	34.845	0.000
		Higher	51,74	48,26		
	Know_Wind_Energy	Younger	43,32	56,68	23.389	0.000
		Older	23,55	76,45		
	Know_Solar_Energy	Younger	65,70	34,30	21.223	0.000
		Older	45,95	54,05		
	Know_Geo_Energy	Younger	59,93	40,07	30.655	0.000
		Older	35,91	64,09		
	Know_Bio_Energy	Younger	56,99	43,01	30.269	0.000
		Older	33,20	66,80		
	Env_Fr_Wind_Energy	Lower	43,32	56,68	23.389	0.000
		Higher	23,55	76,45		
	Env_Fr_Solar_Energy	Lower	65,70	34,30	21.223	0.000
		Higher	45,95	54,05		
	Env_Fr_Bio_Energy	Lower	56,99	43,01	30.269	0.000
		Higher	33,20	66,80		
	RES_Electr	Lower	60,65	39,35	3.568	0.050
		Higher	68,48	31,52		
	Annoy_Hydro	Lower	39,75	60,25	7.653	0.006
		Higher	52,30	47,70		
Income	Electr_Prod	Lower	71,43	28,57	6.087	0.014
		Higher	60,38	39,62		
Political Beliefs	Env_Probl	Left-wing	31,01	68,99	5.147	0.023
		Right-wing	17,71	82,29		
	RES_Electr	Left-wing	56,59	43,41	3.857	0.050
		Right-wing	69,47	30,53		

Table 12.3 The impact of the socio-demographic characteristics on public's perceptions. (*Cont.*)

Socio-demographic characteristics	Variable's name		No (%)	Yes (%)	Pearson Chi-square	*p* value
	Install_Govern	Right-wing	68,99	31,01	6.664	0.010
		Left-wing	52,08	47,92		
	Install_Investors	Right-wing	21,71	78,29	10.769	0.001
		Left-wing	42,11	57,89		
	Manag_Design	Right-wing	43,41	56,59	18.061	0.000
		Left-wing	71,88	28,13		
	Manag_Constr	Right-wing	37,21	62,79	26.499	0.000
		Left-wing	71,88	28,13		
	Manag_Expoit	Right-wing	22,48	77,52	22.557	0.000
		Left-wing	53,13	46,88		
	Manag_Prices	Right-wing	34,11	65,89	20.515	0.000
		Left-wing	64,58	35,42		

5 Environmental sensitivity

The questions about the environmental sensitivity of the respondents showed that the environmental problems significantly affect the daily lives of respondents (73% much and very much). Women and respondents with higher education and with left-wing political beliefs declare to be more affected by the environmental problems in their daily lives. The high agreement with this view and the strongest support among women are consistent with the results of the recent Special Eurobarometer survey conducted in December 2019 in relation to the attitudes of European citizens toward the environment [26]. In our work, the air pollution is considered by the respondents to be the most important environmental issue, while water pollution is considered the second most important one. Climate change and ecosystem disruption is found in the third place, followed by solid waste management and noise pollution. The results of the Eurobarometer survey are slightly different. Air pollution is considered as the second more important environmental issue for the Europeans and Greek citizens while climate change comes in the first place [26]. In our work, the opinion of the respondents regarding the actions taken in Greece for the protection of the environment is very bad, as the 79.7% of the respondents consider these actions as ineffective. These findings are in line with those of the Eurobarometer survey, as the majority of the Europeans claim that their national government is not doing enough for the environmental protection [26].

6 Opinions and knowledge about the RES

According to our results, RES are not widely known in Greece, as only the one third of the respondents declare that they have only a moderate knowledge about RES. Men, younger respondents and respondents with higher education are more familiar with RES. These findings are compatible with a previous survey conducted in Greece (in Greek island of Crete), where one in three respondents stated the low level of knowledge about RES [27]. It seems that there is room for further information and public awareness about RES. Concerning the main source of information about RES, the respondents mentioned the TV/radio and internet (45.7%), followed by family/social environment (30.6%), while there is an important absence of information from the education system: only the 20.2% declare that they obtain information on RES from the education system. The significant role of television and Internet in informing citizens about environmental issues has been also indicated in the Eurobarometer survey concerning the attitudes of European citizens toward the environment [26].

Concerning the forms of RES, the respondents declare that they know more about solar energy (43.8%, sum of "much" and "very much"), with wind energy (32.8%, sum of "much" and "very much") and hydroelectric energy (30.4%, sum of "much" and "very much") coming next. On the contrary, the respondents declare having a very low knowledge on geothermal energy (48.2% of "not at all" knowledge) and biomass (45.4% "not at all" knowledge). Younger respondents and respondents with higher education are more informed about RES (wind, solar geothermal, biomass). It is worth noting that when the question focuses on the knowledge of respondents about specific forms of RES and not on the concept of RES in general, a higher percentage of respondents who seem to know specific forms of RES is recorded. This fact reveals that citizens may not be as familiar with the term RES as they are with specific forms of RES, such as solar and wind energy, as these forms of RES are related to natural resources that are easier to be understood/perceived. The gap of knowledge related to geothermal energy and biomass as well as the great familiarity of the respondents with solar and wind energy has been also highlighted in previous works [17,27].

Furthermore, the majority of the respondents consider that solar energy and wind energy is environmental friendly (72.8% and 70.7% respectively of “very much”). On the contrary, they do not consider that biomass is environmental friendly (42.1%, sum of “little” and “not at all”). Men are more informed about geothermal and hydroelectric energy and consider them environmentally friendly, while respondents with higher education consider solar energy, wind energy and biomass environmentally friendly. The environmental friendliness of solar and wind energy has been also indicated to previous works [28]. In addition, the respondents' view is probably related to past experience from RES projects that have been implemented in Greece, where the most of them concern the exploitation of solar and wind energy (photovoltaic and wind farms). It is also indicative that biomass, that is considered less environmentally friendly, is the form of RES where the greatest lack of knowledge of the participants is recorded.

Taking into consideration the hypothetical installation of a RES in a region, the respondents believe that it would create new job positions (47.6%, sum of “much” and “very much”) and would help the economic development of the region (53.9%, sum of “much” and “very much”), while it would reduce the dependence of the region on fossil fuels (54.8%, sum of “much” and “very much”). On the contrary, they believe that large lands areas will be reserved for the installation of RES (44.7%, sum of “much” and “very much”) and a high initial investment cost will be necessary (62.1%, sum of “much” and “very much”), but they do not consider that a RES can be the origin of an environmental degradation of this region (66.2%, some of “little” and “not at all”). The reduction of oil dependence and the creation of new jobs have been also reported among the main advantages of the expansion of RES in previous works [27].

Furthermore, the respondents believe that the production of electricity from RES is cheaper than conventional sources (64.4%, sum of “cheaper” and “much cheaper”) and they are very positive with the development of RES for electricity production (81%, sum of “much” and “very much”). In particular, women, respondents with higher education and those having left-wing political beliefs consider that electricity production from RES is cheaper than from conventional sources, while younger people and respondents with relatively high income (>10,000) have a higher acceptance of RES for electricity production. The price of energy is an important issue, especially during the current period of economic crisis. According to the Eurobarometer survey related to public opinion about the energy technologies, knowledge, perception and measures, the energy prices is the first energy related issue that concerns one-third of the European citizens [29].

Regarding to who should participate in the decision to install a RES, the respondents consider very important the participation of expert scientists (89.6%), of investors (68.5%), of the local government (64.7%) and of the residents of the area (61.9%), while consider the participation of the central government less important (41.7%). Furthermore, the respondents with left-wing political beliefs consider that investors should not be involved, but the central government should be involved. The elderly and married people prefer the participation of the government, the local government, and the residents of the area.

Finally, the majority of the respondents believe that priority should be given to solar energy (70.3%) and then in wind energy (17.6%). This may be due to the fact that Greece is a sunny country.

7 Hypothetic RES installation scenario

In the hypothetic scenario of installing a RES at a short distance from the respondents' residence, is seems that they would not be against the installation of a photovoltaic system/park (68.7%, sum of "little" and "not at all"), a system for the production of geothermal energy (66.8%, sum of "little" and "not at all") and the installation of a wind turbine/wind farm (63%, sum of "little" and "not at all"), while they would be against the installation of a biomass project (35.6%, sum of "much" and "very much") and the installation of a hydroelectric station/dam (31.1%, sum of "much" and "very much"). However, the respondents, although they generally support the development of energy projects near their residence, they set an average minimal distance of 10km. Younger respondents are less against the installation of all type of RES in a short distance from their home. On the contrary, men, those who have children and respondents with lower education level declare that they would be against the installation of a hydroelectric station/dam a short distance from their home.

Regarding the management of RES, the respondents prefer the private bodies in terms of design (45%) and construction (51%), while they prefer a public body in terms of exploitation of RES (31.9%). This public body should also fix the price of the electricity produced (38.4%). Elderly people, married and people with children prefer the design of RES to be done by a public body. This may be due to the fact that they consider the design of RES more expensive by a private entity. Furthermore, respondents with left-wing political beliefs prefer all RES management (design, construction, exploitation, and setting prices) to be done by a public body.

Finally, the majority of the respondents is not willing to pay for the installation of RES (71.6%, sum of "little" and "not at all"). Mostly men are the ones who are less willing to pay for RES installation, while younger people and people without children are less negative to pay for RES. It is remarkable that the respondents declare that they are not willing to pay any additional amount of money (they declare 0€) to the electricity bills either in case the management of RES facilities was public or private: 50.3% and 49.6%, respectively.

8 Conclusions

This work shows the results of the acceptance of RES in Athens, Greece, using a structured questionnaire. The first finding is that the notion of RES is not widespread. Men and those with a higher level of education seem to be more informed. Solar energy is the most popular form of RES with wind energy following. On the contrary, a significant knowledge gap is identified for both biomass and geothermal. Respondents' information and knowledge about RES primarily derived from the Internet. The lack of information on RES from the education system was strongly highlighted by all age groups. Solar and wind energy are evaluated as the most environmentally friendly forms. However, the environmental friendliness of RES is mainly under doubt by older people and those having lower level of education. The majority of the respondents argue that the implementation of a RES project has generally positive consequences in terms of local economic prosperity, independence from fossil fuels and the possibility of creating new jobs. As significant disadvantages are reported the high cost required for the initial investment, as well as the commitment of large areas of land. The vast majority of citizens are positive about the development of RES projects for the production of electricity, promoting as an incentive the cheaper price of energy produced from RES compared to conventional sources. However, there is great distrust toward the government for actions to promote and develop RES projects. The participation of experts in the decision to install a RES project is considered by most respondents to be absolutely necessary. Nevertheless, the importance of the active participation of the residents in the decision-making process for the installation of an energy project is recognized.

Although citizens are generally positive about the development of RES projects, they are very reluctant to the development of a RES project near their residence. The stronger objections/reservations are expressed for the installation of a biomass combustion plant. For the management of both the design and the construction of RES projects, mainly private bodies are preferred, while on the contrary, for the exploitation and the determination of their prices, the preference is obvious to the public body. As regards to their willingness to pay for the expansion of RES, the majority of the respondents declared their unwillingness.

Overall, the above findings can be used as a tool to enhance public acceptance of renewable energy investments and programs. They can also be used to intensify information activities and awareness campaigns for specific form of RES such as biomass and geothermal energy.

References

[1] M. Pascesila, S.G. Burcea, S.E. Colesca, Analysis of renewable energies in European Union, Renew. Sustain. Energy Rev. 56 (2016) 156–170.

[2] L. Brennan, P. Owende, Biofuels from microalgae: a review of technologies for production, processing, and extractions of biofuels and co-products, Renew. Sustain. Energy Rev. 14 (2010) 557–577.

[3] International Energy Agency (IEA), 2016, Renewables: About renewable energy. Available from: http://www.iea.org/topics/renewables.

[4] J. Twidell, T. Weir, Renewable and Sustainable Energy Reviews, Taylor & Francis, (2006).

[5] Centre for Renewable Energy Sources and Saving (CRES), 2019. Renewable Energy Sources. Available from: http://www.cres.gr/kape/energeia_politis/energeia_politis.htm.

[6] European Commission, 2020. Special Eurobarometer 501, Attitudes of European citizens towards the Environment (Fieldwork December 2019, Publication March 2020). Available from: https://ec.europa.eu/commfrontoffice/publicopinion/index.cfm/Survey/getSurveyDetail/instruments/SPECIAL/surveyKy/2257.

[7] A. Ernst, H. Shamon, Public participation in the German energy transformation: Examining empirically relevant factors of participation decisions, Energy Policy 145 (2020) 111680.

[8] Md. Moula, J. Maula, M. Hamdy, T. Fang, N. Jung, R. Lahdelma, Researching social acceptability of renewable energy technologies in Finland, Int. J. Sustain. Built Environ. 2 (2013) 89–98.

[9] J. Firestone, A. Bates, L. Knapp, See me, feel me, touch me, heal me: wind turbines, culture, landscapes, and sound impressions, Land Use Policy 46 (2015) 241–249.

[10] J. Zoellner, P. Schweizer-Ries, C. Wemheuer, Public acceptance of renewable energies: Results from case studies in Germany, Energy Policy 36 (2008) 4136–4141.

[11] Y. Guo, P. Ru, J. Su, L. Anadon, Not in my backyard, but not far away from me: Local acceptance of wind power in China, Energy 82 (2015) 722–733.

[12] F. Ribeiro, P. Ferreira, M. Araújo, A. Braga, Public opinion on renewable energy technologies in Portugal, Energy 69 (2014) 39–50.
[13] G. Assefa, B. Frostell, Social sustainability and social acceptance in technology assessment: A case study of energy technologies, Technol. Soc. 29 (2007) 63–78.
[14] S. Karytsas, H. Theodoropoulou, Socioeconomic and demographic factors that influence publics' awareness on the different forms of renewable energy sources, Renewable Energy 71 (2014) 480–485.
[15] N. Zografakis, E. Sifaki, M. Pagalou, V. Nikitaki, K. Psarakis, K.P. Tsagarakis, Assessment of public acceptance and willingness to pay for renewable energy sources in Crete, Renewable and Sustainable Energy Reviews 14 (3) (2010) 1088–1095.
[16] J. Paravantis, Multivariate analysis of attitudes of elementary education teachers towards the environment, computers and e-learning. Special issue on environmental sustainability and business, Int. J. Business Studies, A publication of the Faculty of Business Administration, Edith Cowan University 18 (1) (2010) 55–72.
[17] E. Savvanidou, E. Zervas, K. Tsagarakis, Public acceptance of biofuels, Energy Policy 38 (2010) 3482–3488.
[18] J.K. Kaldellis, Social attitude towards wind energy applications in Greece, Energy Policy 33 (2005) 595–602.
[19] D. Papoulis, D. Kaika, Ch. Bampatsou, E. Zervas, Public perception of climate change in a period of economic crisis, Climate 3 (3) (2015) 715–726.
[20] Ch. Vogiatzi, G. Gemenetzi, L. Massou, S. Poulopoulos, S. Papaefthimiou, E. Zervas, Energy use and saving in residential sector and occupant behavior: A case study in Athens, Energy Buildings 181 (2018) 1–9.
[21] E. Drimili, Z. Gareiou, A. Vranna, S. Poulopoulos, E. Zervas, An integrated approach to public's perception of urban water use and ownership of water companies during a period of economic crisis. Case study in Athens, Greece, Urban Water J. 16 (5) (2019) 334–342.
[22] E. Drimili, R. Herrero-Martin, J. Suardiaz-Muro, E Zervas, Public views and attitudes about municipal waste management: Empirical evidence from Athens, Greece, Waste Manage. Res. 38 (6) (2020) 614–625.
[23] C.R. Kothari, , Research Methodology: Methods and Techniques (second revised edition).New Age International Publishers, New Delhi, 2004.
[24] J. Eng, Sample size estimation: How many individuals should be studied?, Radiology 227 (2) (2003) 309–313.
[25] A.G. Kalamatianou, Social Statistics: One Dimensional Analysis Methods, The Economic Publications, Athens, (1994).
[26] European Commission (EC), 2020. Press release, 2020. Commission proposes a public loan facility to support green investments together with the European Investment Bank. Available from: https://ec.europa.eu/commission/presscorner/detail/en/ip_20_930.
[27] N. Zografakis, E. Sifaki, M. Pagalou, G. Nikitaki, V. Psarakis, K.P. Tsagarakis, Assessment of public acceptance and willingness to pay for renewable energy sources in Crete, Renew. Sustain. Energy Rev. 14 (2010) 1088–1095.
[28] J.K. Kaldellis, M Kapsali, El Kaldelli, Ev Katsanou, Comparing recent views of public attitude on wind energy, photovoltaic and small hydro applications, Renewable Energy 52 52 (2013) 197–208.
[29] European Commission (EC), 2006. Special Eurobarometer survey 262, Energy, technologies, knowledge, perception, measures. Available from: http://ec.europa.eu/research/energy/pdf/energy_tech_eurobarometer_en.pdf.

13

Sustainable site selection of offshore wind farms using GIS-based multi-criteria decision analysis and analytical hierarchy process. Case study: Island of Crete (Greece)

Pandora Gkeka-Serpetsidaki, Theocharis Tsoutsos

Renewable and Sustainable Energy Lab (ReSEL), Technical University of Crete, School of Environmental Engineering, University Campus, Chania, Greece

Chapter outline

1 Introduction to our work

There is an increasing need for the development of Renewable Energy technologies globally, in order that the nations could meet the goals related to climate change combat. The long-term strategy for the year of 2050 is an economy with net-zero greenhouse gas emissions; the target for 2030 is to reduce at least at 40% the

Low Carbon Energy Technologies in Sustainable Energy Systems. http://dx.doi.org/10.1016/B978-0-12-822897-5.00013-4

greenhouse gas emissions (compared to 1990 levels), according to the European Commission [1].

In this respect, offshore wind turbines are gaining more and more ground, because they can mainly avoid land-use conflicts, in relation with the onshore ones, they could also combine numerous marine activities and they could effectively contribute to the reduction of greenhouse gases emissions, due to high amounts of energy that are produced from their functionality [2].

As a consequence, there is a demand for deploying a methodological framework for evaluating available marine areas for siting installations of Offshore Wind Farms (OWFs), in an undoubtedly sustainable way.

Our method is based on Analytical Hierarchy Process (AHP) and Geographic Information Systems (GIS), as well as the relative weights, are derived from an extensive survey, including the experts' opinions. Also, it has to be noted that the criteria identified from the global literature and they were adapted along with the Greek legislation [3,4]. An effort became that the criteria selected could cover a wide range of categories such as environmental, technical, economic, cultural, safety, legislative, and criteria related with the potential disturbance from the OWFs. This combination allows us to acquire an overall scope on this subject and we could finally result in the more suitable marine areas for their deployment. Crete island was selected as our case study, due to its high offshore wind potential, its need to be energy independent and because this method could be applied to other similar insular environments with comparable electricity demands [5,6].

In addition to this, it is highlighted that in order to meet the future scale of installed renewable generation infrastructure, offshore wind farms have to be among the main priorities. In this context, the proper regulatory framework for the site selection and consequently, the operation of these projects should also be developed [7,8]. This study aims to provide a fundamental basis on which the most sustainable scenarios, concerning the installation of OWFs should be adapted.

2 Introduction to the offshore wind energy sector

2.1 Worldwide current status

It is an indisputable fact nowadays that there is a continuous need globally of developing low carbon technologies, in order to meet the goals of the nations in relation with the combat towards climate change. In this respect, the offshore wind sector is acknowledging a considerable blooming due to unlimited amounts of offshore wind worldwide and due to its rapid maturity [2].

According to [9] the installations of offshore wind farms globally dispose today a total cumulative capacity of more than 29 GW, representing 4.5% of total cumulative capacity. Until 2018 [2] the 80% was pertained to Europe and was constituted only 0.3% of global electricity generation. However, this technology is considered that has the potential to become the keystone of the world's power supply. The offshore wind sector is anticipated to increase up to 15-fold its capacity over the next 2 decades and it is worth noting also that becomes the leading source of electricity in Europe. Additionally, it is worthy of mention that the total amount of electricity demanded worldwide today is about 23,000 TWh, whereas the most suitable offshore wind sites (less than 60 m deep and within 60 km from shore) globally could potentially produce 36,000 TWh [2].

One of the advantages of this offshore wind is that the capacity factors are more valuable and more stable than onshore wind and up to 2 times than the solar PV, due to the fact that this technology can generate electricity during all hours of the day and especially in winter months. Moreover, the hourly variability of the offshore wind varies up to 20%, compared with solar PV, which fluctuates up to 40% [2]. Additionally, its high availability and seasonality, as well as the security of energy supply, could positively contribute to the needs of energy systems in relation with other renewable energy technologies. Finally, the development of this sector could also offer an alternative to the crucial existing problem of social acceptance because of the avoidance of conflicts about land uses [2].

Finally, the environmental benefit of this technology, in the condition that it gains ground the next 2 decades, is that 5–7 Gigatons of CO_2 emissions from the power sector globally could be avoided [2].

2.2 The situation in Europe

Concerning Europe's targets, the contribution of renewable energy is at least 32%, in terms of final energy consumption, until 2030, according to the 2030 climate and energy framework [1]. OWFs are recognized as very important clean energy generation parks; they experience rapid growth in recent years, especially in north-western European countries, such as the United Kingdom, Germany, Belgium, Denmark, and the Netherlands. There are already 110 offshore wind farms in 12 European countries [10]. It is worth mentioning that Europe had a total installed offshore wind capacity of 18,498 MW by the end of 2018 [11], whereas 22,072 MW at the end of 2019 [10]. This means that 502 new offshore wind turbines connected totally to the grid, across 10 offshore wind farms projects. These installations have been placed across the marine areas of 12 European countries [10].

As shown in Table 13.1, the United Kingdom has the largest amount of offshore wind installed capacity in Europe, with 9,945 MW, corresponds to the percentage of 45% of all installed offshore wind farms in Europe. Germany is second with 34% (7,445 MW), followed by Denmark (8%) (1,703 MW), Belgium (7%) (1,556 MW) and the Netherlands with (5%) (1,118 MW). The five countries mentioned above represent 99% of the installed capacity, whereas the other countries include Spain, Finland, France, Sweden, Norway, Ireland, and Portugal constitute together a percentage of 1% of the installed capacity [10].

Cumulatively, the North Sea accounts for 77% of all offshore wind capacity in Europe. The Irish Sea (13%), the Baltic Sea (10%), and the Atlantic Sea (<1%) follow this rank.

The average distance to shore (59 km) and water depth (33 m) continue to increase even though most wind farms are bottom-fixed.

Table 13.1 Installed cumulative capacity (MW) of grid-connected offshore wind power projects during the last three years [10–12].

Year/country	2017	2018	2019
United Kingdom	6,835	8,183	9,945
Germany	5,355	6,380	7,445
Denmark	1,266	1,329	1,703
Belgium	877	1,186	1,556
Netherlands	1,118	1,118	1,118
Sweden	202	192	192
Finland	92	71	70.7
Ireland	25	25	25.2
Spain	5	10	5
Portugal	0	0	8.4
Norway	2	2	2.3
France	2	2	2
Total	15,779	18,498	22,072

2.3 The case of Greece

Subsequently, as for the energy sector in Greece, it is noted, that the offshore wind energy technology is underdeveloped, contrary to the onshore, which has made considerable progress, having a total installed capacity over than 3.5 MW at the end of 2019, increased up to 25% relatively with the end of 2018 [13].

3 Case study—the island of Crete

3.1 Characteristics of the area

The island of Crete is the largest and most populous—it accounts over 600,000 permanent habitats, which are doubled during summertime [14]—of the Greek islands, the 88th largest island in the world and the fifth-largest island in the Mediterranean Sea, after Sicily, Sardinia, Cyprus, and Corsica. It is surrounded by the Libyan Sea and the Sea of Crete the southern border of the Aegean Sea. Crete lies approximately 160 km south of the Greek mainland. The land consists of an area of 8,336 km^2 and a coastline of 1,046 km. The maximum length of the island is 269 km and the maximum width 60 km [15,16]. The economy of the island relies mainly on trade, tourism and agriculture, being more developing. This is translated into a continuously increased energy demand [17].

3.2 The energy system

The energy system of Crete constitutes an off-grid electricity production system, where at the end of 2019, the overall installed capacity determined to 1076.70 MW. As shown in Table 13.2, the installed capacity on the island of Crete consists of 796.82 MW (74%) thermal energy, 200.29 MW (18.6%) of onshore wind and 78.29 MW of solar PV (7.27%). Small percentages of the installed capacity are the 0.30 MW (0.02) from small hydroelectric power plants and 0.99 MW (0.09%) from a biogas installation by the end of 2019 [17].

The maximum annual (2018) electricity demand for the island amounted to 707 MW. The whole generation of electricity (2019) was defined at 3115 GWh, where 2438 GWh were produced from thermal power plants and 677 GWh from renewables [17].

Consequently, it is mainly based on three principal conventional electricity power plants Linoperamata (Heraklion), Xylokamara (Chania), and Atherinolakkos (Lassithi) that they are burning fossil fuels [18].

Table 13.2 Allocation of total installed capacity for the island of Crete [17].

	Installed capacity on Crete (MW)					
Year	Onshore wind parks	Solar PV installations	Small hydroelectric power plants	Biogas installations	Thermal power plants	Total
2019	200.29	78.29	0.30	0.99	796.82	1076.70

Hence, the more and more increasingly energy demand of the island of Crete especially in the extended tourist season (April to November), as well as the European and national goals for the promotion of low carbon technologies have led to an increasing interest in investments in Renewable Energy Technologies.

Crete is sited at a lucrative geographical position that favors the deployment of renewables due to high volume, especially of solar and wind potential [19]. So, in order to overcome the land-use conflicts and the social unacceptance of existing parks and, at the same time, taking advantage of the huge number of shoreline kms, with the remarkable wind potential [20], the sustainable development of OWFs is necessary. Undoubtedly, taking into consideration the protection of all environmental areas and sensitive ecosystems.

Additionally, the first phase (Crete—Peloponnese) of the interconnectivity of the island is planned to be fulfilled in the current year, as well as the second phase (Crete—Athens), according to the preliminary plan of the 10-year program of development of transport of energy [21]. The arrangement of this interconnection is represented in Fig. 13.1. Each one part is comprised of a submarine cable, 132 km the first and 328 km the second [21]. This installation would surely offer incomparable benefits for the island such as the unlimited penetration of renewables, reduction of fossil fuels, security of energy supply (in cases of power system failures) reduction of cost of electricity, improvement of environmental conditions for wildlife and natural ecosystems, adequate coverage of high demand of electricity especially in the tourist season (April–November), savings of energy and resources, aid to achievement the National/European goals, related to climate crisis and augmentation of opportunities of employment for the residents [22].

The combination of all those mentioned earlier along with the high energy offshore potential of the island of Crete imposes a necessary need of deploying a methodological framework of evaluating the potential sustainable sites of OWFs installations.

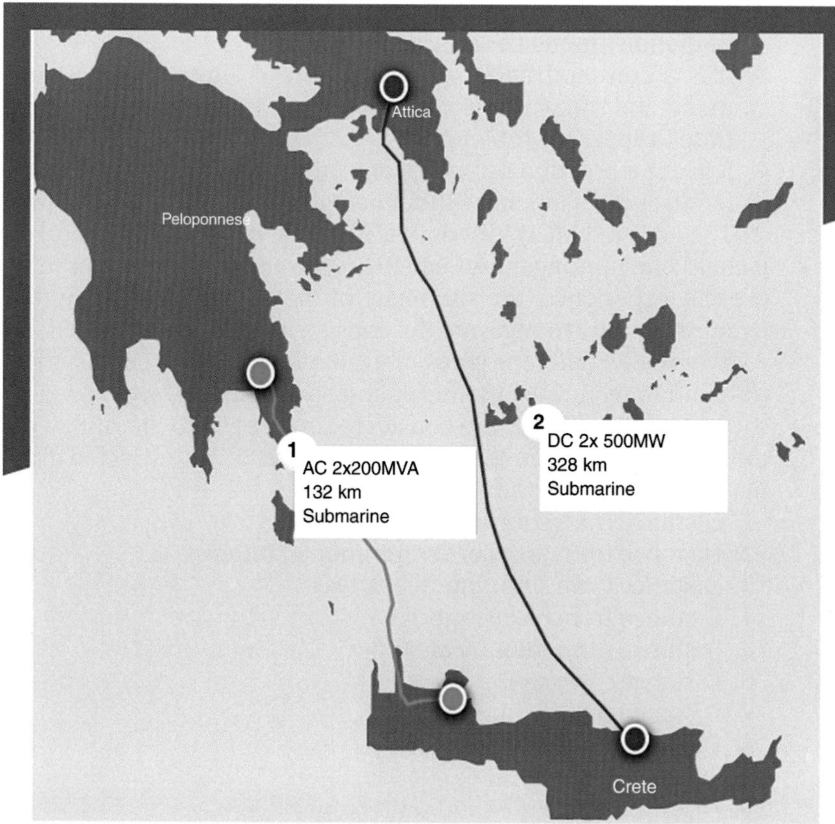

Figure 13.1. Scheduled interconnectivity of Crete. Modified from Ref. [23].

4 Methodology

It is remarkably known from the global literature that Multi-Criteria Decision Analysis (MCDA) techniques are widely used in the spatial planning of RES, especially of Photovoltaics (PV), biomass installations, onshore and offshore wind farms [19,24–26]. The more common methods used in the literature are: AHP, ANP (Analytic Network Process), TOPSIS (Technique for Order Preference by Similarity to Ideal Solution), ELECTRE (Élimination et Choix Traduisant la Réalité), PROMETHEE (Preference Ranking Organization Method for Enrichment Evaluations), and certain out of them are combined with other mathematical methods, as for example, the fuzzy logic [27].

Concerning this work, the utilization of MCDA was selected, because of the following advantages [28]:

1. It permits the integration of numerous factors/criteria, quantitative or even, and qualitative.

2. It copes with the complexity of the problem, as well as the contradiction among the selected criteria.
3. It is a common and applied method of alternatives' assessment, and there are also several research studies that concentrate to specific problems.
4. It is a method that promotes objectivity and integration of different sorts of factors without being energy and cost-intensive.

The adopted MCDA method is AHP due to the advantages that it could offer among others which are the capability of integration experts' experience, the simplicity of the method and the effectiveness in order to measure the consistency of the results [29].

Eventually, different types of evaluation criteria such as environmental, techno-economical, and socio-political were used in this study, after taking into consideration the global literature and the specific characteristics of our study area [5]. The selected evaluation criteria were the following:

1. Distance from submerged cables
2. Distance from areas of environmental interest
3. Distance from shipping/air routes
4. Distance from electric grid
5. Distance from military areas
6. Distance from shore
7. Distance from fishing areas
8. Distance from road network
9. Distance from heritage sites
10. Distance from cities and settlements
11. Distance from mining areas and activities
12. Wind resources
13. Water depth
14. Seabed substrate
15. Noise level/Acoustic disturbance
16. Optical disturbance

Additionally, the relative weights are resulted from the responses of different stakeholders' groups, after having completed the relevant questionnaires. The next step is the calculation of relative weights from the completed matrices (pairwise comparisons of the selected evaluation criteria), for each evaluation criterion, using the method of AHP.

The completion of questionnaires was realized based on this Table 13.3 according to the preferences of the stakeholders among the evaluation criteria. Subsequently, the consistency of the completed matrix is checked in accordance with the following equations [19,26]:

$$CI = \frac{\lambda_{\max} - n}{n-1} \quad (13.1)$$

Table 13.3 The fundamental scale according to Saaty [30].

Intensity of importance	Definition
1	Equal importance
3	Moderate importance of one over another
5	Essential or strong importance
7	Very strong importance
9	Extreme importance
2,4,6,8	Intermediate values

$$CR = \frac{CI}{RI} \quad (13.2)$$

Where CI, consistency index; n, the size of the matrix ($n \times n$); λ_{max}, the maximum eigenvalue of the comparison matrix; RI, random index values.

In our case the matrix is 16×16 due to the number of criteria selected to be used, so the RI for $n = 16$ is about 1.597 according to [31].

CR; consistency ratio

With the condition that:

$$CR < 0.1 \quad (13.3)$$

If this is not possible, the matrix has to be adjusted until it will be achieved.

Afterward, the priority vectors of the experts are aggregated with the help of geometric or arithmetic mean average according to whether they express their own interests in the respect of their organization that they work for or not, respectively [19,26].

Subsequently, the exclusion criteria are selected according to the legislative, technoeconomical, and environmental constraints. In the next step of methodology, GIS are also used, so as to be produced/developed the exclusion and subsequently, the evaluation maps. The relative weights derived from the experts are incorporating in this step of methodology, into the production of evaluation maps. The adopted methodological framework of this work is briefly presented in Fig. 13.2. Finally, after being processed the combination of the two maps with the suitable tools in the GIS environment, the final priority map, with the integration of the aforementioned relative weights is derived.

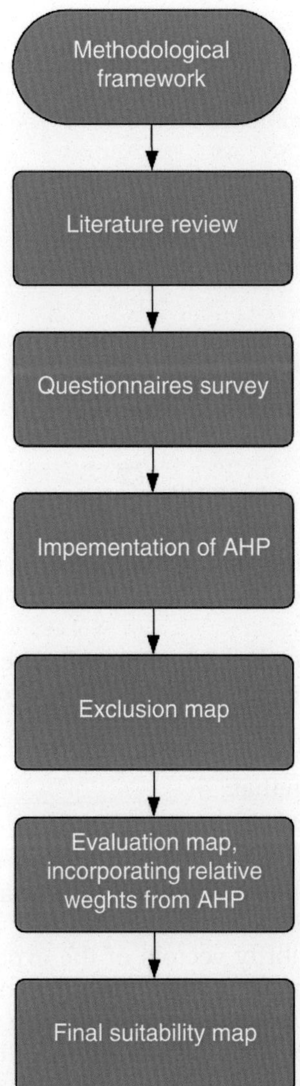

Figure 13.2. Methodological framework followed by this study.

5 Results and conclusions

Subsequently, an example of the criterion-related to the environmentally protected areas is given. This criterion is an exclusion and an evaluation criterion at the same time in terms of our study. The map (Fig. 13.3) was produced with the aid of the software ArcGIS 10.5 and it illustrates with different colors three environmental restrictions which are NATURA 2000 areas (green color),

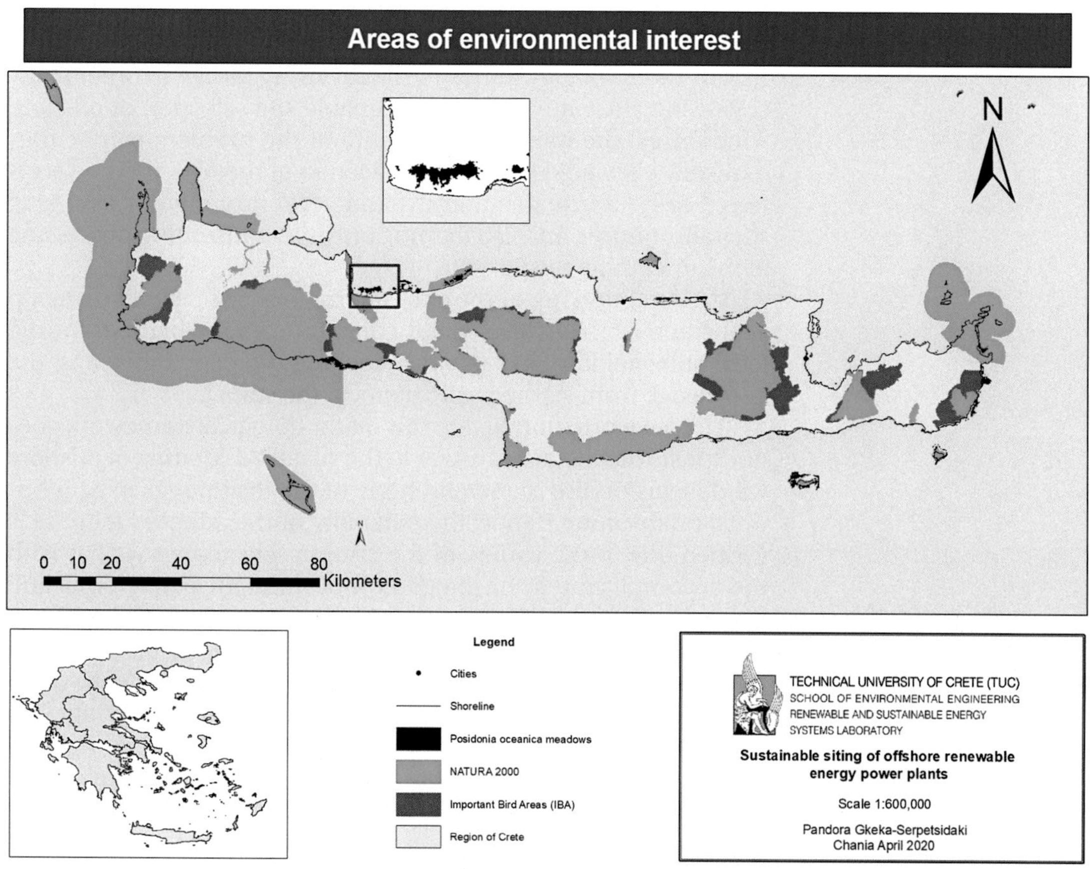

Figure 13.3. A combination of three exclusion criteria—NATURA 2000, Important Bird Areas, and Posidonia oceanica meadows—concerning the protection of the environment.

Important Bird Areas (pink color) and Posidonia oceanica meadows (dark green color), concerning the island of Crete.

More and more researchers and scientists are occupied with the problem of RET (Renewable Energy Technology) siting nowadays. The principal obstacle of this problem is that a sustainable siting implies a certain complexity and contradiction of numerous factors [28]. For example, an environmental criterion, such as the safe distance from NATURA 2000 (protected environmental areas) which has to be maximized, in some cases may oppose with the criterion of the depth of the sea, which in any case is a techno-economic criterion and the target is to be as much as possible minimized [32]. As a result, the main purpose of the OWFs site selection studies is to find the most suitable locations, in order to achieve the minimization of the impacts of the environmental and socio-political criteria, whereas at the same time the maximization of impacts of the techno-economic criteria.

Our methodology indicates that by using MCDA in order to assess siting alternatives for a sustainable site selection of offshore wind farms, the multicriteria nature of the problem can be dramatically alleviated [28]. The application of the AHP method is attested to be a valuable and dynamic tool not only for assessing sites alternatives but also for the non-time-consuming processing of the upcoming results [28].

The final ranking of the sites alternatives, which will come up at the end of this work, would constitute a valuable tool for the local stakeholders that it would include the experience and the knowledge from all the participants in the survey.

It is also worth noting that this methodological framework does not constitute the final answer to the sustainable siting of offshore wind farms on the island and none of the techniques used is optimal or absolute. Hence, the reliability of the adopted method is verified due to the variety of the criteria selected, as well as with the accomplishment of integration of different perceptions and judgments of multiple experts into the equation of the problem [28]. Consequently, this study provides a dynamic methodology that could be used for other RET and in combination with other different study areas, too. Finally, the followed methodological framework could be adopted by all stakeholders and competent authorities as a dynamic decision-making tool, since they are responsible for the legislative planning of OWFs at a national and/or regional level.

References

[1] European Commission. 2030 climate & energy framework. [Online] [Cited: 18 April 2020.]. Available from: https://ec.europa.eu/clima/policies/strategies/2030_en.

[2] International Energy Agency (IAE), Offshore Wind Outlook, World Energy Outlook Special Report. [Online] November 2019. [Cited: 21 April 2020.]. Available from: https://www.iea.org/reports/offshore-wind-outlook-2019.

[3] Ministry of Environment, Energy and Climate Change (MEECC). Specific Framework for Spatial Planning and Sustainable Development for Renewable Energy Sources. JMD 49828/2008, OGHE B' 2464/03-12-08. [Online] 3 December 2008. [Cited: 10 April 2020.].

[4] Ministry of Environment, Energy and Climate Change (MEECC). Accelerating the Development of Renewable Energy Sources to Deal with Climate Change and Other Regulations Addressing Issues under the Authority of the Ministry of Environment, Energy and Climate Change. Law 3851/2010, OGHE A' 85/04-06-10. [Online] 4 June 2010. [Cited: 10 April 2020.].

[5] M. Christoforaki, T. Tsoutsos, Sustainable siting of an offshore wind park a case in Chania Crete, Renew. Energy 109 (2017) 624–633.

[6] V. Kotroni, K. Lagouvardos, S. Lykoudis, High-resolution model-based wind atlas for Greece, Renew. Sustain. Energy Rev. 30 (2014) 479–489.
[7] The Ministry of the Environment and Energy (Greece). National Energy and Climate Plan (NECP). [Online] November 2019. [Cited: 1 March 2020.]. Available from: http://www.opengov.gr/minenv/?p=10155.
[8] Hellenic Wind Energy Associaion (HWEA). Workshop for Floating Offshore Wind in Greece. [Online] 12 April 2019. [Cited: 1 March 2020.]. Available from: https://eletaen.gr/en/workshop-for-floating-offshore-wind-in-greece-2/.
[9] Global Wind Energy Council (GWEC). Global Wind Report. [Online] 2020. [Cited: 21 April 2020.]. Available from: https://gwec.net/global-wind-report-2019/.
[10] Wind Europe. Offshore Wind in Europe, Key Trends and Statistics, [Online] February 2020. [Cited: 18 April 2020.]. Available from: https://windeurope.org/data-and-analysis/product/?id=61#presentations, 2019.
[11] Wind Europe. Offshore Wind in Europe, Key Trends and Statistics, [Online] February 2019. [Cited: 1 March 2020.]. Available from: https://windeurope.org/wp-content/uploads/files/about-wind/statistics/WindEurope-Annual-Offshore-Statistics-2018.pdf, 2018.
[12] Wind Europe. Offshore Wind in Europe, Key Trends and Statistics, [Online] February 2018. [Cited: 18 April 2020.] https://windeurope.org/data-and-analysis/product/?id=39, 2017.
[13] Hellenic Wind Energy Association (HWEA). Hellenic Wind Energy Association (HWEA). [Online] [Cited: 21 April 2020.]. Available from: https://eletaen.gr/.
[14] G.P. Giatrakos, T.D. Tsoutsos, N. Zografakis, Sustainable power planning for the island of Crete, Energy Policy 37 (2009) 1222–1238.
[15] Malagò, Anna, et al. Regional scale hydrologic modeling of a karst-dominant geomorphology: The case study of the Island of Crete, J. Hydrol. 540 (2016) 64–81.
[16] Radek Tichavský, et al. Flash flood reconstruction in the Eastern Mediterranean: Regional tree ring-based chronology and assessment of climate triggers on the island of Crete, J. Arid Environ. 177 (2020) 104135.
[17] HEDNO (Hellenic Electricity Distribution Network Operator). "HEDNO - Monthly Reports of RES & Thermal Units in the non-Interconnected. [Online] December 2019. [Cited: 21 April 2020.]. Available from: https://www.deddie.gr/el/themata-tou-diaxeiristi-mi-diasundedemenwn-nisiwn/agora-mdn/stoixeia-ekkathariseon-kai-minaion-deltion-mdn/miniaia-deltia-ape-thermikis-paragogis/minaia-pliroforiaka-deltia-paragogis-2019/.
[18] T. Tsoutsos, et al. Sustainable siting process in large wind farms case study in Crete. Renew. Energy 75 (2015) 474–480.
[19] M. Giamalaki, T. Tsoutsos, Sustainable siting of solar power installations in Mediterranean using a GIS/AHP approach, Renew. Energy 141 (2019) 64–75.
[20] T. Soukissian, et al. Offshore wind climate analysis and variability in the Mediterranean Sea. Int. J. Climatol. 38 (2018) 384–402.
[21] Independent Power Transmission Operator (IPTO). [Online] December 2019. [Cited: 22 April 2020.]. Available from: http://www.admie.gr/uploads/media/DPA_2021-2030_Prokatarktiko_Schedio_01.pdf.
[22] S. Biza, et al. Crete—Peloponnese 150kV AC interconnection simulation results for transient phenomena in main switches, Energy Proc. 157 (2019) 1366–1376.
[23] Independent Power Transmission Operator (IPTO). [Online] July 2018. [Cited: 22 April 2020.]. Available from: http://www.admie.gr/fileadmin/groups/ADMIE_DADY/Periodiko_energwn/energo6.pdf.
[24] M. Mahdy, A.S. Bahaj, Multi criteria decision analysis for offshore wind energy potential in Egypt, Renew. Energy 118 (2018) 278–289.

[25] M. Vasileiou, E. Loukogeorgaki, D.G. Vagiona, GIS-based multi-criteria decision analysis for site selection of hybrid offshore wind and wave energy systems in Greece, Renew. Sustain. Energy Rev. 73 (2017) 745–757.
[26] T. Höfer, et al. Wind farm siting using a spatial analytic hierarchy process approach: A case study of the Städteregion Aachen, Appl. Energy 163 (2016) 222–243.
[27] Y. Wu, et al. A decision framework of offshore wind power station site selection using a PROMETHEE method under intuitionistic fuzzy environment: A case in China, Ocean Coast. Manage. 184 (2020) 105016.
[28] T. Tsoutsos, et al. Sustainable energy planning by using multi-criteria analysis application in the island of Crete, Energy Policy 37 (2009) 1587–1600.
[29] T.L. Saaty, Décider face à la complexité:Une approche analytique multicritère d'aide à la décision. s.l.: ESF Editeur, 1984. p. 231. ISBN-10: 2710104911, ISBN-13:978-2710104919.
[30] T.L. Saaty, How to make a decision: The analytic hierarchy process. Eur. J. Oper. Res. 48 (1990) 9–26.
[31] J.A. Alonso, T.M. Lamata, Consistency in the analytic hierarchy process: a new approach, Int. J. Uncertain. Fuzz. Knowledge-Based Syst. 14 (4) (2006) 445–459.
[32] E. Loukogeorgaki, D.G. Vagiona, M. Vasileiou, Site selection of hybrid offshore wind and wave energy systems in greece incorporating environmental impact assessment, Energies 11 (8) (2018) 2095.

14

Accounting and Sustainability

Sofia Asonitou
University of West Attica, Aegaleo, Athens, Greece

Chapter outline

1 Introduction

According to NASA [1] "global warming is the long-term heating of Earth's climate system observed since the pre-industrial period (between 1850 and 1900) due to human activities, primarily fossil fuel burning, which increases heat-trapping greenhouse gas levels in Earth's atmosphere." Globalization and increased resource consumption have been the cause of serious problems and dangers for the environment and for the provision of resources in the future. Policy makers, businesses, and educators are required to find ways for more sustainable development in the future [2].

The 2030 Agenda adopted 17 Sustainable Development Goals (SDGs) (The Global Goals) that are intended to "stimulate action over the next 15 years in areas of critical importance for humanity and the planet" [3]. The SDGs have rapidly attracted the attention and the consistent support of several public and private sector organizations, NGOs, business and professional bodies, including the accounting and financial sector.

Accountants have a critical role in supporting the realization of SDGs since their expanded role and skills in the last decades allow them to participate in the management teams and act as advisors to their employers adding value to business output [4–6]. This study refers to the EU strategy to create a roadmap designed to make the EU economy sustainable through the promotion of the non-financial reporting directive (NFRD) and the response of the accounting profession to this endeavor.

Low Carbon Energy Technologies in Sustainable Energy Systems. http://dx.doi.org/10.1016/B978-0-12-822897-5.00014-6

2 Sustainability and EU strategy

The Intergovernmental Panel on Climate Change (IPCC) was founded by the United Nations Environment Programme (UN Environment) and the World Meteorological Organization (WMO) in 1988, and embraces 195 Member countries. The IPCC was created to provide governments and other interesting parties with systematic scientific reports on the dangers from the climate change, as well as activities that could diminish future problems and risks. Climate-related risks to a range of areas such as food and water supply, health, livelihoods, human security, and economic growth are expected to increase with global warming of 1.5°C and increase further with 2°C. The difference of 0.5°C seems to have great impact on sustainable development and extinction of poverty according to the report of IPCC [7]. The estimation of IPCC is that temperature increase of only 1.5°C, can be achieved if human activity attains net-zero CO_2 emissions until 2050 and neutrality for all other greenhouse gases at the end of the century.

UN [3] comprised a list of 17 goals, which serve the aims of terminating poverty, protecting the planet and supporting all human beings to live fulfilling lives. This framework of goals are related to social, ecological and economic activities, which if adapted consistently with limiting global warming to 1.5°C or 2°C, can balance social well-being, economic prosperity and environmental protection. SDGs refer to foundational elements for human well-being such as poverty, hunger, education, gender equality, access to clean water and sanitation, decent work and economic growth, climate action, affordable and clean energy, conservation of oceans, build resilient infrastructure, responsible production and consumption, promotion of sustainable use of terrestrial ecosystems, peace, justice and strong institutions, and revitalizing the global partnership for sustainable development. 193 UN Members States have committed to fulfilling the SDGs by 2030 [8].

Among the above goals climate change and regulation of gas emissions constitute an urgent challenge. European citizens seem to have realized the threats from climate change and desire increased action by policy makers. Citizens are informed and sensitized on the severe impacts on the planet eco-systems and biodiversity, health and food systems on earth. Reversing the impacts and limiting global warming to 1.5°C requires the adoption of coordinated governance measures, mobilization of financial actions, and changes in human behavior and lifestyles.

According to IPCC [7] finance can play significant role in this endeavor. Financing investments in infrastructure and mobilizing private funds and investment banks, may enhance the effectiveness of other public policies and provide additional resources. Average investment needs on a global scale toward controlling temperature increase up to 1.5°C are expected to be around 2.4 trillion USD until 2035, corresponding to 2.5% of the world GDP. These figures reflect the requirement for synergies between public and private funding, the banking system and policy tools to mobilize incremental resources and global investments to maximize co-benefits.

EU responding to the call for collective action, presented on 11 December 2019, the European Green Deal [9] as the roadmap designed to make the EU economy sustainable. The ultimate goal of the Green Deal was to create a clean, circular economy, to support biodiversity, by the effective use of global resources. According to the Green Deal, climate changes and environmental degradation should be turned into opportunities for all policy areas where citizens have equal access. Europe is adopting a new growth strategy through endorsing a resource-efficient and competitive economy with zero emissions of greenhouse gases by 2050. Europe aspires to decouple economic growth from the use of resources and become the first climate-neutral continent by 2050. For this a tight timeline was set out indicative to the importance given by EU on the above aims. According to the timeline the following steps were planned: "1) Presentation of the European Green Deal Investment Plan and the Just Transition Mechanism (14 January 2020) 2) Proposal for a European climate law to ensure a climate neutral European Union by 2050 (4 March 2020) 3) Public consultation on the European Climate Pact bringing together regions, local communities, civil society, businesses and schools (open until 17 June 2020) 4) Adoption of the European Industrial Strategy, a plan for a future-ready economy (10 March 2020) 5) Proposal of a Circular Economy Action Plan focusing on sustainable resource use (11 March 2020) 6) Presentation of the 'Farm to fork strategy' to make food systems more sustainable (20 May 2020) 7) Presentation of the EU Biodiversity Strategy for 2030 to protect the fragile natural resources on our planet (20 May 2020)."

The Strategic Agenda of EU for the years 2019–2024 [10], included the vision of climate-friendly, green, fair and social Europe, declaring climate change as an existential threat. EU recognized that to achieve the above targets by 2050, according to the Paris Agreement, will require significant public and private investments. In this framework EU will affirm that its legislation has no climate and environmental consequences and that it is aligned with the target of mitigating global warming to under 1.5°C. Transport, energy, agricultural, trade and infrastructure investment policies should be reformed in order not to cause any biodiversity loss. Between 1990 and 2018, EU policies contributed to a reduction of gas emissions by 23%, while the economy grew by 61%. If EU follows the current policies greenhouse gas emissions will be reduced by 60% by 2050 and this is not enough to reach climate neutrality. Therefore additional action needs to be taken to transform the economy and every sector should contribute toward climate neutrality. The transformation of the economy needs to be done in a way that will reaffirm EU's commitment to transparency and accountability, business and job creation thus ensuring and sustaining prosperity for European citizens.

3 Sustainability and the Accounting Profession

EU recognized the importance of the climate change and issued directive 2014/95/EU—the non-financial reporting directive (NFRD)—which set the regulations on how the large companies should disclosure non-financial information. This directive intended to improve the previous accounting directive 2013/34/EU. According to the last issued guidance 2014/95/EU companies are required to include non-financial information in their annual reports after 2018 [11].

The way that large companies treat the environment and other social issues like gender, age and human rights constitute important indicators of how these companies approach business. Investors, consumers, and policy makers should be aware of companies' policies as these could affect their financial decisions. EU requirements can help companies recognize the importance of the limited resources on the planet and the impact their operations may have on them. The companies which should comply with this directive are large public-interest companies which employ 500 employees or more. 6000 large companies in the EU have the above characteristics, comprising listed companies, banks and insurance companies as indicated by member states.

NFRD states that large companies are obliged to publish reports which present their practices to a series of issues such as "environmental protection, social responsibility and treatment of employees, respect for human rights, anti-corruption and bribery and diversity on company boards (in terms of age, gender, educational and professional background)" [11]. For their reports, businesses may rely on the UN Global Compact guidelines, the OECD guidelines for multinational enterprises or the ISO 26000. EU currently is expected to issue a new directive to improve NFRD, in order to reinforce its sustainability strategy.

Issuance of central laws, amendments, guidelines and reports, prove that the environmental issues have become part of the core business subjects and companies are obliged to analyze their impact on the environment, the business risks, the threats and the opportunities they face and report back to the public. The accounting profession very early realized its role within the SDG framework. Financial and accounting organizations have taken the lead in recent years in cooperation with EU and UN, to support SDGs. Accounting professional bodies have been developing commitments to sustainability and keep pursuing substantive practices in furthering their support. The financial sector takes action toward involving in this process and concentrating experience and knowledge to support businesses and influence them to adapt and change toward cleaner environment.

Indicative is the new executive role that Ernst and Young (EY) created, that of the *global sustainability leader*, with the responsibility to produce and guide practices to help EY become a world leader in sustainable business strategies. Another distinguished role is that of UN special envoy for climate change and finance. The focus of this position will be to implement actions and changes to attain the 1.5°C goal of the Paris Agreement, by fostering guidance for financial reporting and risk management of private financial decision-making [12].

Consumers, investors, and citizens seek information regarding the holistic value of organizations and the impact their activities have on environment and their overall conduct. Integrated Reporting can be one way to provide financial and non financial information about organizations in a manner, which is closely aligned with the EU Non-financial Reporting Directive. Integrated Reporting provides information based on the value that a company creates over time. This value is not defined in monetary terms, rather it is seen through the formation of financial, human and intellectual capital, providing Key Performance Indicators and emphasizing on ethical behavior, environmental impacts, and human rights. In the framework of Integrated Reporting, King and Atkins [13] proposed the creation of a new role that of Chief Value Officer (CVO) to drive forward integrated thinking and reporting and enhance sustainable value creation. This proposal redefines the role of accountant, transforming it in a profession that creates long-term value for society, giving a wider and integrated scope in his tasks [12]. The Institute of Chartered Accountants in England and Wales (ICAEW) considers the support to SDGs as part of their commitment to serving the public interest and contend that "… a successful economy depends upon and interacts with a successful society and a successful environment" ([14], p. 40).

Professional accountants are well positioned to deliver information and analysis that connect financial and non-financial information further supporting the targets of EU and businesses to better perform financially and comply with rules. Furthermore the accounting profession can become part of the process in helping translate and adapt the government policies into businesses actions. In countries with a high percentage of accountants in the workforce, research has shown a lower percentage of the global Corruption Percentage Index as indicated by Transparency International. This is of critical importance considering that fraud and corruption globally amounts at least $2.6 trillion that could be redirected to supporting SDGs [15]. The accounting profession at a global level has recognized its aim to consistently support eight of the 17 SDGs. The eight SDGs include “quality education, gender equality, decent work and economic growth, industry, innovation and infrastructure, responsible consumption and production, climate action, peace and justice and strong institutions and partnerships for the goals” [16].

IFAC very recently proposed the creation of a new sustainability standards board that would operate alongside the International Accounting Standards Board (IASB) under the International Financial Reporting Standards (IFRS) Foundation. This board would fill the gap for a coherent reporting system that provides consistent and comparable data relevant to sustainable development and stakeholder demands. A cohesive approach to reporting, innovation and proper technology can support high quality reporting and governance. Executive leaders from various professional financial organizations such as Deloitte, IFAC, IIRC (International Integrated Reporting Council), IOSCO (International Organization of Securities Commissions), AICPA (American Institute of Certified Public Accountants) and WEF (World Economic Forum), agree that this is the right momentum to align forces with the backing of public authorities, to achieve transparent measurement and disclosure of sustainability performance. A report of these standards would become an essential part of effective business management, would facilitate the necessary linkages to break down silos and to restore trust. The opportunity seems ideal to accelerate progress and gather all the relevant standards under one roof bridging the gap between the two worlds of financial reporting and sustainability reporting.

The importance of sustainability is emphatically shown in The World Economic Forum [17], which restated the original Davos

Manifesto of 1973 about the Universal Purpose of a Company in the Fourth Industrial Revolution. The WEF contends that the mission of a company includes more than the generation of wealth. Companies are more than economic organizations, they aim to satisfy other human and societal purposes. Therefore their performance measurement should embrace shareholders earnings, in parallel to data about ecological, social, and decent governance aims.

The WEF/IBC with the cooperation of the four largest accounting firms—Deloitte, EY, KPMG, and PwC issued a proposal about the use of a core set of global metrics for NFI (non-financial information) alongside to conventional financial reports. The objective is to report NFI in such a way that would satisfy stakeholders' concerns about the rigor treatment of sustainability issues and in relation to issues of transparency and comparability [18].

In parallel to NFI reporting, an important issue concerns the speed of energy transition from fossil fuels to sustainable energy systems. Will this be gradual or rapid? The decision and actions taken to implement specific directions, will impact the climate of the earth in the future. Also this decision will have implications for industrial production, technology, governments, energy producers and private consumers. Adopting a rapid transition will provide the opportunity to meet the goals of the Paris Agreement on the target of mitigating global warming to under 1.5°C. With the adoption of gradual transition, humanity will miss this opportunity because the energy systems of tomorrow will remain almost the same as they are today. The decision between the gradual and the rapid transition will have great impact on the global energy industry and consequently this will impact all business sectors on a global scale. Clearly, investors and business executives will base their current decisions, on the evolution of the energy systems. Researchers concentrate on the year 2030 as a base year that will show whether the rapid scenario can possibly be achieved. The following pointers for 2030 reveal the steps to be taken during the next decade: (1) The price of solar electricity in 2030 should fall to the $20–30 level, (2) solar capacity installations in 2030 should rise to over 200 GW per year, (3) electric vehicles market share in 2030 should reach over 30%, and (4) according to World Bank total global carbon taxes were at the level of $44 billion in 2018. Under the rapid scenario, there should be a major increase in the level of taxation, approximately the level of action that would be needed is around half of emissions being taxed at average tax rates of around $20 per tons taxed [19].

4 Sustainable Finance and Circular Economy

Sustainable finance refers to "how finance (investing and lending) interrelates with economic, social, and environmental issues (ESG)" [20]. Sustainable finance is about incorporating ESG factors into financial decision-making, but this process has many different approaches between countries and between industries. Although there is no unique path to disclosure of ESG factors, this has become the source of innovative practices, challenges and new investment strategies. Each economic sector has different approach toward ESGs. Policy makers, regulators, entrepreneurs and civilians are gradually recognizing climate-related risks as a source of financial risk that can affect businesses, whole economic sectors and even more the stability of the financial systems. IOSCO [21] states that the transition to more sustainable economic models will provide investment opportunities and firms should start disclose material risks and related information. The financial system can support the transition into the long-term sustainable development of economies acting as an intermediator of risks. Also securities regulators can contribute to necessary reforms by promoting transparency and allowing participants to identify and assess risks related to sustainability issues through proper disclosure [22]. The importance of the role of the financial system becomes obvious when considering that the gap in financing to achieve the Sustainable Development Goals (SDGs) is estimated at $2.5 trillion per year in developing countries. Also UN has estimated that attaining the SDGs could create US$ 12 trillion of market opportunities [23]. According to UN Biodiversity Finance Initiative (Biofin) $ 440bn will be needed in order to protect nature.

Accountancy Europe [24] makes proposals on how the accounting profession can contribute to EU strategy to achieve climate-neutrality. The organization ascertains that the financial system will take on a critical role during the transition to a low-carbon, climate-protective society. The financial system has the potential to contribute to environmental and social long-lasting prosperity by concentrating on three pillars: corporate reporting to accommodate long-term horizons, disclosure requirements to address investors' needs and accounting frameworks to be in line with long-term investments. High quality information on environmental, social and governance (ESG) issues is important for sustainable decision-making. Accountants and financial professionals can help to upgrade the relevance, quality and comparability of the reported information to meet stakeholders' demands for assessment of ESG risks and opportunities. In this framework the Sustainability Accounting Standards Board Foundation has established the Sustainability Accounting Standards Board that sets sustainability disclosure standards that are industry-specific and connected to the notion of materiality to investors. The standards are focused to sustainability matters that are financially material and may influence financial performance [25]. Disclosing the financial performance of a company is not enough anymore as it has become apparent that the market value of a company consists of tangible and intangible capitals such as financial assets, intellectual capital, customer relationships, and brand value, environmental, social, and human capital. SASB's Sustainable Industry Classification System® (SICS®) categorizes firms under a sustainability lens analyzing their business model, their resource intensity and sustainability impacts, and their sustainability innovation potential [25].

A part of the actions taken by EU, within the Green Deal initiative, is the enhancement of the Circular Economy model. It is estimated that the increase in annual waste and the consumption of biomass, metals and fossil fuels, will reach 70% by 2050. These numbers indicate the intense extraction of the limited natural resources of our planet, and a significant source of greenhouse gas emissions and loss of biodiversity. European Green Deal enforces a strategy to replace the linear economic model with a circular economic model, which will be climate-neutral and resource efficient. To achieve the economic benefits out of this strategy, more economic players need to participate and contribute to this endeavor. The overall target of the circular economic model is to keep the resource consumption within limits and double circular material use rate in the next ten years. New business opportunities can appear within the framework of creating sustainable products that last longer and come from reusable material, especially for SMEs. The circular economy model may provide an additional increase of EU GDP by 0.5% by 2030 and create around 700,000 new jobs. In recent years, cyclical activities, such as repair, reuse, and recycling have created additional value of about 147 billion euros while the value of related investments reached 17.5 billion euro. Recycling of wastes in cities during the years 2008–16 increased considerably, however, recyclable material covers only 12% of material demand within Europe. A full model of circular economy is applied only to 9% of global economy, indicating that there are margins for further improvement [26]. Digital technologies such as the Internet of things, Block chain, artificial intelligence, and big data, have an important role in the framework of circular economy. They can support the transition to a cleaner industrial production, can accelerate circularity, and promote the sharing and collaborative business models which will make industries less dependent on natural resources [27].

5 Conclusions

In the new era, investors and consumers demand to have a clear picture of the companies' ability to create sustainable value in the long run. Economic players need to have trusted information not only about companies' financial performance but also broader information related to environmental, social and governance factors. Accounting and finance professionals have a decisive role to play in the process toward climate neutrality and the transition to Circular Economy in order to reassure investors, regulators, and other stakeholders that they can have higher-quality information and insights about company performance, risks, and opportunities.

Professional accountants with their technical and professional skills are ideally positioned to support EU strategy on the transition to the circular economy and the mitigation to global warming to under 1.5°C. They are the intermediaries between businesses and EU strategies. They support their employers to create and enable their strategic plans. Accountants, working either as employees or practitioners in a firm, perform a plethora of activities directly related to the economy. Among other activities, accountants "measure and manage performance, implement reporting and internal control systems, analyze information, and develop governance and risk management policies" [15].

Research shows that there is further need to raise awareness of the SDGs among accounting academics to help not only in the research development of related areas but also to initiate the expansion of related accounting topics in the education of prospective accountants and business executives [28]. Accounting and accounting education have a critical role in furthering sustainable development, in applying related concepts and methods and in establishing a solid perception on the importance of sustainable development in business and accounting graduates [29,30]. Accounting professionals and accounting academics need to meet the challenges of the new era by keeping abreast with the evolutions, the strategies set by EU and the demands of employers, consumers and policy makers in order to defend the only planet we have.

References

[1] NASA. Overview: weather, global warming and climate change. (2020). https://climate.nasa.gov/resources/global-warming-vs-climate-change/.

[2] United Nations World Commission on Environment and Development (UNWCED). Our Common Future (The Brundtland Report), United Nations World Commission on Environment and Development, Oxford University Press, Oxford, 1987.

[3] UN. Transforming Our World: The 2030 Agenda for Sustainable Development. United Nations, New York, NY, 2015. Available from: https://sustainabledevelopment.

[4] S. Schaltegger, I. Etxeberria, E. Ortas, Innovating corporate accounting and reporting for sustainability—attributes and challenges, Sustain. Dev. 25 (1) (2017) 113–122.

[5] M. Storey, S. Killian, P. O'Regan, Responsible management education: mapping the field in the context of the SDGs, Int. J. Manage. Edu. 15 (1) (2017) 93–103.

[6] S. Asonitou, T. Hassall, Which skills and competences to develop in accountants in a country in crisis?, Int. J. Manage. Edu. 17 (3) (2019) https://doi: org/10.1016/j.ijme.2019.100308.

[7] IPCC. Summary for policymakers. In: V. Masson-Delmotte, P. Zhai, H.-O. Pϕrtner, D. Roberts, J. Skea, P.R. Shukla, A. Pirani, W. Moufouma-Okia, C. Pɩan, R. Pidcock, S. Connors, J.B.R. Matthews, Y. Chen, X. Zhou, M.I. Gomis, E. Lonnoy, T. Maycock, M. Tignor, and T. Waterfield (Eds.), Global Warming of 1.5°C. An IPCC Special Report on the impacts of global warming of 1.5°C above pre-industrial levels and related global greenhouse gas emission pathways, in the context of strengthening the global response to the threat of climate change, sustainable development, and efforts to eradicate poverty, 2018.

[8] UN. The Sustainable Development Goals Report, United Nations, New York, NY, 2016.

[9] EU. The European Green Deal (2019b), https://ec.europa.eu/info/sites/info/files/european-green-deal-communication_en.pdf.

[10] EU. A new strategic agenda for the EU 2019–2024 (2019a), https://www.consilium.europa.eu/en/eu-strategic-agenda-2019-2024/.

[11] EU. Directive 2014/95/EUof the European parliament and of the Council, amending Directive 2013/34/EU as regards disclosure of non-financial and diversity information by certain large undertakings and groups (2014), https://eur-lex.europa.eu/legal-content/EN/TXT/?uri=CELEX%3A32014L0095.

[12] AIA. Climate action and the finance profession (2020), https://www.accountingcpd.net/ accessed 28/6/2020.

[13] M. King, J. Atkins, Chief Value Officer: Accountants Can Save the Planet, Greenlead Publishing, UK, 2016 .

[14] P. Wilson, Goal setters, Economia 57 (2017) 40–44.

[15] IFAC. Speaking out as the global voice (2020), https://www.ifac.org/what-we-do/speak-out-global-voice.

[16] IFAC. The 2030 Agenda for Sustainable Development: A Snapshot of the Accountancy Profession's Contribution, International Federation of Accountants, New York, NY, 2016.

[17] WEF. Toward common metrics and consistent reporting of sustainable value creation (2020), https://www.weforum.org/reports/measuring-stakeholder-capitalism-towards-common-metrics-and-consistent-reporting-of-sustainable-value-creation.

[18] Accountancy Europe. Interconnected standard setting for corporate reporting (2019), https://www.accountancyeurope.eu/publications/interconnected-standard-setting-for-corporate-reporting/.

[19] WEF. The speed of the energy transition gradual or rapid change? (2019), https://www.weforum.org/whitepapers/the-speed-of-the-energy-transition.

[20] D. Schoenmaker, Schramade S W, Principles of Sustainable Finance, Oxford University Press, (2019).

[21] IOSCO. Objectives and principles for securities regulation (2017), https://www.iosco.org/library/pubdocs/pdf/IOSCOPD561.pdf.

[22] IOSCO. Sustainable finance and the role of securities regulators and IOSCO final report (2020), https://www.iosco.org/library/pubdocs/pdf/IOSCOPD652.pdf.

[23] UN. United Nations Secretary-General's, Roadmap for Financing the 2030 Agenda for Sustainable Development 2019–2021 (2019), https://www.un.org/sustainabledevelopment/wp-content/uploads/2019/07/UN-SG-Roadmap-Financing-the-SDGs-July-2019.pdf; un.org/content/documents/21252030%20Agenda%20for%20Sustainable%20Development%20web.pdf.

[24] Accountancy Europe. Accountancy Europe leads the non-financial reporting debate (2020), https://www.accountancyeurope.eu/reporting-transparency/accountancy-europe-leads-the-non-financial-reporting-debate/.

[25] SASB. Sustainability accounting standards board, mission (2020), https://www.sasb.org/governance/.

[26] National Documentation Centre. Circular Economy: A new business model for sustainable development (2019) (in Greek), https://www.ekt.gr/el/magazines/features/23377.

[27] EU. A new circular economy action plan, for a cleaner and more competitive Europe, Brussels, 11.3.2020, Brussels, 11.3.2020, COM (2020) 98 final, https://eur-lex.europa.eu/legal-content/EN/TXT/?qid=1583933814386&uri=COM:2020:98:FIN.

[28] G. Kyriakopoulos, S. Ntanos, S. Asonitou, Investigating the environmental behavior of business and accounting university students, Int. J. Sustain. Higher Edu. 21 (4) (2020) 819–839 https://doi.org/10.1108/IJSHE-11-2019-0338.

[29] F. Annan-Diab, C. Molinari, Interdisciplinary: practical approach to advancing education for sustainability and for the Sustainable Development Goals, Int. J. Manage. Edu. 15 (1) (2017) 73–83.

[30] J. Bebbington, C. Larrinaga, Accounting and sustainable development: an exploration, Acc. Org. Soc. 39 (6) (2014) 395–413.

PART 3

Conclusions and future research

15

Should low carbon energy technologies be envisaged in the context of sustainable energy systems?

Grigorios L. Kyriakopoulos
School of Electrical and Computer Engineering, Electric Power Division, Photometry Laboratory, National Technical University of Athens, Athens, Greece

Chapter outline

1 Introduction

Public concerns about the adverse impacts of energy systems, coupled with technological changes that have made alternative energy sources increasingly more affordable, they are driving notable shifts toward the use of renewable energy. During the period 2008–17 the global installed capacity of renewable energy sources (RES) increased from 1.06 TW to 2.18 TW [1]. Besides, in 2017 the worldwide share of renewable energy generation in whole power generation sector was 25%. At the same decade, 2008–17, among the total installed renewable energy capacity: (1) hydropower capacity amounted to 1.27 TW, (2) wind capacity increased more than fivefold, from 0.1 TW to 0.51 TW, (3) photovoltaics (PV) capacity

Low Carbon Energy Technologies in Sustainable Energy Systems. http://dx.doi.org/10.1016/B978-0-12-822897-5.00015-8

increased more than 20 times, from 0.015 TW to 0.39 TW [1]. The main driver behind the steady increase in installed capacity of PV and wind turbines has been a significant price drop. At this decade of referencing, 2008–17, the price of PV dropped by 80%, while the price of wind turbines reduced by 50% over the same period. It is expected that these technologies will continue to dominate the renewable energy installation portfolio in the future [1].

Since the industrial revolution humans well being is depended heavily on coal, oil, and other fossil fuels to energize everything from light bulbs to factories. Fossil fuels are involved in almost any activity, and consequently, the greenhouse gases emitted from the combustion of those fuels have today reached an alarming level. As these gases capture heat in the atmosphere, the average temperature on the surface of the earth is increasing. The continuous temperature rise has led to the phenomenon known as global warming, a signal of climate change, which in turn affects the planet's weather and climate. Climate change, however, involves acute dangers like extreme weather events, negative changes in wildlife populations and habitats, rising sea levels, and many other adverse impacts [2].

The steady increase for energy production during the last decades worldwide is caused by the incremental growth of the global population and the overall stress to support high living standards and humanitarian well-being. However, the intensification of energy production is stressing the environmental carrying capacity of energy sources, thus, drawing up urgent policies and the actions that have to be prioritized, accordingly. Besides the environmental dimension, the optimization of energy systems sustains the economics dimension, under which the mismatch between the curves of energy demand and energy supply is predominately characterizing as a necessity to control resource wastes and to investigate better ways on the efficient and cost-effective exploitation of operated technologies. Such an approach of energy systems sheds light on energy systems dynamics, the interoperability behavior, the functionalities, the embedded technological abilities, and the social participation of varied stakeholders to regional or national energy schemes. In this respect, it is noteworthy current research to be oriented to unveil trends and state-of-the-art perspectives in analyzing the recent technological advances on key issues of energy systems, as well as in assimilating low carbon economy, strategies and future perspectives [3]. A typical scheme of energy systems' investigation should contain the following dimensions [3]:

- Energy systems and low carbon technologies
- Supply chain, utilization, and harvesting

- Storage mechanisms and cycling
- Applications and case studies

Among the energy systems of sustainable orientation and low carbon footprint are that of renewable-based structure. Indeed, renewable energy sources, such as solar and wind, seem to be a reasonable alternative to fossil fuels and their planet-warming effects, as they emit much less carbon dioxide or other harmful greenhouse gases. However, as with all new technologies, the deployment of renewable energy sources encounters many challenges with the most important being its high cost which renders it uncompetitive compared to fossil-based electricity [2]. Particularly, in order to facilitate the transition to a renewable-based energy system, it is critically important that policymakers take into account the attitude of the public toward renewables [4]. Experience has shown that the public reacts differently to proposed renewable energy installations and, if these reactions are negative, the opposition can lead even to the cancellation of the project altogether. At the same time, frequent cancellations of energy projects are not translated only into losses for energy companies but also into the shrinkage of the renewable sector. For this reason, knowing how the public feels and reacts to renewable energy enables developers and policymakers to plan projects which resonate with the public and are more likely to be accepted by the it. In addition, insights into public attitudes can guide efforts to design effective strategies intended to shape favorable attitudes to renewable energies [4]. From this perspective, it may be stated that the public attitude is a determinant of widespread implementation of renewable energies to the same degree that the technological development of the sector is. Acknowledging the strong influence of attitudes, researchers have conducted multiple studies and employed several methodologies in order to capture attitudes. Due to the abundance of these works, a literature review may serve as a valuable tool to draw important conclusions and provide clues on the directions of future studies. Hence, the aim of this chapter is to review recent empirical research which was carried out in countries of the Europe Union during the period 2015–20. In other words, the attention is placed on the recent period that followed the Paris Agreement and on the countries of the European Union as this is where the most ambitious renewable targets were set [4]. At the following Table 15.1 a literature framework of energy systems and models' devoted to low carbon energy technologies, is given.

Based on the aforementioned literature outline, Table 15.1, it is critical to note that carbon capture storage (CCS) technologies will influence the long-term energy system of developed economies, taken into consideration those primary energy supplies and

Table 15.1 Literature framework of sustainable energy systems, models, and technologies.

References	Field of analysis	Methodology modeling simulation	Research outcomes
[5]	In China, coal-to-gas, carbon capture, and renewable energy utilization are important measures among various low-carbon plans in the energy sector. However, technological constraints against the development of these measures are that of inadequate gas supply, unclear utilization methods of captured carbon dioxide, and curtailment of excessive wind and solar power.	The methodology of this study included a coupling model considering power to gas (P2G) technology and carbon capture power plants (CCPP), being extended into an integrated energy system (IES) that includes electricity, thermal and gas. An economic and environment scheduling model was proposed at a wind power penetration level.	The simulation results stressed out that the ability of the system to absorb renewable energy is significantly enhanced with the presence of CCPP and P2G.
[6]	Nanofluids play a determining role for the development of efficient and competitive heat transfer fluids. Therefore, a research of the thermal properties for developing water and Therminol oil-based nanofluids was conducted to operate at high pressure and temperature. The outcomes of this research can support a concrete understanding for nanofluids based heat transport phenomena.	Simulations were carried out on parabolic trough collector solar thermal system, while using nanofluid properties and system advisor model software.	Simulation results showed enhancement in energy production along with 10.37% reduction in real power purchase agreement price as well as 21% reduction on the thermal energy storage volume.
[7]	Along with the introduction of its energy concept in 2010, the German government set ambitious targets for its national energy and climate policy. The economic and the technological feasibility toward the decarbonization of the German energy system until 2050, they were investigated.	The analysis is based on an hourly simulation model EnergyPlan, under three scenarios developed for possible application at the German energy supply system until 2050, while forecasting that greenhouse gas (GHG) emissions will have to be reduced by 80% by 2050.	Massive decarbonization of the German energy system until 2050 seems feasible in case that: a) smart grid costs are disregarded and b) this sustainable energy transformation is accompanied by determining political willingness and public acceptance to materialize the setting goals.

[8]	Hydrogen source plays a key role through its versatile production methods, end uses and as a storage medium for renewable energy at a future energy system of low-carbon. The usage of a global model cognizant of energy policy, technology learning curves, and international carbon reduction targets, they were deployed to optimize the future energy system in terms of cost and carbon emissions up to 2050.	A joint approach of four exploratory and modeled scenarios incorporated hydrogen city gas blend levels, nuclear restrictions, regional emission reduction obligations for carbon capture, and storage deployment timelines. Hydrogen has proven a feasible energy source to supply approximately 2% of global energy needs by 2050.	Irrespective of the quantity of hydrogen produced, regions of reference and scenarios modeled, the transport sector and the passenger fuel cell vehicles are consistently a preferential end use for future hydrogen. Besides, a shift toward renewable energy and the significant role for carbon capture and storage, they are identified to underpin carbon target by 2050.
[9]	The cost-effectiveness of operation strategies, which can be used to abate carbon dioxide emissions in a local multi-energy system, it was carried out at a case study using data from a real energy system that integrates district heating, district cooling, and electricity networks at Chalmers University of Technology. The optimized abatement strategies include increased usage of biomass boilers, substitution of district heating and absorption chillers with heat pumps, and higher utilization of storage units.	The operation strategies at this study were developed using a mixed integer linear programming multi-objective optimization model with a short foresight rolling horizon and a year of data. The cost-effectiveness of different strategies was evaluated across different carbon prices.	Insights into developing abatement strategies for local multi-energy systems could be used by utilities, building owners, and authorities. The emission abatement cost of all strategies was ranged at 36.6–100.2 (€C/tCO_2), which meets the setting estimates of carbon prices if the Paris agreement target is to be achieved.
[1]	The futured 100% renewable energy systems have to integrate different sectors: heating, cooling, and transport, all including power provision. Such energy systems are needed to mitigate the negative impacts of economic development based on the use of fossil fuels and relied on variable renewable energy resources. The decarbonization of energy systems is imperative to combat the climate change. Integrating future energy systems with carbon dioxide capture and utilization technologies can support deep decarbonization.	A review outline regarding the state-of-the-art literature on carbon dioxide captures and utilization technologies were conducted while emphasizing on the potential of technologies integration into a (low-carbon and high-renewables) penetration grid.	The captured carbon dioxide can be either utilized as a feedstock for various value-added applications in the chemical industry and related sectors such as the food and beverage industries. Such a technologies can be used to balance the grid to allow for high levels of variable renewable energy in the power mix. The potential market size for carbon dioxide as raw material can be also investigated.

(Continued)

Table 15.1 Literature framework of sustainable energy systems, models, and technologies. (*Cont.*)

References	Field of analysis	Methodology modeling simulation	Research outcomes
[10]	Energy system decarbonization and changing consumer behaviors can create and destroy new markets in the electric power sector. This means that the energy industry will have to adapt their business models in order to capture these pools of value. Recent work explores how changes to the utility business model that include digital, decentralized or service-based offers could both disrupt the market and accelerate low carbon transitions.	This study signified the need for comprehensive demonstration trials, which can iteratively combine and test information and communications technology (ICT) solutions. This aspect of innovation support requires a new approach to energy system trials.	The study unveiled those technologies that may hinder electricity business model innovation and where more research or development is necessary. Regardless the fact that none of the business models that are compatible with a low carbon power sector are facing technology barriers that cannot be overcome, work still has to be done in the domain of system integration, while it is also unclear whether these business models are technologically feasible.
[11]	Low-carbon technologies will play a vital role in the realization of environmentally sustainable economies. However, uncertainties upon the feasibility of their development and implementation necessitate possible scenarios for the potential of these technologies to be considered, enabling flexible decision-making with respect to long-term energy strategies in Japan.	The role of hydrogen in future energy systems in Japan was examined by using a MARKet ALlocation (MARKAL) model. A range of uncertainties was considered for nuclear power generation and carbon capture and storage (CCS) from fossil power generation.	An 80% reduction of carbon dioxide emissions from the 2013 level by 2050 requires emissions from the electricity sector to decrease to nearly zero. Hydrogen power can play a functional role in future energy systems in Japan, but its contribution should also depend on nuclear power and CCS.
[12]	China is experiencing a transition to low-carbon economic development. At this study literature search on the potentials of and barriers to China's transition to low-carbon development was conducted, identifying promising fields of action and suggesting a research agenda to systematically address research constraints and shortcomings.	A broad literature review was conducted based on the selection of three main research areas of interest: low-carbon cities, low-carbon technologies and industries, and the transition of China's energy system. Summary of some specific issues have been discussed more in Chinese-language journals, rather than of English language ones.	The research outcomes signified that specific elements of a more comprehensive research agenda, they can improve researchers' understanding of China's ability to enter a low-carbon development pathway.

[13]	Differently from typical analysis focusing on technology portfolio for each route, authors of this study analyzed the deployment of each technology across policy routes, for optimizing technology through R&D. Such a R&D priority should be given to those less-policy-sensitive technologies that are rapidly deployed across modeled-time horizon (e.g., PV), but also to those deployed up to their technical potentials, being typically prone to less sensitivity on exogenous policy routes.	The optimization energy system model JRC-EU-TIMES was used to support energy technology R&D design by analyzing power technologies deployment up to 2050 and their sensitivity to different decarbonization exogenous policy routes. Such policy routes are based on scenarios of the EU Energy Roadmap 2050 combining energy efficiency, renewables, nuclear, or CCS.	Cost-effective technologies should be very sensitive to the policy routes (e.g., CSP and marine), while R&D efforts are putting forward to improve their techno-economic performance. In general, such policy decisions for the configuration of the low carbon power sector, they should be a certain research objective, especially on nuclear acceptance and available sites for new RES plants.
[14]	Focus on energy affordability and environmental impact is driven to the potential value of low carbon technology (LCT) interventions in buildings and district energy systems. Relevant interventions may include improvements of the insulation levels and installation of low carbon and renewable generation technologies (e.g., combined heat and power, photovoltaics, heat pumps). These LCT interventions for both electricity and heat supply can, in principle, reduce energy costs and carbon dioxide emissions for final energy consumers.	While coupling and optimizing multiple energy vectors from traditionally independent systems (e.g., electricity, heat and gas) a transition to distributed multi-energy systems can be achieved. This argument introduces complex physical and commercial interactions between the different energy vectors. Such interactions are overly different at different aggregation levels (e.g., premises, district, and commercial level).	A techno-economic assessment framework of business cases of LCTs, it systematically models the physical and commercial multi-energy flows at the premises, grid connection point, and commercial levels. This is particularly important given the commonly asymmetrical nature of various energy price components and associated actors (e.g., retailers and energy service companies). Various energy system actors, including policy makers and regulators aim to facilitate the uptake of low carbon multi-energy technologies.

(*Continued*)

Table 15.1 Literature framework of sustainable energy systems, models, and technologies. (*Cont.*)

References	Field of analysis	Methodology modeling simulation	Research outcomes
[15]	Low carbon energy technologies have to consider the social interactions upon materialization. In particular, there is a variety of complex ways in which people understand and engage with these technologies and the whole changing energy system. The role of the public socio-environmental sensitivities to low carbon energy technologies, as well as public responses to energy deployment in planning decarbonization pathways to 2050, they cannot be undermined. Resistance to certain resources and technologies based on particular socio-environmental sensitivities would alter the portfolio of options available which could shape the decarbonization path: How the energy system achieves decarbonization as well as affects the cost and the achievability of decarbonization.	A series of three modeled scenarios illustrated the way that a variety of socio-environmental sensitivities could impact the development of the energy system and the decarbonization pathway. The scenarios represented risk aversion (DREAD), which avoids deployment of potentially unsafe large-scale technology, local protectionism (NIMBY) that constrains systems to their existing spatial footprint, and environmental awareness (ECO) where protection of natural resources is of paramount importance.	The deployment of these three-modeled scenarios identified very different solutions for all three sets of constraints. In particular, the DREAD scenario seemed slightly implausible, while all three scenarios showed increased cost, especially that of ECO.

electricity generation in 2050 under different settings of carbon dioxide capture efficiency, and the upper limit of annual carbon dioxide storage [1,5,9–12]. Among the most prevailing energy uses worldwide, the energy share of RES technologies to electricity generation it is apparently influenced by the assumed parameters of CCS from fossil power generation. Depending on the share of electricity generated from fossil fuels with CCS, the share of RES-based power generation in 2050 can vary differently, being accompanied by fluctuating amounts of imported RES to meet the energy demands among those countries that are planning to meet their energy needs from RES, accordingly [7,8,11,13,15].

Another significant finding of the aforementioned literature framework, is the need to evaluate the uncertainties associated with the development and implementation of low-carbon technologies in the future, while simulations were performed based on multiple scenarios, which assumed different carbon dioxide emissions targets and CCS, under an joint power generation from, each case different, RES and fossil power generation [11]. Simulation analyses suggest that the carbon dioxide emissions from the electricity sector should be reduced to nearly zero by 2050, especially at developed economies, if these economies can achieve the long-term target of reducing carbon dioxide emissions by considerable amounts -up to 80% [11], or even more at a scenarios-based analysis [7]- from today levels by 2050. In this context, it is noteworthy to stress out that the vital role of any RES technology to low carbon power source in the future energy system should depend on not only its own specifications, but also its interaction with other technologies, as with other studies based on energy modeling [7,8,11,13,15].

The interrelation of these technologies has not been fully considered in national energy plans even among the developed economies. Other factors that can affect the deployment of RES technologies by 2050 include renewable energy capacity and the efficiency of end-use technologies. The effects of these parameters on energy systems can be evaluated to better understand the contribution of hydrogen energy technologies and enhance the utilization of domestic energy sources, thus increasing energy self-sufficiency among, the most energy-intensified, developed economies [7,8,11,13,15].

2 Methods

At this concluding chapter of this book, the main contributors of low carbon contexts and the materialization of sustainable energy systems, they have been gathered and classified into two dimensions: technological and social. Regarding the technological dimension, the examined aspects of investigation were that of:

- Mining marketplace and power consumption
- Resource recovery technologies in reducing use of fossil fuels and fossil-based mineral fertilizers
- Renewable energy sources and energy crops
- Offshore wind farms using GIS-based multi-criteria decision analysis and analytical hierarchy process

Besides, the main examined aspects of the social dimension were that of:

- Household sector
- Education
- Bibliometric analysis on energy, sustainability, and climate change
- Public acceptance of energy systems based on renewables

A detailed analysis of the contained methodological approaches is developed in the following Section 3.

3 Results

As aforementioned, the two research dimensions of analysis were that of low carbon contexts and the materialization of sustainable energy systems, as these were investigated in terms of their technological and social dimensions of analysis. The main results of the research are presented further.

3.1 Low carbon energy: the technological dimension

The maturity assessment of technologies is a key enabler for low-carbon business model innovation in the electricity system sector. The outline of technology readiness levels for energy system technologies, it can be organized into the following three different groups [10]:

- Supply-side technologies: Stationary batteries; Micro combined heat and power (CHP); Solar PV; Solar thermal; Heat storage; Fuel cells; Carbon capture and storage; Synthetic fuels; District heat networks; Gas fired power plants; Diesel generators; Nuclear, Wind, Biomass supply chain; Hydrogen storage.
- Demand-side technologies: Heat pumps; Remote controlled electric storage heaters; Energy efficient lightning; Intelligent heating controllers; In-home displays; Smart appliances; Smart EV chargers; Hydrogen-hybrid-electric vehicles.
- Technologies that are needed in the domain of data, communication and integration: Smart meter technologies; Home energy management systems (HEMS); Building energy management

systems (BEMS); Demand-side response; Sensors (IoT); Vehicle-to-Grid communication; Communication for wholesale market; Cyber security; Blockchain; Peer-to-peer communication; Reactive power control; Local network balancing; Market/Trading platform; Trading optimization; Advanced distribution management system (ADMS); Machine-Learning; Machine-to-Machine communications; Generation optimization; Wide-Area energy management systems (WAEMS); Virtual power plant (VPP).

The aforementioned grouping of energy system technologies with environmental orientation signifies that the entry of new business models into an established market with substantial incumbent advantage puts technological niches at a price and scale disadvantage. Therefore, there is an opportunistic need of technological trials, which focus less on individual engineering innovation and more on combining and recombining ICT and systems integration tools, enabling new business models to enter the energy market [10].

At another study of technological background on energy system technologies with environmental orientation, the direct utilities and uses of carbon dioxide can be framed below [1]:

- Solvent
- Refrigerant
- Protecting gas
- Enhanced oil and gas recovery
- Cleaning and extracting agents
- Enhanced coal bed methane
- Techniques for reversible adsorption and assimilation of carbon dioxide
- Impregnating operator
- Acidity controller for aqueous solution
- Use in food and beverage industries

Besides, indicative reduction reactions of carbon dioxide can be framed below [1]:

- Chemical (Fischer Tropsch; Reverse water gas shift)
- Electrochemical (Solid oxide fuel cell; Platinum; Nickel and its complexes; Palladium and its complexes)
- Photochemical (Transition metal complex; Cascade metal; Semiconductor; Metallic molecular catalyst)
- Bioelectrochemical (Microbial fuel cell with microbial electrolysis cell; Stainless steel electrode with bacteria)

From the aforementioned groups it can be stressed out that there are technologies like: communication for wholesale markets, peer-to-peer trading agents, generation optimization and virtual power plant, all of which need a minimum scale to reach a higher maturity. In this respect, the creation of either a set of large-scale demonstrators, or one major demonstrator, it is recommended; where these technologies are deployed on a regional level. However, there are other technologies that do not require significant further R&D investment by policy makers [10]. It is, therefore, important to denote that further basic research and development in these technologies is not always critical to the transition toward sustainable future energy systems. The aforementioned twofold statements are differently supporting energy systems innovation. It cannot be argued that further types of development, as at offshore wind efficiency or solar cell development, they are unlikely to be made, but the fact that the returns on public innovation support in these areas may be diminished or at least outstripped by the potential for systems innovation that support the consumer end of the supply chain, based on specific business models. These customer-centered innovations may include peer-to-peer contracts and incentives for flexibility, requiring a different innovation ecosystem developed by iterative trials and target at system optimization [10].

3.1.1 Mining marketplace and power consumption

In the contemporary entrepreneurial environment of energy market the conditions for the economic motivation to use resources in a cost-effective manner are a prime requirement if an organization is to boost the efficiency of energy-saving technologies implementation, regardless of its purpose and type of activity. Therefore, the policies of energy saving should be applicable at both the technical and the systemic levels of analysis. The reasoning of energy saving is the systematic implementation of a set of technical and technological measures, which should be preceded by the optimization of organizations' energy consumption at the system level. The target of money savings is earmarked for paying for consumed energy resources, which is obtained by organizational measures by developing scientifically sound preconditions, like targeted energy inspections, along with implementation of technical measures for energy conservation [16].

At the systemic level, it is significant to improve the energy efficiency of mining enterprise units, which is carried out by using the optimal energy management methodology and by following procedures to create a database, interval estimation, forecasting and rationing of energy consumption. This methodology is based on a dynamic model that allows policymakers to simulate the process of power consumption of the coal production units for a time interval of up to 5 years or more [16].

The urgency of developing a dynamic adaptive model of energy consumption by adding the actions of external control has generated a new theoretical direction, focused mainly on study of the processes of technocenoses energy consumption at the so-called bifurcation stages. In generalizing the prospects of this model, the optimal management of energy consumption are driven to the creation of specialized software and hardware systems, which could be used among various stakeholders, such as regional energy grid, energy service and energy retail companies, as well as energy plans of electricity generation at regions, cities, municipalities, enterprises, and organizations. These key-aspects are framing a foreseeable future development [16].

3.1.2 Resource recovery technologies in reducing use of fossil fuels and fossil-based mineral fertilizers

Nowadays, it is generally accepted that the waste streams -solid, liquid, and gaseous- should be considered as sources of embodying invaluable resources: energy (mainly in the forms of heat and electricity) and nutrients. Resources' recovery can efficiently produce energy by mitigating pollution and improve a sustainable approach in operation. In particular a plethora of technologies established and integrated into a sustainable waste cycle. Such a cycle can balance out resource scarcity and demand for primary materials, reducing energy consumption and, subsequently, carbon dioxide emissions. Among the most important procedures of waste management are that of: (1) anaerobic digestion of sludge with biogas recovery, (2) co-digestion, incineration, and co-incineration with energy recovery, and (3) pyrolysis, gasification, wet-oxidation. These processes can be utilized to manufacturing and fabrication of materials. Moreover, the production of biofuels is gaining importance as a promising alternative energy source to fossil fuels [17].

Wastewater treatment concerns about disinfection of hazardous material, recovery and retreatment of valuable wasted materials, and the overall protection of human health and environmental protection from the so-named: mixed gray- and black-water; otherwise termed as storm-water or combined sewer system. The majority of such wastewater flows is still unexploited, where typical projects derive secondary resources and products from unsegregated urban wastewater, including: reclaimed fertigation and irrigation water (water and nutrients), phosphorous (P)-rich sludge, biopolymers, alginates, cellulose, construction material, and energy in the forms of biogas, biofuel, electricity, and heat [17].

Among the aforementioned nutrients' met at water sources, phosphorus is an essential fertilizer, but the accessible reserves of phosphorus can be depleted in the next 50 years, making utmost priority its recovery. Again, nitrogen and struvite recovery is associated with the bioelectrochemical system (BES) that possess realistic economic potential, while, future research can be further directed to confront the constraints of: minimizing methane production in anodes as well as hydrogen and electrochemical losses [17].

An important consideration of addressing the shortage of resources availability worldwide, the recovery of nutrients from wastewater treatment plants (WWTPs) is of utmost important priority. In the case of nitrogen, ammonia recovery can be achieved since ammonium is a widely used fertilizer, and it has been conventionally obtained by the energy intensive Haber-Bosch process. The energy costs associated to both Haber-Bosch process and nitrification-denitrification treatments could be minimized, while ammonia is recovered from WWTPs in the form of urine and utilized as fertilizer [17].

In a similar study, it has been reported that carbon dioxide can be captured in algal systems to produce algal-bio-fertilizer that can, then, be used to improve rice quality and simultaneously increasing its yield. From a climate viewpoint this alternative is more environmental friendly, comparing to synthetic chemical fertilizers. Among various technologies of utilization carbon dioxide in multiple field trials, algal fertilizer has proven highly efficient to greatly achieve emissions reduction [1]. In particular, the integration of chemicals and energy production in large-scale industrial algal biofarms has conceptualized the "algal biorefinery" process. The advantage of this process is the large intake or consumption of carbon dioxide for algal biomass production, since production one ton of dry algal biomass production requires about 1.8 tons of carbon dioxide. Typical products from algal biorefinery include bioenergy, carbohydrate, protein extract, as well as a variety of organic chemicals for cosmetic, pharmaceutical, and nutraceutical industries [1].

3.1.3 Renewable energy sources and energy crops

Increase of human population worldwide leads in intensification of land uses for food and energy, by simultaneously generating huge volumes of residues and wastes, especially among developed economies. In this context by-products and residues of agri-food processing are incurring high management expertise of costs and environmental remediation caused by untreated residues from the food chain [18]. Such residues and wastes should be valued as raw materials for energy production, emboding value-added assets of natural antioxidant, antimicrobial, and bioactive importance. Therefore, substances that can benefit food-producers, consumers and environment, they should prioritize EU policies to support a sustainable agriculture vision [18].

RES are considered as alternative proposals that offer environmental, energy, economic, and social benefits. Nevertheless, these benefits should not lead to unconditional acceptance of their technical exploitation projects. Contrarily, the construction of energy plants and sometimes the use of RES are raising environmental, economic and social issues. Among RES that are essential for human well-being and improve the quality of life, biomass (as one of the most prevailing RES) is expected to triple the energy capacity, from 1313 Mtoe in 2010 to 3271 Mtoe in 2040 [18].

Besides to heat and electricity production, biomass can be used for the production of biofuels, mainly focused on the production of bioethanol by fermenting sugars, starch, cellulose, and semicellulose derived from various biomass-rich plant species. Among this wide variety of cultivated and wild plant species these can produce not only edible products but also residues at the stages of cultivation and harvesting that could be used to produce biomass energy for electricity, heat, and biofuel purposes. Such energy crops are commonly observed in the spacious geographical areas of USA, Canada, Mexico, China, India Thailand, Malaysia, and Brazil. Among energy crops, perennial species in tropical and subtropical regions are highly productive in energy production. Furthermore, some energy crops, due to their resistance to adverse climatic and soil conditions; they can be used to support agricultural income in remote areas with non-fertile soils. Agricultural specifications of crop characteristics, economic importance, geographical distribution, and climate adaptability, are jointly considered to support energy crops to be used in crop rotation systems, enabling the: pesticides control, improvement of soil fertility, maintenance of long-term land productivity, and subsequently, increasing yields and profitability of farms [18]. Certain key-aspects of energy crops to embody low carbon context, they are framed as follows [18]:

- Crop residues for energy purposes offset the increase in energy demand by enhancing clean energy production.
- During biofuels production and through photosynthesis, significant amounts of carbon dioxide are re-bounding.
- Among developing countries the agricultural crops are regulated according to the national independence from imports of fossil fuels such as oil, the prices are also bounded on unstable political situations in producer countries.
- Ongoing demands for bioenergy are indirectly related to land use changes to support domestic production of biomass from energy crops, thus, causing soil degradation, nitrate fertilizer pollution, deterioration of the greenhouse effect, the reduction of land used in production of food crops, rising prices, as well as soil and water acidity.

Social norms and attitudes, in alignment with public interest toward different RES, they are all determining the feasible use of renewable energy among large energy-producer countries [4]. Such socio-cultural features are differently exhibited among European countries, thus, it is rather problematic an overall positive attitude to renewable energy sources, whereas it is significantly more familiar and knowledgeable energy policy makers to be specialized on localized renewable types, than for others. These multileveled preferences and knowledge are bounded on particular factors which exert great influence individuals and can contribute at shaping, either positive or negative, their attitudes. These attitudes are expected to support greater acceptance of renewables among individuals. However, better understanding of policy attitudes in a comprehensive and farsighted way may prerequisite the examination of other renewable types at future research works, enabling researchers to apply innovative methodologies [4].

Political efforts to promote renewable energy technologies should help EU members to develop a "green" electricity market. Such a full development of the renewable energy market relies on whether the electricity consumers are willing to pay (WtP) for renewables [2]. From this perspective, WtP for renewable energy is proven an important tool to design effective marketing and promotion strategies, to successfully increase payments for renewables, and to accelerate their marketable diffusion [2]. The consumerism behavior in favor of WtP for renewable energy is widely varying across countries. In a systematic approach of determining the influential key-drivers and key-barriers (buying unwillingness) of WtP, it is important energy policy makers to specify marketing and managerial prospects of those green electricity schemes devoted to the strategic energy planning on renewables [2].

From a joint approach, technological advancements and increased efficiencies that could help curbing GHG emissions in the agriculture sector, they are that of: nitrification inhibitors, feed additives for ruminants, as well as covering and flaring of slurry facilities [1]. From another viewpoint it has been also denoted that the currently plausible technologies will not be enough to meet the climate targets and that the development of methane inhibitors are needed [1]. Besides, more efficient use of nitrogen fertilizers are prerequisited, as N_2O emissions grow exponentially and nitrogen inputs exceed crop needs, no linearly as it could assume. Such a reduction of GHG emissions from agriculture sustains several difficulties, mostly linked to quantification of actual levels of emissions [1].

3.1.4 Offshore wind farms using GIS-based multi-criteria decision analysis and analytical hierarchy process

Offshore wind farms globally dispose today a total cumulative capacity representing 4.5% of total cumulative capacity and only 0.3% of global electricity generation. However, the wind technology sustains the potential to become a certainly supportive RES technology to meet the global power supply. A 15-times increase of today capacity of offshore energy production is forecasting within the next two decades, making it a protagonistic RES technology to the electrification of Europe. From a social viewpoint offshore wind has the advantages to prevent either social conflicts from competitive land uses, or social unrest issues, which other RES technologies are confronting. From an environmental viewpoint the wind farm technology is prone to future development within the next two decades, reaching at 5–7 Gtonnes of carbon dioxide emissions prevention from the power sector worldwide [19]. It is noteworthy that offshore wind farms can certainly play a decisive role to meet a future scaling-up of installed infrastructure for renewables-based era of the future. In this respect, proper regulatory framework should consider the site selection and, consequently, and the rules of operation for these wind energy-driven projects [19].

At the study of Gkeka-Serpetsidaki and Tsoutsos [19] the authors provided a fundamental basis on which the most sustainable scenarios, concerning the installation of offshore wind farms (OWFs) should be adapted. The selected methodology and evaluation was based on the surrounding marine areas of the island of Crete (Greece), being shown as the most sustainable scenarios' running regarding the installation of OWFs [19]. Different types and evaluation criteria, such as environmental, techno-economical and socio-political, they were jointly addressed while the methodology included questionnaires. These questionnaires were fully consisted of pairwise comparisons among the aforementioned criteria, being gathered from eight different groups of stakeholders/experts and then the relative importance of the selected criteria was measured with the aim of the Analytical Hierarchy Process (AHP). The proposed analysis utilized Geographic Information Systems (GIS) to deploy a site spatial assessment of the area studied. Additionally, the relative weights had been derived from the responses given by different stakeholders who participated in the survey [19].

Such a developed methodology can be adopted by all stakeholders and competent authorities, as a decision-making tool, since municipal authorities are responsible for the legislative planning of OWFs at a regional level of analysis [19]. Among others, specific consideration can be taken to the protection of areas of environmental interest and the accommodated sensitive ecosystems, fostering the sustainable development of the offshore wind farms' installed. Such an installation of wind farms it is anticipated to offer incomparable benefits for island contexts, including: unlimited usage of wind-RES, usage reduction of fossil fuels, energy supply security, reduction of electricity cost, environmental conditions' improvement for wildlife and natural ecosystems, autonomous electricity at high demanding periods, especially at tourism seasons, and summer times, savings of energy and resources, supportiveness to achieve national and/or European energy goals' related to climate crisis, as well as employment opportunities for the local citizens [19].

On the other hand, the sustainable prospects of wind-farms siting imply certain complexities and contradictions among numerous factors, including regulatory and techno-economic constraints. Therefore, the suitable site selection of OWF by energy policy makers should co-evaluate various environmental and socio-political criteria, toward minimization of the OWF impacts and maximization of the techno-economic benefits [19]. The reliability and verification of the adopted method are mainly based on the criteria selected, while planning considerations refer also to the successful integration/synthesis of different perceptions and judgments among all experts involved. These characteristics are proving that the followed methodology should be considered as a dynamic decision-making tool, where all stakeholders are responsible for the regulatory preparedness and the legislative planning of OWFs at a regional, and wider national, level of analysis [19].

3.2 Sustainable energy systems: the social dimension

3.2.1 Household sector

At the relevant literature, the introduction of Sustainable Energy Systems (SES) encompasses energy homeostasis and systemic resiliency that recursively operate in association to each other in a wide spectrum of environmental challenges and operational issues [20]. In such an analysis of electrical power control and energy management in the household sector, the perspectives of renewable energy sources' integration by local electric utilities has been investigated [20]. In this analysis, a set of strategies for the coordination and control of the electricity supply versus demand was tested and adapted to customers' needs and the infrastructure of the network. SES can be fully operative with the grid, aiming at maximizing the supply of green energy supply versus demand and systems' capacity. The simulated scenarios are precisely reflecting proposed strategies of real-word conditions in terms of different tariff options, systems capacity, and energy storage alternatives. Such proposed electric tariffs and energy management strategies could proven advantageous for the integration of distributed generation systems in the context of the smart grid transformation, under the specific research focus on ENEL in Chile [20–23]. The research framework of SES in the household sector is determined as follows:

- Resiliency and Energy Homeostasis: How to engineer resiliency in SES, focuses on the need Smart Energy Systems to incorporate homeostasis-based control systems in the design of SES [20].
- Grid-tied microgrids with and without energy storage: Answering the questions of when, where and how to apply each case [20].
- The conceptualization of Smart Energy Systems, Energy Sustainability and Grid Flexibility in the Smart Grid Agenda of Electric Utilities, like ENEL Distribution in Chile [21].
- The role of tariffs' differentiation, frequency footprint, possible rewards system for green energy integration, and the concept of tariffs' analysis with energy sharing innovations [21].
- Energy homeostasis and homeostatic control strategies to enhance and accommodate different communal lifestyles and scenarios of energy consumption needs to the integration and expansion of green energy systems tied to the grid [21].
- The key role of energy storage and the reasons of its importance for electric utilities to advance the distribution of electricity sector [22].
- Insights and reflections from industry and academia, approaching the issues of future holds with climate change and severe drought scenarios: an urgent call for action on the part of government and electric utilities [22].
- The general concepts of exergy, thriftiness and resiliency in electric power systems and on-grid distributed generation, which are referring to: How to engineer them toward the integration of green energy systems in buildings and condos [23].
- The concept and role of "prosumers" and the concept of the sustainable block, referring to applications in residential systems [23].

3.2.2 Education

In envisaging low carbon energy technologies among sustainable energy systems, it is crucial researchers to determine the drives and the barriers of such transition to low carbon energy society and the attribution of economic value to natural sources [24]. Among the main determinants of such a transition to knowledge management, education plays a decisive role [25]. Among all educational fields, it is noteworthy that accounting professionals are committed to sustainability and keep adopting substantive practices to support sustainability through accounting education.

In parallel a reformulation of accounting education should depart from knowledge dissonance to weak or strong sustainability, in terms of cultural and political paradigms of shift. From an economics viewpoint the financial sector can involve the accumulated experience and knowledge, in order to support businesses to adapt and change toward cleaner environment. From a technological viewpoint educational policies should interrelate various digital technologies -such as the internet of things, blockchain, artificial intelligence, and big data- enabling the transition to a cleaner industrial production of less dependence on natural resources and simultaneous acceleration of circularity and promotion of sharing collaborative business models.

Accounting education can support an integrated perception of the sustainable development while applying related educational concepts and teaching methods among business and accounting graduates [26]. Accounting professionals and academics need to meet the challenges of the new era of liquefied economic growth by keeping abreast with the evolutions, the EU strategies, and the demands of employers, consumers, and policy makers [26][27].

It can be signified that accounting and finance professionals should play a crucial role in the process of climate neutrality, through the transition to circular economy and the valuation of depreciated waste material. Such a transition can reassure investors and policy makers to retrieve higher-quality information and insights about the performance, the risks and the opportunities offered to companies, SMEs, and maker spaces communities. Therefore, approaches of low carbon energy technologies in the context of sustainable energy systems are envisaged through the educational reforms of environmental and accounting education among business schools and higher education institutions [25,28,29].

3.2.3 Bibliometric analysis on energy, sustainability, and climate change

Such bibliometric type of analysis aim at investigating and analyzing the scientific production related to energy, sustainability, and climate change. The literature search is based on specific key word to retrieve published papers in the topics of energy, sustainability, and climate change within specific period of analysis. The bibliographic domains of data retrieval include that of: scientific field, source type, document type, journal, country and institution, in association with the publication year [30].

For the scope of this book, the bibliometric research is optimistically served by the keywords “climate change,” “sustainability,” and “energy.” It has been reported a remarkable increase of the annual number of publications for all aforementioned three keywords, from the year 1990 onwards. It has been specifically shown that literature production of climate change is mainly in the field of "Environmental Science," followed by "Earth and Planetary Sciences" and "Agricultural and Biological Sciences." Sustainability is mainly if the field of "Environmental Science," followed by "Social Sciences" and "Engineering." Energy, is mainly in "Physics and Astronomy," followed by "Engineering" and "Materials Science" [30].

Besides, among those countries reported, USA, UK, and China were the top-three productive countries for the "climate change" research, USA, UK, and Australia, for the "sustainability" research and USA, China, and Germany for the "energy" research. For all three keywords, USA and other western countries showed a constantly decreasing profile, whereas the percentage contribution of China is constantly increasing. The most productive institution referring to all three keywords: "climate change," "sustainability," and "energy" was the Chinese Academy of Sciences, followed by the CNRS Centre National de la Recherche Scientifique. At a general context, the institutions with the highest productivity come from USA or China [30].

In focusing on a typical research paper regarding the joint approach of the aforementioned keywords with the themes developed at this book, it has been denoted that the increasing penetration of variable renewable energy sources (VRES), it has shown the increasing need for a flexible energy system. As old fossil fuel plants are gradually closing, flexible power plants can be built to supply electricity when VRES is not available. VRES penetration is especially needed in certain time periods where there is an excess of electricity that is not exportable, and the spot market electricity prices become very low or even negative. Such electricity surplus is subject to economically valuable utilization that represents an important opportunity of development. Among the most promising and appropriate solutions for such electricity surplus is energy storage. However, two crucial aspects are that conversion systems must be market controlled, and that different types of stakeholders are shown participatory interest in that energy market [1].

The operational growth of the energy market it has been long debated, if carbon capture is a sustainable solution. Carbon capture from stationary carbon dioxide emitting installations is a technology that can drastically reduce the emissions to the atmosphere [1]. In this respect, it is assumed that fossil power plants with carbon capture and storage (CCS) will become available at the next decade. A typical range of CCS parameters at a national context is including: Cost for carbon dioxide capture [euros/t-CO_2]; Cost for carbon dioxide transport and storage [euros/t-CO_2];

power generation efficiency loss [%]; carbon dioxide capture efficiency [%]; upper limit of annual carbon storage in 2050 [Mt-CO_2/year]. For the cost for carbon dioxide capture and power generation efficiency loss, different values were applied for each power generation method: supercritical pulverized coal (SCPC), integrated coal gasification combined cycle (IGCC), integrated coal gasification fuel cell combined cycle (IGFC), and natural gas combined cycle (NGCC) [11].

In the relevant literature, there are two ways in which the captured carbon dioxide can be handled [1]. One is to geologically store it via the aforementioned CCS, and the other is to reuse it as a new resource via carbon capture and utilization (CCU). It is arguable that CCS as a technology can enable the prolonged use of fossil fuels. Besides, CCS is not very suitable for a 100% RES energy system that is based on VRES, since these plants operate as baseload production, requiring increased fossil fuel consumption for capture processes. Compared to CCS, CCU can provide different and more sustainable products and/or services, and integrates different systems using waste, as a resource, from one to the other. CCU can be valued as key-component and enabler technology for future 100% RES carbon-free energy systems [1].

Two carbon capture (CC) technologies are particularly interesting for integration with energy systems; post-combustion and pre-combustion CC technology. As pre-combustion CC technology includes gasification process of the fuel, there is a demand for electricity for the gasification process. The gasification process can run flexibly, while it does not need to match the exact timing of the power plant electricity generation, which could be beneficial for balancing the supply and demand on the power grid. On the other hand, the post-combustion capture removes carbon dioxide from flue gases, after combustion occurred, and is more mature and established technology [1].

Regarding the operation strategies which can be utilized for CC and carbon emission abatement in a local multi-energy system, the investigation of the cost-effectiveness against carbon prices in carbon markets, carbon tax schemes, and cost-effectiveness of similar pilot projects, they should be all considered [9] The results of the case study show that, by utilizing all the abatement strategies, almost 20% emission reduction can be achieved with a 2% increase in cost. The abatement strategies' proposed by the optimization results included: more usage of biomass boilers in heat production, substitution of district heating and absorption chillers with heat pumps, as well as higher utilization of storage units [9].

3.2.4 Public acceptance of energy systems based on renewables

Conventional energy sources are finite and under depletion, whereas RES are constantly appearing in the natural environment, mainly in the forms of solar energy, wind energy, hydroelectric energy, geothermal energy, and biomass. The ongoing adoption of RES to national energy mixes allows many countries worldwide to become energy independent. Therefore, acceptance or rejection of RES by citizens plays a determinant role, as hardly any new RES-technology can be effectively implemented without its social acceptance. In the relevant literature, the investigation of the acceptance or the rejection of RES has been examined using a structured questionnaires delivered at urban areas [31,32]. It has been revealed that most of the participants are in favor of the development of RES, declaring themselves supporters to these investments, but stressing out that they are reluctant to the development of RES projects near to their residential areas, and they are not particularly willing to contribute financially to these projections' development [32]. At a similar survey made 2 years ago, the respondents were more flexible toward the RES perceived advantages and WtP for renewable energy, being positively associated with education, energy subsidies, and state support [31]. The today rigid and social behavior and "distant" attitude against RES technologies at the Greek context should be certainly related to the ongoing economic recession and the deep affection of national economic growth (in Greece) by governmental measures taken for pandemic confrontation (let alone the unconfordable therapeutic conditions of medical treatment and patients' recovery).

Focusing on the study of [32] the term and the operability opportunities of "RES" are not widespread. In particular, men and highly educated participants seem to be more informed, while the solar energy is the most popular RES, followed by the wind energy. Contrarily, significant knowledge gap was identified for biomass and geothermal. The main sources of RES knowledge are primarily derived from the internet and relevant electronic platforms, being accessible and interested in all age groups. Regarding the environmental friendliness, solar energy and wind energy are considered the most environmentally friendly RES, though under doubt by older people and those having lower level of education. Other advantages reported from RES projects are related to local economic prosperity, independence from fossil fuels, and creation of new jobs. Among the reported disadvantages are that of high cost required for the initial investment, and the engagement of large land areas. Besides, the majority of participants are shown distrust toward governmental actions that promote the development of RES projects. However, more trustworthy appreciation was reported to the participative role of experts, being considered absolutely necessary in the decision to install RES projects, where residents can be actively participated in the decision-making process of such RES projects' installation and operation [32].

Other reservations from participants are expressed for the installation of a biomass combustion plant, while private funding bodies are preferred to the design and the construction of RES projects, whereas public bodies are more suitable to the exploitation and the tariff/pricing determination of such RES projects [32]. Overall, it is noteworthy the highly liquefied and transitional attitudes prevailing from time to time and from place to place, regarding the public acceptance of energy systems based on RES. However, these social appreciation or disprove can be used as a feasible managerial tool to enhance public acceptance of RES technologies and projects, supporting the gathering of informative activities and awareness campaigns for specific mainly suspective forms of RES, such as biomass and geothermal energy. Subsequently policy makers can better design effective energy policies toward the feasible penetration of RES in national or international energy mixes [32].

4 Discussion and current research considerations

A wide spectrum of low-carbon technologies should have functional roles in decarbonizing the national energy systems by 2050. However, each low-carbon technology has some level of uncertainty associated with its development and implementation, which can affect the long-term energy transition of national and international contexts of analysis. Multiple possible scenarios and models' developed for low-carbon technologies should support flexible decisions given these uncertainties, which have not yet been fully accounted for in the 2030-target energy strategies among, mainly, developed economies [11].

In the European context during the last decade there is an ongoing research activity for energy management, being especially focused on energy management from various types of renewable energy sources [33,34], while earlier studies (dating back a decade ago) had been focused on small hydro stations and woodfuel-based projects for energy production [35–37]. Among the recently published studies, the role of a New Ecological Paradigm Scale (NEP) has been examined as a unidimensional measure of environmental attitudes, in order to measure the overall relationship between humans and the environment. NEP also correlated a statistically significant level with respondents' willingness to pay for the expansion of renewable energy [31].

Regarding the energy-consuming applications at the built environment the main environmental considerations of contemporary energy systems include, among others, multicriteria decision analyses of energy consuming domestic applications [38,39], the recycling of energy-consuming devices at the household sector, as well as the redesigning of urban landscapes and protected areas from light pollution [40–42].

Combustion of fossil fuels to generate electricity and to provide heating, cooling, and transportation services are the major contributors to GHGs emissions and air pollution. In this context, energy systems account for about two-thirds of global GHG emissions. While carbon dioxide is responsible for 77% of radiative forcing, other GHGs are also generated by energy systems, such as that of methane (CH_4) and nitrous oxide (N_2O) that may be generated at various points -resource extraction, processing, transportation, storage, and combustion- jointly to fossil energy supply chains [1]. In approaching the key-aspects of energy systems with low carbon perspectives in an integrated manner, the main outcomes are presented in the following Table 15.2.

Table 15.2 Key-aspects impacting on energy systems of low carbon orientation.

References	Research field Technological application	Key-considerations and challenges
[1]	• Electrification of transport	• Electric motor is much more efficient than the internal combustion engine, since electric cars are 3.5 times more efficient, and electric buses are 2.5 times more efficient than their internal combustion counterparts. • The advantage is also compounded because electricity can be generated using various low-carbon sources. • Not all transportation modes can be directly electrified by the currently available technology. Indeed, for long distance transportation, heavy-duty transportation, marine, and air transportation, direct electrification is not a feasible solution with currently available technology, different alternatives are more suitable to the future transport energy systems. For the remainder of the transport sector that is not suited to electrification, several proposed alternatives contain biofuels, hydrogen, electrofuels, or a combination of these technologies

(Continued)

Table 15.2 Key-aspects impacting on energy systems of low carbon orientation. (*Cont.*)

References	Research field Technological application	Key-considerations and challenges
[7]	• Improvements in energy efficiency	• Consistent buildup of RES systems is required, especially in focusing on the reduction in conventional power capacity due to the nuclear phase-out by 2022. • Fossil-fueled power plants have to be significantly reduced through the implementation of CCS technology and the replacement of coal-fired power plants by gas-fired plants or CHP plants. • Measures need to be implemented to stabilize the price of carbon dioxide certificates at a much higher level, while financial incentives for the transition to RES technologies have to be also offered.
[1]	• Improvements of process efficiency and energy efficiency in the industry.	• Process efficiency and energy efficiency in the industry are important transitional steps of the industry toward a more sustainable and low-carbon future. • Emissions reduction from industry is challenging due to the existence of many different subsectors. • Electrification of the industrial sector and production of hydrogen and hydrocarbons with power-to-liquid and power-to-gas technologies could be feasible transitional pathways. • Single measures have a limited impact on the decarbonization of the industrial sector, but the integrated approach can make a greater difference.
[7]	• Energy sector interconnection • High energy conservation potential	• It reduces primary energy demand and, consequently, GHG emissions. • Challenges are related to interconnecting different energy sectors, managing energy supply and demand, and the temporary storage of energy. • Challenges of interconnecting energy sectors are achieved through CHP systems, heat pumps, and electric vehicles. These technologies can make a major contribution to overall energy efficiency and to grid stability. • Because of renewable intermittent behavior, flexibility options such as storage systems should be proposed. • The proposed measures will pay off in the long run if a concerted effort is made by all actors involved to successfully implement and complete this historic transition. • The estimates for renewable energy systems can indicate lower costs than the costs of today's • energy system. Therefore, policy makers ought to create the correct innovation and investment incentives, in order to advance the development of RES projects.

Table 15.2 Key-aspects impacting on energy systems of low carbon orientation. (*Cont.*)

References	Research field Technological application	Key-considerations and challenges
[8]	• Global linear optimization model of hydrogens role in the future energy system. • Hydrogen for industry is linked to chemical feedstock syngas and it can be used to produce methanol, ammonia, and hydrocarbons through the Fischer-Tropsch process.	• Cognizant of policy, technology learning curves and carbon targets. • Hydrogen is estimated to account for approximately 2% of energy needs by 2050, playing a prominent role in the transportation sector. • Ongoing research can determine the appropriate production and use of hydrogen methodologies, fuels, and blends toward industrial applications. • While considering the production of hydrogen it should be denoted that approximately 98% of global hydrogen production is directly or indirectly bounded on fossil fuels.
[8]	• Hydrogen for storage • Electrolyzer and hydrogen for storage and production of energy, in terms of energy efficiency. • The use of hybrid renewables and hydrogen are complimented in terms of improving overall system efficiency.	• It is noteworthy that without a carbon tax, which would reduce fossil based generation, cost reductions through concentrated research and development are required for hydrogen to improve renewable penetration levels and meet seasonal storage needs. • The incorporation of hydrogen production and storage into renewable energy and traditional energy systems was proven feasible, albeit more expensive, than battery bank choices. • Surplus energy to hydrogen was identified as a possible solution for the generation of environmentally friendly fuel and its storage.
[1]	• Increase in power generation demand under targeted policy scenarios. • Implementation of different policy measures.	• Power generation demand is forecasting to increase by more than 25% until 2040, according to the International Energy Agency (IEA). • Achievement of 100% renewable energy systems can be achieved by the year 2050. Use of 100% renewable energy for all sectors, and not only for that of renewable power, are proven technically and economically feasible among distinct countries and regions, in South-east Europe, the EU, and the United States.

5 Conclusions and future research orientations

5.1 Challenges of carbon abatement based on energy systems

From modeling viewpoint, an integrated model should contain outputs that include the change in energy system cost (fixed and variable), energy supply and consumption, CCS utilization levels, energy system structure, the quantum of hydrogen introduced into the energy system, production methods, end uses and geographic origins [8]. From a technological viewpoint, renewable energy sources include wind, solar PV, geothermal, hydropower, and biomass. Within biomass, several sources are considered including energy crops, forestry biomass, log residues, black liquor, and wastepaper, sawmill residues, harvest residues from crops, sugar cane residue, bagasse, household waste and feces. Secondary energy carriers should include hydrogen, methane, methanol, dimethyl ether (DME), oil products, carbon monoxide, and electricity. The final energy demand should incorporate solid fuels, liquid fuels, gaseous fuels and electricity, cognizant of daily load cures and seasonal variations [8]. Energy conversion methods account for efficiency, capacity factor, plant lifetime, annual expense rates, and reducing costs or efficiency improvements over time (learning curves) [8].

Future energy demands can potentially be met with 100% renewables provided that they are conceptualized and planned using a system-level integrated approach. Use of CCU technologies in the future energy system could play an important role as it can help to increase the flexibility of renewable energy systems by indirectly storing variable electricity into a wide range of fuels and chemicals. In a recent report, the European Commission acknowledged that the CCS's role could only be in the short/medium term for industrial processes, and to generate net-negative emissions from biomass combustion [1]. In this research orientation, integrated value chains of CCU can help the integration of technology with energy system and better support the integration of renewable energy with different end-use products. Carbon capture from more distributed generation systems in the future will not necessarily prove non-feasible, especially in the case that the transportation and storage demand is minimized [1].

The great challenge of integrating carbon capture in the energy systems is the high and long-term investments and finding the most suitable end-utilization of the collected carbon. Integrating different industrial actors to invest and to use CCU or CCS technologies is remaining challenging. The low carbon pricing and carbon policy is not supporting implementation of these technologies. Carbon capture has also a technical challenge of collection and purification of CO_2 from different sources and intermittent operation of the capture itself is not yet fully investigated nor demonstrated on the large scale. Other unresolved issues include storage safety and leakage of stored carbon that are still debated, in the light of careful selection of storage sites and carbon dioxide monitoring. One of the critical obstacles for implementing CCU in the energy systems is a lack of clear system boundaries for the emission assessment and in some cases system analysis, coupled with lacking policy incentives [1].

5.2 Policies and implications

From a policies viewpoint, modeled parameters and constraints are commonly set to estimate the societal penetration of RES for the first-half of 21st century, period 2000–50. Following the modeling of the initial and/or basic scenarios, sensitivity analyses should be undertaken through testing of societal RES penetration under key exploratory scenarios to assess the targets and impacting on policy, technology and carbon abatement [8].

From an economics viewpoint, energy system investments are linking several issues that have to be identified, since they impede the low-carbon transition. Among them, the pronounced role of the continued growth of coal-fired power plants (particularly in Asia), they are necessary to be phased out rapidly to meet Paris Agreement goals. Besides, profitable gas, oil, and coal projects are funded in preference to energy efficiency and renewable energy projects, while nations are not spending enough on research and development for public energy. The continuation of such trends means that the Paris Agreements may not be unanimously met [8].

From a technology viewpoint, a technology which presents challenges to future research work is CCS, currently in the pilot phase in a lot of case-studies, implying an imperative need to move rapidly toward large scale deployment, in order to satisfy carbon emission reduction requirements. The imperative need to apply CCS not only to fossil fuel power plants -already pursuing their commercial viability- but also to negative emission technologies which incorporate biomass energy (i.e., BECCS) faces significant challenges that may limit their contribution to Paris Agreement targets [8].

In terms of policy implications, policy makers are challenged to develop suitable tools toward an "ideal" energy system at regional, national, or international levels by identifying technology, carbon target and policies which are primarily impact in on energy supply mix, hydrogen production and consumption, CCS deployment, and system cost parameters. This research approach should identify the most effective measures to cost effectively increase of RES-based energy production while meeting carbon reduction goals [8].

Finally, it is noteworthy that the pre- and post-combustion carbon capture technologies are especially promising for integration with energy systems. Their inherent flexibility can allow these two carbon capture technologies to be used for balancing the electricity supply and demand in the power grid. In such systems, techno-economic penalties relative to steady-state capture systems can be offset by variations in electricity price in a renewable-intensive power grid. This is significant since, in future energy systems, there will be a need for fast dynamic response technologies, taken into consideration that it needs to be emphasized that there is a large difference in scale (by orders of magnitude) between the potential market for carbon dioxide as feedstock, and the amount that needs to be removed to be significant for climate stabilization purposes [1]. In the future, further large-scale projects should be inseparable tools to fully deploy fossil power generation systems with CCS globally, thus achieving the annual carbon dioxide storage by 2050 [11].

References

[1] H. Mikulčić, I. Ridjan Skov, D.F. Dominković, S.R. Wan Alwi, Z.A. Manan, R. Tan, et al. Flexible carbon capture and utilization technologies in future energy systems and the utilization pathways of captured CO_2, Renew. Sustain. Energy Rev. 114 (2019) 109338 DOI: 10.1016/j.rser.2019.109338.

[2] E. Karasmanaki, Understanding willingness to pay for renewable energy among Europeans during the period 2010–2020. In: Low Carbon Energy Technologies in Sustainable Energy Systems. Elsevier, 2020 (Chapter 5).

[3] V.C. Kapsalis, Multi Energy Systems of the Future, in: Low Carbon Energy Technologies in Sustainable Energy Systems. Elsevier, 2020 (Chapter 10).

[4] E. Karasmanaki, G. Tsantopoulos, Public attitudes towards the major renewable energy types in the last five years: A scoping review of the literature, in: Low Carbon Energy Technologies in Sustainable Energy Systems. Elsevier, 2020 (Chapter 4).

[5] X. Zhang, Y. Zhang, Environment-friendly and economical scheduling optimization for integrated energy system considering power-to-gas technology and carbon capture power plant, J. Clean. Prod. 276 (2020) 123348 DOI: 10.1016/j.jclepro.2020.123348.

[6] T. Singh, I. W. Almanassra, A. Ghani Olabi, T. Al-Ansari, G. McKay, M. Ali Atieh, Performance investigation of multiwall carbon nanotubes based water/oil nanofluids for high pressure and high temperature solar thermal technologies for sustainable energy systems, Energy Convers. Manage. 225 (2020) 113453 DOI: 10.1016/j.enconman.2020.113453.

[7] S. Kumar, M. Loosen, R. Madlener, Assessing the potential of low-carbon technologies in the German energy system, J. Environ. Manage. 262 (2020) 110345 DOI: 10.1016/j.jenvman.2020.110345.
[8] A. Chapman, K. Itaoka, H. Farabi-Asl, Y. Fujii, M. Nakahara, Societal penetration of hydrogen into the future energy system: Impacts of policy, technology and carbon targets, Int. J. Hydrogen Energy 45 (7) (2020) 3883–3898 DOI: 10.1016/j.ijhydene.2019.12.112.
[9] N.M. Alavijeh, D. Steen, Z. Norwood, L.A. Tuan, C. Agathokleous, Cost-effectiveness of carbon emission abatement strategies for a local multi-energy system - A case study of chalmers university of technology campus, Energies 13 (7) (2020) 1626. DOI: 10.3390/en13071626.
[10] C. Mazur, S. Hall, J. Hardy, M. Workman, Technology is not a barrier: A survey of energy system technologies required for innovative electricity business models driving the low carbon energy revolution, Energies 12 (3) (2019) 428 DOI: 10.3390/en12030428.
[11] A. Ozawa, Y. Kudoh, A. Murata, T. Honda, I. Saita, H. Takagi, Hydrogen in low-carbon energy systems in Japan by 2050: The uncertainties of technology development and implementation, Int. J. Hydrogen Energy 43 (39) (2018) 18083–18094 DOI: 10.1016/j.ijhydene.2018.08.098.
[12] C. Wang, A. Engels, Z. Wang, Overview of research on China's transition to low-carbon development: The role of cities, technologies, industries and the energy system, Renew. Sustain. Energy Rev. 81 (2018) 1350–1364 DOI: 10.1016/j.rser.2017.05.099.
[13] S. Simoes, W. Nijs, P. Ruiz, A. Sgobbi, C. Thiel, Comparing policy routes for low-carbon power technology deployment in EU – an energy system analysis, Energy Policy 101 (2017) 353–365 DOI: 10.1016/j.enpol.2016.10.006.
[14] N. Good, E.A. Martínez Ceseña, L. Zhang, P. Mancarella, Techno-economic and business case assessment of low carbon technologies in distributed multi-energy systems, Appl. Energy 167 (2016) 158–172 DOI: 10.1016/j.apenergy.2015.09.089.
[15] B. Moran Jay, D. Howard, N. Hughes, J. Whitaker, G. Anandarajah, Modelling socio-environmental sensitivities: How public responses to low carbon energy technologies could shape the UK energy system, Sci. World J. 2014 (2014) 605196 DOI: 10.1155/2014/605196.
[16] M.V. Zinovevich, D.V. Antonenkov, E.Y. Sizganova, D.V. Solovev, Increasing efficiency of mining enterprises power consumption, in: Low Carbon Energy Technologies in Sustainable Energy Systems. Elsevier, 2020 (Chapter 2).
[17] M. Zamparas, The role of resource recovery technologies in reducing the demand of fossil fuels and conventional fossil-based fertilizers, in: Low Carbon Energy Technologies in Sustainable Energy Systems. Elsevier, 2020 (Chapter 1).
[18] S.V. Leontopoulos, G. Arabatzis, The contribution of energy crops to biomass production, in: Low Carbon Energy Technologies in Sustainable Energy Systems. Elsevier, 2020 (Chapter 3).
[19] P. Gkeka-Serpetsidaki, T. Tsoutsos, Sustainable site selection of offshore wind farms using GIS-based multi-criteria decision analysis and analytical hierarchy process. Case study: Island of Crete (Greece), in: Low Carbon Energy Technologies in Sustainable Energy Systems. Elsevier, 2020 (Chapter 13).
[20] F. Yanine, A. Sanchez-Squella, A. Barrueto, A. Parejo, H. Rother, Linking energy homeostasis, exergy management and resiliency to develop sustainable grid-connected distributed generation systems for their integration into the distribution grid by electric utilities, in: Low Carbon Energy Technologies in Sustainable Energy Systems. Elsevier, 2020a (Chapter 6).

[21] F. Yanine, A. Sanchez-Squella, A. Barrueto, A. Parejo, H. Rother, Smart energy systems: the need to incorporate homeostatically controlled distributed generation systems to the electric power distribution industry: an electric utilities' perspective, in: Low Carbon Energy Technologies in Sustainable Energy Systems. Elsevier, 2020b (Chapter 7).
[22] F. Yanine, A. Sanchez-Squella, A. Barrueto, A. Parejo, H. Rother, Grid-tied distributed generation with energy storage to advance renewables in the residential sector: tariffs analysis with energy sharing innovations, in: Low Carbon Energy Technologies in Sustainable Energy Systems. Elsevier, 2020c (Chapter 8).
[23] F. Yanine, A. Sanchez-Squella, A. Barrueto, A. Parejo, H. Rother, Integrating green energy into the grid: how to engineer energy homeostaticity, flexibility and resiliency in energy distribution and why should electric utilities care, in: Low Carbon Energy Technologies in Sustainable Energy Systems. Elsevier, 2020d (Chapter 9).
[24] G. Kyriakopoulos, G. Kyriakopoulos. Ecosystems services valuation (ESV) then and now: a review, in: M.J. Acosta (Ed.), Advances in Energy Research, vol. 27, Nova Science Publishers, 2017, pp. 1–61 (Chapter 1).
[25] S. Asonitou, Accounting and Sustainability, in: Low Carbon Energy Technologies in Sustainable Energy Systems. Elsevier, 2020 (Chapter 14).
[26] G. Kyriakopoulos, S. Ntanos, S. Asonitou, Investigating the environmental behavior of business and accounting university students, Int. J. Sustain. Higher Edu. 21 (4) (2020) 819–839 https://doi.org/10.1108/IJSHE-11-2019-0338.
[27] G. Kyriakopoulos, V. Kapsalis, K. Aravossis, M. Zamparas, A. Mitsikas, Evaluating circular economy under a multi-parametric approach: A technological review, Sustainability 11 (21) (2019) 6139, doi: 10.3390/su11216139.
[28] G. Kyriakopoulos, Secondary education at the dawn of the twenty-first century, in: Secondary Education, Nova Science Publishers, 2016a, pp. 1–32 (Chapter 1).
[29] G. Kyriakopoulos, The applicability of the science, technology, and society (STS) concept towards the pedagogic didactics of natural sciences. In: Progress in Education, 45, 35–60. editor: Roberta V. Nata. Nova Science Publishers, 2016b (Chapter 3).
[30] Th. Kalyvas, E. Zervas, Bibliometric analysis of scientific production on energy, sustainability and climate change. In: Low Carbon Energy Technologies in Sustainable Energy Systems. Elsevier, 2020 (Chapter 11).
[31] S. Ntanos, G. Kyriakopoulos, M. Chalikias, G. Arabatzis, M. Skordoulis, Public perceptions and willingness to pay for renewable energy: A case study from Greece, Sustainability 10 (3) (2018) 687 DOI: 10.3390/su10030687.
[32] Z. Gareiou, E. Drimili, E. Zervas, Public acceptance of renewable energy sources, in: Low Carbon Energy Technologies in Sustainable Energy Systems. Elsevier, 2020 (Chapter 12).
[33] G. Arabatzis, G. Kyriakopoulos, P. Tsialis, Typology of regional units based on RES plants: The case of Greece, Renew. Sustain. Energy Rev. 78 (2017) 1424–1434 DOI: 10.1016/j.rser.2017.04.043.
[34] S. Ntanos, G. Kyriakopoulos, M. Skordoulis, M. Chalikias, G. Arabatzis, An application of the new environmental paradigm (NEP) scale in a Greek context, Energies 12 (2) (2019) art. no. 239, DOI: 10.3390/en12020239.
[35] K.G. Kolovos, G. Kyriakopoulos, M.S. Chalikias, Co-evaluation of basic woodfuel types used as alternative heating sources to existing energy network, J. Environ. Protect. Ecol. 12 (2) (2011) 733–742.
[36] G. Arabatzis, C. Malesios, An econometric analysis of residential consumption of fuelwood in a mountainous prefecture of Northern Greece, Energy Policy 39 (12) (2011) 8088–8097.

[37] G. Arabatzis, D. Myronidis, Contribution of SHP Stations to the development of an area and their social acceptance, Renew. Sustain. Energy Rev. 15 (8) (2011) 3909–3917.

[38] C.A. Bouroussis, F.V. Topalis, Assessment of outdoor lighting installations and their impact on light pollution using unmanned aircraft systems - The concept of the drone-gonio-photometer, J. Quant. Spectr. Radiat. Transfer 253 (2020) 107155 DOI: 10.1016/j.jqsrt.2020.107155.

[39] E.-N. Madias, L.T. Doulos, P.A. Kontaxis, F.V. Topalis, Multicriteria decision aid analysis for the optimum performance of an ambient light sensor: methodology and case study, Operat. Res. Int. J. (2020) DOI: 10.1007/s12351-020-00575-5.

[40] A. Papalambrou, L.T. Doulos, Identifying, examining, and planning areas protected from light pollution. the case study of planning the first National Dark Sky Park in Greece, Sustainability 11 (2019) 5963.

[41] C. Grigoropoulos, L.T. Doulos, S. Zerefos, A. Tsangrassoulis, P. Bhusal, Estimating the benefits of increasing the recycling rate of lamps from the domestic sector: Methodology, opportunities and case study, Waste Manage. 101 (2020) 188–199.

[42] O. Ardavani, S. Zerefos, L.T. Doulos, Redesigning the exterior lighting as part of the urban landscape: The role of transgenic bioluminescent plants in mediterranean urban and suburban lighting environments, J. Clean. Prod. 242 (2020) 118477.

Index

Note: Page numbers followed by "f" indicate figures, "t" indicate tables.

E

Printed in the United States
By Bookmasters